Compendium of Bioenergy Plants
CORN

Compendium of Bioenergy Plants

Series Editor

Chittaranjan Kole
Vice-Chancellor
Bidhan Chandra Agricultural University
West Bengal
India

Books in this Series:

Published or in Press:

- Stephen L. Goldman & Chittaranjan Kole: *Corn*
- Hong Luo, Yanqi Wu & Chittaranjan Kole: *Switchgrass*
- Eric Lam, Helaine Carrer, Jorge da Silva & Chittaranjan Kole: *Sugarcane*

Compendium of Bioenergy Plants
CORN

Editors

Stephen L. Goldman
Plant Science Research Center
The University of Toledo
Ohio
USA

Chittaranjan Kole
Vice-Chancellor
Bidhan Chandra Agricultural University
West Bengal
India

CRC Press
Taylor & Francis Group
Boca Raton   London   New York

CRC Press is an imprint of the
Taylor & Francis Group, an **informa** business

A SCIENCE PUBLISHERS BOOK

CRC Press
Taylor & Francis Group
6000 Broken Sound Parkway NW, Suite 300
Boca Raton, FL 33487-2742

© 2014 Copyright reserved
CRC Press is an imprint of Taylor & Francis Group, an Informa business

Cover Illustration: Halftone figure: Reproduced by kind courtesy of Dr. Roman Brunecky (author of Chapter 2).

Library of Congress Cataloging-in-Publication Data

Compendium of bioenergy plants : corn / editor: Stephen L. Goldman.
     p. cm. -- (Compendium of bioenergy plants)
  ISBN 978-1-4822-1058-3 (hardcover : alk. paper) 1. Corn--Genetic
engineering. 2. Energy crops. I. Goldman, Stephen L. II. Title:
Corn. III. Series: Compendium of bioenergy plants.

  SB191.M2C66 2014
  631.5'233--dc23

                     2013028936

Visit the Taylor & Francis Web site at
http://www.taylorandfrancis.com

CRC Press Web site at
http://www.crcpress.com

Science Publishers Web site at
http://www.scipub.net

Preface to the Series

The need for sustainable energy is growing at an increasing rate with the alarmingly high rate of increase in population coupled with the fast growth of urbanization. By 2050 the world population is estimated to be seven billion computed at a conservative rate of growth. By 2100, the number is projected to be over ten billion by another estimate. The source of fossil fuels being predominantly used over time will face depletion around the end of this century unless non-conventional energy sources are put in place. Besides depletion, fossil fuel use is constrained by geo-political issues and threat of greenhouse gas emission. Among the alternative energy sources, bioenergy is emerging as the most promising as compared to atomic, solar, and wind. Bioenergy including bioethanol and biodiesel can be produced from cellular biomass, starch, sugar, and oil derived from several plants and plant products in huge amounts once the required strategies and technologies are formulated and validated for commercialization in cost-effective ways.

Scientific exploratory research conducted during the last few years has identified a large number of plants as potential sources of bioenergy. These include maize, sorghum, switchgrass, canola, soybean, and sugarcane among field crops; eucalyptus and poplar among forest trees; and jatropha, oil palm and cassava among plantation crops. Several other promising field crops including Brachypodium, minor oilseeds, sugarbeet, sunflower, and sweetpotato; forest trees including diesel trees and shrub willow; plantation crops such as Paulownia; many lower plants; and even vegetable oils, organic farm waste and municipal sludge have been found to be promising. Therefore, 'fuel' has made its place in the list of principal agricultural commodities along with food and fiber.

Significant studies have also been conducted in natural and social sciences to facilitate utilization of plants and plant products as the most potential source of bioenergy. In bioenergy crops, research has been carried out on genetics, genomics and breeding for relevant traits employing traditional and molecular breeding, genomics-assisted breeding, and genetic engineering. Physiological works have been done for *in planta* production of cell-degrading enzymes and enzymatic conversion of cell walls into biofuels. Significant advancement has been made on the works on post-harvest technologies and chemical engineering, fuel quality, and greenhouse gas impacts of bioenergy. Most importantly, economics, public policies, and perceptions have also been critically examined.

There are, at present, only a few books on bioenergy crop plants available. I have myself edited a book recently with two other co-editors for the CRC Press of the Taylor and Francis Group. This book entitled 'Handbook of Bioenergy Crop Plants' elucidates on the general concepts of and concerns about bioenergy crop production, genetics, genomics and breeding of commercialized bioenergy crop plants, and emerging bioenergy crops or their groups besides deliberations on unconventional biomass resources such as vegetable oils, organic waste and municipal sludge.

As expected, there is also an array of research and review articles on the basic concepts, strategies and means of utilization of bioenergy crop plants and their products in scientific journals, web sites, newsletters, newspapers, etc. However, there is no endeavor to present any compilation about all the relevant aspects related to particular bioenergy crop plants already commercialized or having potential to be commercialized in near future. The present book series will hopefully fill up that vacuum. This is particularly important as the subject of bioenergy has already occupied its place in academia, research labs, and public life. This was the underlying force behind conception of a book series on 'Compendium of Bioenergy Crop Plants'.

At the outset, I formulated the tentative outline for 15 chapters to maintain more or less uniformity throughout the volumes of the compendium. These included basic information on the crops; anatomical and physiological researches relevant to feedstock; special requirements related to agricultural and industrial infrastructure; elucidation on genetics, genomics and breeding of bioenergy traits; public platforms for sharing results and building initiatives; role of public and private agencies in fostering research and commercialization; regulatory, legal, social and economic issues; general concerns and their compliance; and also future prospects and recommendations. However, the volumes of this compendium are devoted to various crop plants and obviously the concerned volume editors had to improvize on the contents of the respective volumes based on the unique information available and specific requirements. Thus, each volume of this compendium has the 'stand-alone' potential at the same time, thanks to the excellent balancing job performed by the volume editors. Fortunately most, if not all, of the volume editors have long standing association with me as an author of a chapter in some other book, or volume editor of another book series or colleague in a research platform. Therefore, it has been highly comfortable and enriching for me to work with them again for this compendium. I take this opportunity to express my heartiest gratitude to them for offering me this opportunity. The authors of the chapters for each of the volumes have produced high quality deliberations both in terms of comprehensive contents and lucid write-ups. As the series editor, I must join with my volume editors to extend our thanks to the authors of the chapters for their elegant contributions as well as sincere cooperation all along.

This compendium was originally conceptualized by my wife and colleague, Phullara. She had meticulously reviewed the relative importance and quantum of works accomplished in the commercialized and promising bioenergy crops plants and had eventually identified the leading bioenergy crop plants to which the individual volumes of this compendium are devoted to. She was always there for help in editing this compendium similar to several other book series containing over sixty books published or in press. Expressing just thanks will not do justice to her contribution to this book project. I have, therefore, dedicated this compendium to her in recognition of her contributions to this book project and also for all her support, advice and inspiration for all my academic activities besides shouldering most of our domestic loads, taking the major responsibility to navigate our family and nourish our three growing kids, Papai, Titai and Kinai, as that provided me with enough extra time for my book editing jobs in addition to my professional duties.

Chittaranjan Kole

Dedication by Series Editor

Dedicated to
My beloved wife and colleague,
Phullara

The infinite source of support, strength, guidance, and inspiration for my mission to serve science and society.

Preface

In attempting to identify sustainable and environmentally friendly alternatives to fossil fuels President Jimmy Carter's Administration, with prophetic boldness, established a Department of Energy. Its mission was to look at wind and solar as fuel substitutes, in addition to creating federally supported, privately owned, ethanol plants deriving alcohol from corn starch. Despite these forceful measures, it was immediately understood that these infant industries could not even supply the increased energy demands as measured by significant rises in oil imports. As a defensive measure, a petroleum reserve was established in the United States. The wisdom of this decision was immediately apparent as fossil fuel imports rose 65% from 1965 into the early 1970s. Given this increased demand and the continuing and growing political instabilities in the Middle East, the price of fossil fuels would continue to rise significantly. Such a scenario could only compromise or bring to a halt the vigorous Western economies to say nothing of those in the developing world entering their birth pangs.

The 1973 oil crisis loomed large in Carter's mind and so billions of dollars were secured that were to be directed to the manufacture of ethyl alcohol from food grain or sugar beet feed stocks. This funding came as federal loans and loan guarantees for ethanol plant construction. Ethanol subsidies were also used to incentivize the program. This assistance continued under Reagan and his successors, accompanied by subsidy reductions, and now have been largely eliminated. Its long term effect on energy pricing remains to be seen but it is a certainty that the asking price of a bushel of corn has risen and with it the cost of food.

A little more than ten years ago the cost of a bushel of corn fell under two dollars, whereas the cost is now approaching seven dollars (MIT Technology Review, March 2012, www.technologyreview.com; Fortenberry and Park, 2008, www.ideas.repec.org). Indeed the Chicago Board of Trade during the week of July 16, 2012 predicted correctly that the price would increase to $7.49/bushel. These successive cost upsurges parallel exactly the dedication of the corn harvest to bio-ethanol manufacture. In 2005 approximately thirteen percent of the return crop was dedicated to the making of ethanol rising successively in 2006/2007 to 2011 from twenty to forty percent. In keeping with these observations, the food price tag rose between ten to fifteen percent in 2007/2008, a pattern that remains consistent today. To be sure, the FAO reports that food prices have risen consistently during the last eight months. This is not likely to abate. In the United States alone the federal government has mandated that 15 billion gallons of renewable fuel is to be available by 2015 rising to 36 billion in 2022 (Solomon et al. 2009).

Given the national mandate for sustainable fuel manufacture, one cannot begin to anticipate the effect of 2012's unprecedented heat wave and drought on food costs. At the beginning of 2012's growing season, the availability of ample grain supplies was predicted. In keeping with rising demand, Indiana, Iowa, Nebraska and Illinois increased the number of acres dedicated to corn production. Even with this increased harvest potential, the government projected $6.40/bushel up from previous projections ranging from $4.20 to $5.00. Of course these projections have now proven worthless as fields across the Midwest never came to maturity due to the scarcity of water. Should America prove unable to ameliorate the effects of drought, the resulting dietary deficit's impact on rising nutritional costs may well prove incalculable.

The effects of competing for the same energy resources are now beginning to be understood globally. On July 2, 2012, the Wall Street Journal reported that India is now in a large scale energy crisis. Four million rural residents are without power, a tragedy that continues to be compounded by the monthly increases in the number of road vehicles at 1.5 million. The Journal has ascribed this emergency to the fact that India is competing for the identical energy sources with China, Japan, and South Korea. Given this, it should come as no surprise that India reported its lowest growth rate in ten years at 5.3 percent, approximately 40 percent of the country's goal.

The problem is clearly delineated. Mature and emerging economies, in order to meet the demands of growing world-wide populations, must have access to sustainable alternate fuels. Hence, there is a need to identify high biomass plant based alternatives that either produce cellulose which can easily and inexpensively be processed into an alcohol, or, alternatively, seeds synthesizing large amounts of oil that, in turn, can be cheaply converted into biodiesel. The identification of these plants should be a data driven global priority with a set of agreed upon criteria and goals for investment. With such information it becomes more likely that candidate plants will evolve effectively into producers. Ideally, plants so chosen should be able to grow on marginal land or land not dedicated to food production, while at the same reducing the carbon footprint. Moreover as custodians of this planet, the obligation is ours to model potential harm as global soil/water ecosystems are altered to meet the demands of a changing world.

This series, "The Compendium of Bioenergy Crops" was conceived in order to evaluate a number of candidate plants and assess their potential. In addition, a successfully indentified plant energy source requires that the input energy needed to generate the fuel is significantly exceeded by the energy harvested, while heeding social, economic, and political considerations. A caveat needs insertion here. It is common place that investigators and politicians advance THEIR research organism as the one to be developed as an alternate energy source. It is likely that once data is properly vetted, combinations of organisms may prove to be the richest source of energy recovery, an idea not much discussed.

This said, one such plant being considered is maize and our evaluation proceeds from two perspectives. First, can corn be improved to where its biomass is significantly increased, its cellulose-lignin complex more amenable to harvesting and processing and grown in regions not normally associated with its cultivation? Do the breeding protocols and molecular techniques exist to modify this cereal to satisfy increased energy needs? For example, it may not prove possible to alter development and growth habits to the point of profit, but nevertheless the existing molecular tools and value added cloned genes may prove critical to altering other cereals that lack these necessary features. This maize volume looks at corn as a potential bioenergy plant, in addition to its genetic tool box capacity and ramifications on improving additional candidate plants.

With these considerations in mind, a brief review of the contents of this book is in order. In Chapter 1, Professor Solomon and his co-workers give an extensive review of maize, providing information as to why its seed figured so prominently among first generation biofuels. In order to set the tone of this volume, basic information is provided on corn biology and its domestication in addition to its current uses and the genetic tool box available to modify the organism. Indeed a strong case is made that with improvements in technology ranging from corn planting and production along with enhancements in ethanol refining, Net Energy Values (NEV) have increased to where maize may be considered a source of alternate fuel.

This view stands in stark contrast to that embraced in Chapter 11. Professor Pimentel in a powerful polemic makes the case that worldwide starvation has reached epic proportions. Furthermore, he argues that regardless of the biofuel under consideration, the cost of making it exceeds the value of the fuel so manufactured. This is independent of whether the fuel source is either cellulose or oil. The dedication of plants to fuel production becomes more dangerous, in his view, as it has led to poor land management practices which will subsequently do great harm to the soil ecosystem. Clearly the contrasting views expressed on Chapters 1 and 11 need resolution. Indeed, any conclusions that are generated must agree first on a set of principles and premises and, hopefully, it will be the publication of this book that drives this discussion.

In Chapter 2, Professor Bruneky and his coworkers focus on the target energy molecules associated with feed stocks like corn, namely lignocelluloses and their subsequent modification allowing for their entry into energy production. Hence the huge body of literature relative to cell wall development, its particular molecular linkages and its association with lignin is reviewed in detail. Of especial importance, considerable attention is given to the latest developments associated with enzymology of cell wall digestion.

This is critical to the processing of not only the corn seed itself during wet and dry milling but also to the stover in addition to other sources of lignocelluloses. In Chapter 3, Professor Cardonna's group describes the necessary requirements for an effective agricultural/industrial infrastructure. In their analysis, every aspect of the price point is considered, beginning with the seed and soil tillage, and its available methodologies, concluding with a detailed discussion of maize product production using either dry or wet milling. A most interesting aspect of this chapter is the simulation studies devoted to the simultaneous recovery of ethanol and biodiesel.

In Chapter 4, Dr. Tyler and her associates focus on the breeding for bioenergy traits and their genetics, not only on identifying the traits themselves but also on their optimization. This contribution is broad in its delivery. It looks at starch degradation, attempts to increase the molecules production, potential modification of the photosynthetic apparatus and also considers genetic modification of flower development that may result in biomass production. This chapter is valuable in that the maize plant is viewed from the perspective of Bracypodium experts and concludes with a review of the new candidate plant and its likely impact on energy recovery.

In Chapter 5, Drs. Muszynski and Yandeau-Nelson focus on the molecular genetics of bio-energy and bio-energy complex traits. The available maize genetic toolbox is explored with the express purpose of identifying and subsequently cloning those genes impacting the biosynthesis of key biofuel molecules. The identification of these genes and their subsequent modification may lead to enhanced production/purification of lignin biocellulose in addition to the development of more efficient processing technologies. These changes are made possible by the richness of the maize toolbox, including forward and reverse genetics, in addition to targeted and transposon mutagenesis as well as RNAi technologies. Notably this review is in sufficient depth to assist scientists contemplating using maize in their research.

Chapter 6 written by Mr. Brandon Jeffrey and Dr. Thomas Lubberstedt coordinates well with the preceding contribution. Here the focus is on molecular breeding. Particular attention is paid to both mapping populations and methods that lend themselves to the detection of bioenergy traits, the ultimate objective being the use of map based cloning to discriminate among those genes impacting energy recovery.

Chapter 7 by Professors Barros and Morris explores the genomic resources available to maize workers to exploit corn starch as an effective, viable fuel. Conceptually, functional genomics centers around the relationship between genome and phenotype. This requires not only an understanding of a particular gene and its variations with respect to natural alternatives but, in addition, how and in what tissues and under what circumstances these are expressed. A particularly useful resource associated with this contribution is an inclusion of a table giving easy access to Internet-based genomic, transcriptomic, proteomic, and metabolomic reference information.

Chapter 8 written by Dr. Brad Barbazuk and Wenbin Mei describes in detail the tremendous sequencing initiatives that have taken flight since the sequencing of the maize B 73. This whole genome sequencing effort has now opened the possibility of exploring in depth gene regulation, epigenetics, genome structure and evolution. Each of these disciplines either alone, or more powerfully in concert, opens up the option for increased agronomic improvement. Such a research effort holds at its core the possibility of adapting second and third generation candidate biofuel plants to ecosystems to which they are not currently tailored, while at the same time sustaining the integrity of the soil ecosystem.

Professors Msangi and Ortiz explore in Chapter 9 the issues related to the impact of increased plant cultivation for biofuel production on food cost, biodiversity, climate change, and habitat. These investigators concede that, in the short term, shifting 40% of the maize seed harvest to biofuel production has and will continue to swell food prices. Interestingly enough these cost gains may adversely affect the utility of corn as a substitute fuel source. This in turn may give rise to a scenario where continuing improvement in farming methodology might increase food availability at reduced charge. With respect to alterations on biodiversity, land use reassignment, and green house emissions, the long term effects remain to be modeled accurately. It is indisputable that shift of plants, indeed particular plants that are dedicated to cellulose harvest because of their biomass, will significantly tax any ecosystem, while the downstream benefit can at best only be guessed.

Professors Kessan and Slating in Chapter 10 focus on the laws and regulations governing sustainable fuel production as well as the patent mechanism that protects new innovation. The Federal Clean Air Act defines what is lawful with respect to air emissions at biofuel refineries as well as in motor vehicles, and further regulates blending of the fuel components. In addition, incentives and mandates are discussed focusing on the Federal Renewable Fuel Standard (RFS2) and the options available for the protection of new and innovative biofuel technology.

Professor Fletcher in Chapter 12 looks at the economics of corn ethanol production in detail and concludes that its challenges are far from abating. The producers only stay in business as long as income generated is capable of paying both fixed and variable costs. In the short term, this is likely to be quite difficult. Incentivizing this industry has diminished with vanishing subsidies especially in this country. Hence the challenge remains for existing ethanol plants to remain open as profitability is substantially reduced. Given the current state of the industry, it will become increasingly more difficult to raise funds that rely on the ability to convert cellulose to fuel, especially if the projected technological drivers remain in a state of infancy.

Finally, in chapter 13 Professor Moore and his associates provide a vision for the future that addresses all of the biological social, economic and political issues that cloud corn ethanol's future. They rightly point out that the industrial embracing of ethanol as a biofuel was driven by subsidies and blender credits. Second and third generation candidate plants have now also been identified. The need to identify high biomass plants for lingo-cellulose degradation has been driven not by scientific concerns but rather by the Environmental Protection Agency. Specifically, the Renewable Fuel Standard mandate anticipated the production of nine billion gallons by 2008 increasing to 39 billion by 2022. These standards were set independently of any real understanding of the drivers shaping this industry. Where do we go from here as a global community? Surely new advances in technology along with their anticipated deliverables must be the benchmarks whereby investment is governed.

Finally, I would like to express my thanks not only to Professor Chita Kole but also to the contributors who shared with me in the preparation of this volume. In September 2011, I was diagnosed with stage 4 lung cancer that greatly impaired my ability to function. From each contributor, I have known nothing but patience and understanding for which I will be ever grateful. I would like to thank Professor Wlfred Vermerris. Although we have never met, his papers, especially the one that appeared in the Journal of Integrative Plant Biology in 2011, greatly contributed to my understanding of the problem and the shaping of this volume. Finally this project would never have come to completion without the technical assistance of Professor Linda Bowyer and Drs. Paul Hirth and Patsy Scott. These three along with Annmarie Heldt, CB Rimmelin, Bruce and his judge along with the remainder of my much extended family produced an environment where not only could I work but work well. And then too, there is the Lubavitcher Rebe of Holy Memory who taught that when you think good, it is good.

References

Solomon BD, Barnes JR and Halvorsen KE (2009) From grain to cellulosic ethanol: History, economics and policy. In: BD Solomon and VA Luzadis (eds) Renewable Energy From Forest Resources in the United States. Routledge, London, pp 49–66.
Vermerris W (2011) Survey of genomics approaches to improve bioenergy traits in maize,sorghum and sugarcane. J Intergr Plant Biol 53: 105–119.

Stephen L. Goldman
Chittaranjan Kole

Contents

List of Contributors

W. Brad Barbazuk
Department of Biology, and the University of Florida Genetics Institute
The University of Florida, Cancer & Genetics Research Complex, 2033
Mowry Road, PO Box 103610, Gainesville, FL 32610, USA.
Email: bbarbazuk@ufl.edu
Phone: (352) 273-8624

Eugenia Barros
CSIR Biosciences, Meiring Naude Road, Brummeria, Pretoria 0001, South
Africa.
Email: ebarros@csir.co.za
Phone: +27 12 841 3221

James Birchler
University of Missouri, 1-31 Agriculture Building, Columbia, MO 65211,
USA.
Email: birchlerj@Missouri.edu
Phone: (573) 882-4905

Roman Brunecky
National Renewable Energy Laboratory, 15013 Denver West Parkway
Golden, CO 80401, USA.
Email: roman.brunecky@nrel.gov
Phone: (303) 384-6878

Michael N. Burgess
Tower Road East, Blue Old Insectary, room 161, College of Agriculture and
Life Sciences, Cornell University, Ithaca, NY 14853.
Phone: (607) 255-2212

Carlos A. Cardona
Instituto de Biotecnología y Agroindustria, Departamento de Ingeniería
Química, Universidad Nacional de Colombia sede Manizales, Cra. 27
No. 64-60, Manizales, Colombia.
Email address: cardonaal@unal.edu.co
Phone: +57 6 8879300x50199

George Chuck
Plant Gene Expression Center, US Department of Agriculture (USDA),
University of California—Berkeley, Albany, CA 94710, USA.
georgechuck@berkeley.edu
Phone: (510) 559-5710

Bryon S. Donohoe
National Renewable Energy Laboratory, 15013 Denver West Parkway
Golden, CO 80401, USA.
Email: bryon.donohoe@nrel.gov
Phone: (303) 384-7773

Lehman B. Fletcher
Iowa State University, Ames, Iowa USA.
Email: lbf@iastate.edu
Phone: (515) 294-4515

Stephen L. Goldman
University of Toledo, Department of Environmental Sciences, Toledo,
Ohio 43606, USA.
Email: stephen.goldman@utoledo.edu
Phone: (419) 277-0290

Michael E. Himmel
National Renewable Energy Laboratory, 15013 Denver West Parkway
Golden, CO 80401, USA.
Email: mike.himmel@nrel.gov
Phone: (303) 384-7756

Brandon Jeffrey
Iowa State University, Department of Agronomy, Ames, Iowa 50011,
USA.
Email: bjeffrey@iastate.edu
Phone: (515) 294-1360

Douglas L. Karlen
USDA-ARS-NLAE, 2110 University Blvd., Ames, IA, 50011, USA.
Email: doug.karlen@ars.usda.gov
Phone: (515) 294-3336

Jay P. Kesan
Professor and Workman Research Scholar, University of Illinois College
of Law, Program Leader, Biofuels Law & Regulation Program, University
of Illinois Energy Biosciences Institute, Champaign, Illinois 61801, USA.
Email: kesan@illinois.edu

Phone: (217) 333-7887

Kendall R. Lamkey
Department of Agronomy, Iowa State University, Ames, IA 50011, USA.
Email: krlamkey@iastate.edu
Phone: (515) 294-7826

Thomas Lübberstedt
Iowa State University, Department of Agronomy, IA 50011, USA.
Email: thomasl@iastate.edu
Phone: (515) 294-5356

Wenbin Mei
Department of Biology, The University of Florida, Cancer & Genetics
Research Complex, 2033 Mowry Road, PO Box 103610, Gainesville, FL
32610, USA.
Email: wmei@ufl.edu
Phone: (352)392-1175

Kenneth J. Moore
Agronomy Department, 1571 Agronomy Hall, Iowa State University,
Ames, IA 50011, USA.
Email: kjmoore@iastate.edu
Phone: (515) 294-5482

E. Jane Morris
African Centre for Gene Technologies, PO Box 75011, Lynnwood Ridge
0040, South Africa and Department of Biochemistry, University of
Pretoria, South Africa.
Email: jmorris@csir.co.za
Phone: 12-420-6007

Siwa Msangi
International Food Policy Research Institute (IFPRI), 2033 K Street NW,
Washington DC 20006, USA.
Email: s.msangi@cgiar.org
Phone: (202) 862-5600

Michael G. Muszynski
Department of Genetics, Development and Cell Biology, Iowa State
University, Ames, IA 50011, USA.
Email: mgmusyn@iastate.edu

Phone: (515) 294-2496

Carlos E. Orrego
Instituto de Biotecnología y Agroindustria, Departamento de Ingeniería
Química, Universidad Nacional de Colombia sede Manizales, Cra. 27
No. 64-60, Manizales, Colombia.
Email:ppba_man@unal.edu.co
Phone: 01 8000 916956

Rodomiro Ortiz
Swedish University of Agricultural Sciences (SLU), Dept. Plant Breeding
and Biotechnology, Sundsvagen 14, Box 101, Alnarp, SE 23053, Sweden.
Email: rodomiro.ortiz@slu.se
Phone: +018-67 10 00

David Pimentel
Tower Road East, Blue Old Insectary, room 165, College of Agriculture
and Life Sciences, Cornell University, Ithaca, NY 14853, USA.
Email: dp18@cornell.edu
Phone: (607) 255-2212

Michael Resch
National Renewable Energy Laboratory, 15013 Denver West Parkway
Golden, CO 80401, USA.
Email: mchael.resch@nrel.gov
Phone: (303) 384-7854

Luis E. Rincón
Instituto de Biotecnología y Agroindustria, Departamento de Ingeniería
Química, Universidad Nacional de Colombia sede Manizales, Cra. 27
No. 64-60, Manizales, Colombia.

Michael J. Selig
Department of Forest & Landscape, University of Copenhagen,
Rolighedsvej 23, 1958 Frederiksberg C, Denmark.
Email: michaeljselig@yahoo.com

Timothy A. Slating
Regulatory Associate, Biofuels Law & Regulation Project, University of
Illinois Energy Biosciences Institute, Champaign, IL 61820, USA.
Email: slating2@illinois.edu.
Phone: (217) 333-6178

Barry D. Solomon
Michigan Technological University, 1400 Townsend Drive, Houghton, MI
49931, USA.
Email: bdsolomo@mtu.edu

Phone: (906) 487-1791

Ludmila Tyler
Department of Biochemistry and Molecular Biology, University of
Massachusetts–Amherst, Amherst, MA 01003, USA.
Email: ltyler@biochem.umass.edu
Phone: (413) 545-4026

Hui Wei
National Renewable Energy Laboratory, 15013 Denver West Parkway
Golden, CO 80401, USA.
Email: hui.wei@nrel.gov
Phone: (303) 384-6620

Marna D. Yandeau-Nelson
Department of Biochemistry, Biophysics and Molecular Biology, Iowa
State University, Ames, IA 50011, USA.
Email: myn@iastate.edu
Phone: (515) 294-6116

Hugh Young
Department of Plant and Microbial Biology, University of California–
Berkeley, Berkeley, CA 94720, USA.
Email: hugh.young@ars.usda.gov
Phone: (510) 559-5600

Qiong Zhang
University of South Florida, 4202 E. Fowler Avenue, Tampa, FL 33620-
5350, USA.
Email: qiongzhang@usf.edu
Phone: (813) 974-6448

List of Abbreviations

Ac	Activator
ADM	Archer Daniels Midlands
ADP	Adenosine diphosphate
AFEX	Ammonia fiber expansion
AGPase	ADP-glucose pyrophosphorylase
ATP	Adenosine triphosphate
AIL	Advanced intercross lines
AFLP	Amplified fragment length polymorphism
avr	avirulence
AZM	Assembled maize DNA sequences
BAC	Bacterial artificial chromosome
BC	Backcross population
BR	Brassinosteroid
Bt	*Bacillus thurigenesis*
BT2	Billion ton update
CAA	Clean Air Act
Cg1	*Corngrass1*
CBP	Consolidated bio-processing
CeSA	Cellulose synthaseA
CCR	Cinnamoyl-CoA reductase
CGH	Competitive genome hybridization
CIM	Composite interval mapping
CNV	Copy number variation
CODDLe	Codons optimized to detect deleterious lesions
COMT	Caffeic acid *O*-methyltransferase (*Bm3*)
$C_o t$	index of DNA reassociation kinetics
CroPS	Complexity reduction of polymorphic sequences
CSL	Cellulose synthase like protein
DHL	Doubled haploid line
DDGS	Distillers dried grains with solubles
Ds	*Dissociation*
DSB	Double strand DNA breaks
EISA	Energy Independence and Security Act

EMS	Ehyl methanesuilfonate
EPA	(United States) Environmental Protection Agency
E10	Ninety percent gasoline/10% ethanol blend
E15	Eighty five percent gasoline/15% ethanol blend
E85	Eighty five percent ethanol/15% gasoline
EMC	Equiliibrium moisture content
EST	Expression sequence tag
ETBE	Ethyl tertiary butyl ether
ET	Evapotranspiration
FAO	Food and Agriculture Organization
FISH	Fluorescence *in situ* hybridization
FST	Flanking sequence tag
GA	Gibberellic acid
GC-TOF M	GC–time of flight mass spectrometry
GDP	Gross domestic product
GEO	Global environmental outlook
GH	Glycosyl hydrolase
GHG	Greenhouse gas
GLS	Gray leaf spot
GST	Gene-specific tag
GT	Glycosyl transferase
GUS	β-glucuronidase
GWS	Genome wide selection
HAP	Hydroxyapitite chromatography
HapMap	Haplotype map
HFCS	High fructose corn syrup
HR	Hypersensitive response
IBM	Intermated B73xMo17
iLUC	Indirect land use change
IP	Intellectual property
IPPC	Intergovernmental Panel on Climate Change
iTRAC	Isobaric tagging
JGI/DOE	Joint Genome Initiative/Dept. of Energy
LCA	Life cycle analysis
LCFS	Low Carbon Fuel Standard
MAGI	Maize assembled genomic island
MAS	Marker-assisted selection
MDR	Multiple disease resistance
MMS	Methyl methane sulfonate
MTBE	Methyl tertiary butyl ether
MEA	Millenium Ecosystem Assessment
MIM	Multiple interval mapping
MTM	Maize targetted mutagenesis

Mu	*Mutator*
NAM	Nested mapping association
NAD-ME	NAD-dependent malic enzyme
NCBI	National Center for Biotechnolgy Information (NIH)
NCGA	National Corn Growers Association
NER	Net energy ration
NEV	Net energy value
NF-Y	Nuclear factor Y
NGS	Next-generation sequencing
NHEJ	Non-homologous end joining
NLB	Northern leaf blight
NPGI	National Plant Genome Initiative
NSF	National Science Foundation
NUE	Nitrogen use efficiency
OPEC	Organization of the Petroleum Exporting Countries
PacBio	Pacific Biosciences
PCT	Patent Cooperation Treaty
PEPC	Phosphoenolpyruvate carboxylase
PEP-CK	Phosphoenolpyruvate carboxykinase
PPDK	Pyruvate Pi dikinase
PRR	PSEUDORESPONSE REGULATORS
PTO	Patent and Trademark Office
PVPA	Plant Variety Protection Act
QTL	Quantitative trait loci
RFLP	Restriction fragment length poylmorphism
R gene	Resistance gene
RFS	Renewable Fuel Standard
RIL	Recombinant inbred line
RIN	Renewable identification number
RNAi	RNA interference
RSSC	Repeat substraction-mediated sequence capture
Rubisco	Ribulose 1,5-bisphosphate carboxylase oxygenase
RuBP	Ribulose 1,5-bisphosphat
SAM	Shoot apical meristem
SBD	Starch binding domain
sex	Starch excess
SIM	Simple interval mapping
SLB	Southern leaf blight
siRNA	Short interferring RNA
SMAF	Soil management assessment framework
SNP	Single nucleotide polymorphism
SSH	Suppression subtractive hybridization
SSR	Simple sequence repeat

SV	Structural variation
TALEN	Transcription activator-like effector nuclease
TAIL	Thermail asymmetric interlaced
T-DNA	Transfer DNA
TE	Transposable element
TUSC	Trait utility system for corn
2-DE	Two dimensional electrophoresis
UN	United Nations
UNEP	United Nations Environment Programme
UNICEF	United Nations Childrens Fund
UPOV	International Union for the Protection of New Varieties of Plants
VEETC	Volumetric Ethanol Excise Tax Credit
VLCFA	Very long-chain fatty acid
VOC	Volatle organic compounds
WEPS	Wind erosion protection system
WEQ	Wind Erosion Equation
WHO	World Health Organization
WUE	Water use efficiency
XTHs	Xyloglucan endotransglucosylases/hydrolases
ZF	Zinc finger
ZFN	Zinc finge nuclease
ZFP	Zinc finger protein
ZmCAD2	Cinnamyl dehydrogenase2 (*Bm1*)
ZMW	Zero-mode wave guide

CHAPTER 1

Basic Information on Maize

Barry D. Solomon,[1,*] *James Birchler,*[2]
Stephen L. Goldman[3] *and Qiong Zhang*[4]

ABSTRACT

This chapter begins by reviewing the long history of maize as a biofuel crop, with a focus on ethanol production and use in the United States. It then turns to its center of origin in Mexico and domestication in the Western Hemisphere. The next sections focus on its botanical descriptions, and economic importance both domestically and internationally. The following sections describe the use of maize as a model plant for bioenergy research, and studies on its suitability as a bioenergy plant from a variety of perspectives. Finally, the chapter ends with brief conclusions.

Keywords: biofuel, ethanol, gene, genome, greenhouse gas, net energy value

Introduction

Maize (corn) has had a long history in the United States (US) as a bioenergy crop, particularly as source of ethanol. With economic growth increasing rapidly in many developing countries, the question as to whether maize is a legitimate source to meet ever-increasing energy demands requires detailed examination. The ability to manipulate the genetics and molecular biology

[1]Michigan Technological University, 1400 Townsend Drive, Houghton, MI 49931-1295, USA.
[2]University of Missouri, 1-31 Agriculture Building, Columbia, MO 65211, USA.
[3]University of Toledo, Department of Environmental Sciences, Toledo, OH 43606 USA.
[4]University of South Florida, 4202 E. Fowler Avenue, Tampa, FL 33620-5350, USA.
*Corresponding author: bdsolomo@mtu.edu

of maize has increased to a level of sophistication that could not have been anticipated when Watson and Crick first postulated the double helix in 1953. This observation is counterbalanced by the established roles that maize has long played in the food and feed chains, to say nothing of the numerous molecules that are harvested for industrial products. Anything that alters the availability of a natural resource by shifting it away from traditional applications in the supply chain will ultimately affect the comfort and safety of significant numbers of people. Today this change is already being felt in the US, where almost four out of ten plants are being used to extract ethanol from corn starch. This volume is dedicated to the evaluation of maize as a candidate biofuel plant. Thus, the chapter provides background for understanding the basic biology of maize, its historical role in agriculture and industry as well as for biofuels, its social and economic functions and ultimately a context through which the remaining contributions may be viewed.

History as a Biofuel Crop

It has been well established that biofuels, especially the use of ethyl alcohol (ethanol) as an energy source has a long and rich history. For example, the first internal combustion engine invented in the US by Samuel Morey in 1826 could run on ethanol and turpentine; more than 13 million gallons of alcohol were burned for lighting in the US in 1860; and several countries in Europe passed legislation in the late 19th Century to encourage the use of industrial alcohol (Herrick 1907; Cummins 2002; Carolan 2009). By the early 20th Century, both England and especially Germany were starting to rely on alcohol to power internal combustion engines (Carolan 2009). The cost of alcohol production, however, stymied its ability to play a large role in the US energy system at the time. For example, the Internal Revenue Act of 1861 instituted a $2.08/gallon excise tax on industrial alcohol to help pay for the Civil War, the same rate as on alcohol for human consumption, substantially increasing its price (Dimitri and Effland 2007).

In the late 1890s, Henry Ford, Nicholas Otto and other automotive pioneers built car engines that could run on pure ethanol. But even after the tax on industrial alcohol was repealed in 1906, when maize (corn) ethanol was thought to be cheaper than gasoline, the use of ethanol and alcohol fuel blends in the Model T and other vehicles was unable to effectively compete (Baskerville 1906; Kovarik 1998). A key reason for this was that to avoid the alcohol 'sin tax', it had to be denatured by mixing 90% ethanol with 10% methyl (wood) alcohol, which was much more costly. In the early 1900s, maize was only one of several feedstocks used to make ethyl alcohol, along with other sources including other grains, potatoes and sugarbeets.

Despite the dominance of petroleum as an automobile fuel in the early 1900s, substantial interest remained in the use of ethanol as an alternative and eventual replacement fuel. This was because there was serious (though erroneous) concern over an impending shortage and depletion of domestic petroleum resources, which lasted until 1924 when vast new oil reserves were discovered in Texas, Louisiana and California (Yergin 1991). Still, ethanol provided other advantages. It was seen as cleaner burning than oil, had higher octane value and superior antiknock properties, promoted rural agricultural development, and was fully domestic (Kovarik 1998). However, only a small portion of US oil supplies were imported in the early 20th Century and interest in alternative fuels eventually waned, not to be revived until the 1970s. Support for industrial alcohol fuel also suffered during the period of National Prohibition, from 1919 to 1933 (Bernton et al. 2010). In addition, low compression engine automobiles were designed to best run on gasoline, and the oil industry increasingly opposed ethanol throughout the 20th Century (Bernton et al. 2010). Finally, oil prices stayed low and exceeded $30 (real) per barrel only once from 1877 until the 1973–1974 oil embargo (Carolan 2009; EIA 2011a).

Thus, in the first three-quarters of the 20th Century ethanol made from maize was unable to compete effectively with cheap gasoline and the emergence of tetraethyl lead in the 1920s for antiknocking in engines. Exceptions occurred during wartime, however, when resources became scarce and alternatives were needed. For example, during the World War I years of 1917–1919 because of gasoline rationing the need for fuel increased ethanol production to 60 million gallons/year (Songstad et al. 2009). US demand for industrial alcohol spiked even more during World War II, peaking at nearly 600 million gallons in 1944 (Ryder 1944). Alcohol fuel blends had already experienced a brief resurgence in the mid to late 1930s, as falling maize prices encouraged Midwestern farmers to find alternative uses for their crops (Solomon et al. 2007). At this time, alcohol-gasoline blends ranging from 5–17.5% were marketed under trademarked names such as Agrol and Alcolene. Agrol was sold in a 10% blend in around 2,000 retails outlets in eight states from Indiana to South Dakota during 1938, though the Agrol Company suspended operations at its Atchison, Kansas distillery by 1939. At that time, however, fuel alcohol was being produced in 40 nations (Bernton et al. 2010). Despite the inflated demand for alcohol fuels in the US during World War II, afterwards interest in ethanol again waned since leaded gasoline was cheaper and easier to produce while new oil discoveries reduced the urgency of finding petroleum substitutes (Kovarik 1998).

The 1973–1974 embargo by the Arab oil producers of the Organization of the Petroleum Exporting Countries (OPEC) against the United States and later the Netherlands stimulated a revival in the production of fuel alcohol (among other energy alternatives). Fuel alcohol in the US was based on maize while in Brazil sugarcane was used as feedstock. Initially the US farming sector concentrated on production in small-scale distilleries to produce ethanol, though after 1978 the industrial giant Archer Daniels Midland (ADM) came to dominate the industry for the following two decades or so (Federal Trade Commission 2006).

Until the 1980s, US corn ethanol production came exclusively from hydrous (wet grind) mills. As the industry has grown to over 200 refineries more recently most production has shifted to anhydrous (dry grind) mills, as they have become cheaper and more efficient (Solomon et al. 2007). Wet mills require the separation of the grain kernel into its component parts (germ, protein, fiber, starch) before fermentation. This is typically done with a dilute combination of sulfuric acid and water in order to create a slurry. In dry mills, the entire grain kernel is ground up in a hammer mill into a coarse, flour-like consistency ("meal") before it is cooked in a water tank, with enzymes added to produce the resulting slurry mash. Different coproducts are produced from wet and dry milling (see Fig. 1).

The first dry mill to make ethanol was a demonstration plant established by Dr. Paul Middaugh at the South Dakota State University in 1979 (Songstad et al. 2009). The first large-scale dry mill plant was opened by New Energy Corp. in South Bend, Indiana in 1985 (current capacity is 102 million gallons/year, though the plant was idled and the company filed for bankruptcy in November 2012). Not all ethanol plants are large, however, as family farm cooperatives have grown to account for 30–40% of the total commercial US corn ethanol production. At the same time, the ADM share of national output has decreased from a peak of 75% in 1990 down to 13% today, while it has gained large competitors in POET at 12% and Valero Energy at 7% (RFA 2012). Another large company, VeraSun Energy, was for a few years the largest ethanol producer in the US until it went bankrupt in 2008 and sold off its plants in 2009 including some to Valero Energy (Songstad et al. 2009).

The first large support for the modern corn ethanol industry came in the form of a 40 cents/gallon exemption from the excise tax on gasoline for ethanol produced for ethanol/gasoline fuel blends. This provision was included as part of the Energy Tax Act of 1978, part of the broad, five-part National Energy Act passed that year by Congress (Tyner 2008). Later legislation changed this exemption into a Volumetric Ethanol Excise Tax Credit (VEETC). The tax exemption/credit has risen and fallen over time, peaking at 60 cents/gallon for gasoline blenders in the late 1980s, and was eventually allowed to expire at the end of 2011 (Woodyard 2011). Other tax, loan, and other financial support for ethanol production have been approved at both the federal and state levels, including the Small Ethanol Producer Tax Credit of 1990, which has provided small plants (< 30 million gallons/

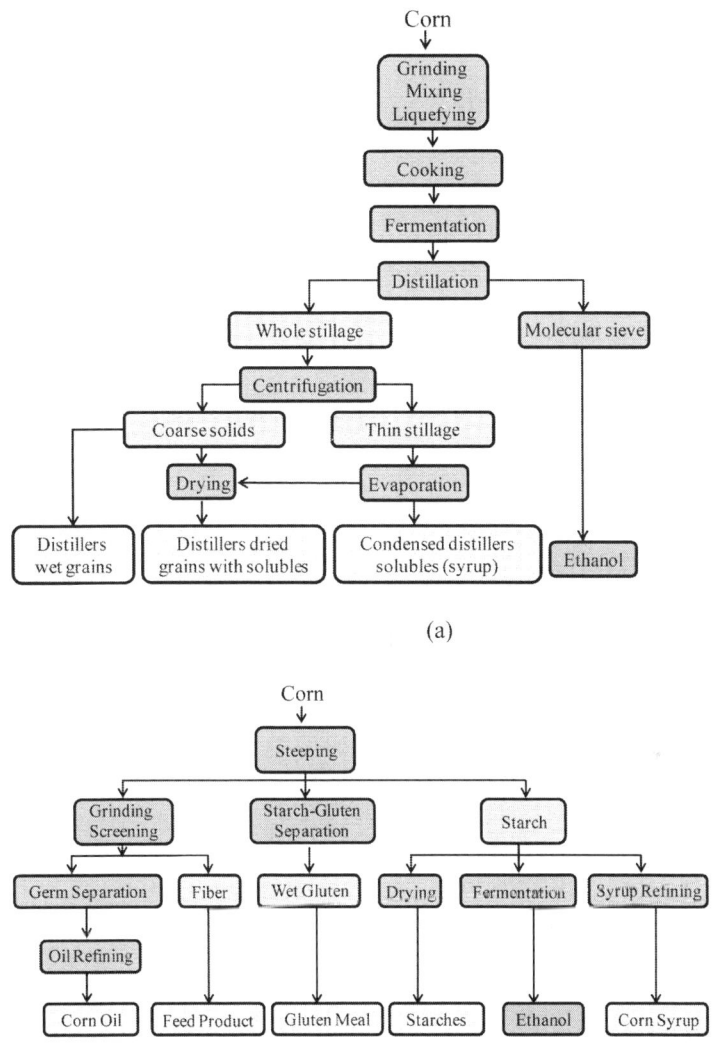

(a)

(b)

Figure 1 Schematic of processes for maize-based ethanol production. (a) dry milling process; (b) wet milling process.

year production capacity) with an additional $0.10 gallon income tax credit for volumes up to 15 million gallons/year (Solomon et al. 2007).

Despite the large government subsidies the corn ethanol industry stagnated in the mid to late 1980s. This occurred when oil and corn prices plummeted, devastating the farm sector. A variety of factors helped to re-stimulate demand for corn ethanol at this time. For example, when the US Environmental Protection Agency (EPA) mandated a phase-out of leaded gasoline by the late 1980s, interest increased in using ethanol as an octane booster and volume extender. However, methyl tertiary butyl ether (MTBE) dominated most oxygenated gasoline markets over ethyl tertiary butyl ether (ETBE) throughout the 1990s. MTBE in fact became the fuel additive of choice under the requirements for 2.0–2.7% oxygen blends in ozone and carbon monoxide nonattainment areas under the Clean Air Act Amendments of 1990, with the exception of the Midwestern maize states that chose ETBE. Corn ethanol production was further stimulated when MTBE bans in California, New York and Connecticut because of groundwater and soil contamination, states that had accounted for over 40% of national MTBE consumption, took effect on January 1, 2004. As a consequence, significant and steady growth in corn ethanol production has occurred since 1996 (Tyner 2008).

The policy makers decided that the existing corn ethanol support mechanisms were inadequate, and enacted a Renewable Fuel Standard (RFS 1) under the Energy Policy Act of 2005. The RFS 1 mandated that 7.5 billion gallons of ethanol would be produced by 2012, a figure that was actually surpassed in 2008 (RFA 2012). The RFS 1 was then replaced with RFS2 under the Energy Independence and Security Act of 2007 (EISA), which requires that corn ethanol production grow to 15 billion gallons a year by 2015 and then be capped at that level, while total renewable fuels production grow to an ambitious 36 billion gallons by 2022, mainly through the use of cellulosic biofuels (Solomon et al. 2009).

Thus, in reflection of broader social debates Congress has decided that there needs to be a limit to the growth in the use of maize as a biofuel crop. The next section of this chapter will consider the center of origin of maize production, its botanical origin and evolution, domestication and dissemination. This will be followed by detailed botanical descriptions of maize. The next section will review the economic importance of the maize crop, its production level and yield, use as a biofuel, as well as use in other markets. We will then consider why maize has been so prominent as a model plant for bioenergy research, along with its drawbacks and limitations. The penultimate section will consider the overall suitability of maize as feedstock for biofuel production from environmental, energy, climate change, and societal perspectives. Finally, some brief conclusions will close the chapter.

Center of Origin and Domestication

There is now considerable evidence that maize (*Zea mays* L., subspecies *mays*) was domesticated from a weedy grass in the Balsas river valley in Mexico (Doebley 2004). This species of teosinte (*Zea mays*, subspecies *parviglumis*), as the commonly known progenitor species, appears to be the major contributor to the genetic constitution of maize although recent evidence shows that another teosinte species (*Zea mays* subspecies *mexicana*) has genetic material that has been introgressed into what is present day maize (van Heerwaarden et al. 2011). During the process of domestication, there are six major genes and multiple minor genes that changed the morphology from the progenitor species (Doebley 2004). One of these has been identified on the single gene level and involves alleles at the *teosinte branched 1* locus. This transcription factor shows complex differences in expression between the progenitor and domesticated maize and affects the level of branching. It acts in a negative fashion such that in maize the branching pattern is suppressed. In teosinte the side branches are extended and end in a tassel. These are greatly reduced in maize and terminate with an ear. Thus, maize has a single stalk for the most part but some varieties have numerous tillers.

In addition to a reduction in overall branching pattern, the other features of morphology that have changed with domestication include a reduction in the fruitcase that surrounds the teosinte kernel (Doebley 2004; Wang et al. 2005). In maize, these form the cob and thus the kernels are exposed. The change involved with the difference in the fruitcase is controlled in large part by the transcriptional regulator, *teosinte glume architecture 1* (Wang et al. 2005). Also with regard to the ear, the development of the spikelets is double in maize compared to teosinte. In the latter, two are initially formed in development but one aborts. This abortion does not occur in maize producing ears with an even number of kernel rows. Also, in teosinte there are two ranks of kernels in the ears. In maize the number of ranks has been increased depending on the variety. Lastly, in teosinte the ear will shatter easily at maturity to facilitate dispersion but in maize the kernels are held together on the ear.

From the humble beginnings in Mexico, maize spread through much of the Americas before contact with Europeans. Further selection for local adaptation occurred as well as for other desired characteristics. Following contact, European explorers carried maize to Europe and eventually it spread worldwide.

Botanical Descriptions

Maize (*Zea mays* L., subspecies *mays*) is within the tribe Andropogoneae, the subfamily Panicoideae and the grass family Poaceae that reside among the monocots and the angiosperms within the plant kingdom (Doebley and Iltis 1980; Kellogg 1998). The species is divided into subspecies *Z. mays, Z. parviglumis, Z. mexicana,* and *Z. huehuetenangensis* of which the first two wild species have contributed to domesticated maize. Other species in the genus include *Z. nicaraguensis, Z. luxurians, Z. perennis* and *Z. diploperennis.*

These other species can be crossed with domesticated maize with relative ease with the appropriate coordination of flowering time but the hybrids show greater sterility than those produced with *Z. parviglumis* and *Z. mexicana*. *Z. perennis* is a tetraploid species and the *Z. diploperennis* species is a highly related diploid form. These two species, as the name implies, have a perennial habit in contrast to the other species in the genus that are annuals.

A related genus of note is *Tripsacum*. There are several species in this genus. They have greater basic chromosome numbers than the genus *Zea*. The species within the genus that has been studied most thoroughly is gama grass, *Tripsacum dactyloides*, of which the diploid form has 18 as a basic chromosome number. Crosses of maize with pollen from *Tripsacum* can succeed in producing viable kernels under the right circumstances but are not routine (deWet et al. 1973).

The maize plant is monoecious in that it has separate male flowers and female flowers on individual plants (see Kiesselbach 1980 for a description of maize anatomy and morphology). The tassel at the top of the plant is the male flower. It contains the anthers in which meiosis occurs before the tassel emerges from the whorl of leaves. Meiosis produces four products in a tetrad (Rhoades 1950). Each of these cells becomes a microspore. Each microspore undergoes two mitoses to produce the male gametophyte or pollen grain. The first division produces the generative and vegetative nuclei. The vegetative nucleus does not divide further and produces the gene products necessary for the remainder of pollen grain development and eventual growth. The generative nucleus divides a second time to produce two crescent shaped sperm that are present in mature pollen grains. When the tassel is mature, the glumes spread open to allow the anthers to dehisce and release the pollen.

The female flowers are the ears. In the young ear shoots, female meiosis occurs to produce megaspores. Usually, the first division of meiosis produces two cells of which the basal one divides again to produce two cells. The basal most megaspore develops into the female gametophyte while the other three products of meiosis degenerate. In the female gametophyte, further mitoses occur to produce an egg cell accompanied by two synergids, two polar nuclei and a variable number of antipodals.

Upon pollination of the silks, the grains germinate and grow the length of the silk to approach and enter the ovule. At entry of one of the synergids, the two sperm are released. One joins with the egg cell to form the embryo and the other sperm joins with the two polar nuclei to form the triploid endosperm. The endosperm encases the growing embryo and persists at kernel maturation with a store of starch and proteins. The storage molecules are broken down and used by the germinating embryo before the leaves emerge from the soil and photosynthesis begins to supply sugars for further growth.

The genome of maize is approximately 2.3 gigabases (Schnable et al. 2009) but varies to some degree by great variability for the heterochromatic knobs that are present at various locations in the chromosomes (Rayburn et al. 1989). The DNA sequence of the inbred line B73 has been determined (Schnable et al. 2009). This genome size is contributed in large part to a proliferation of retrotransposons during the past three million years. They comprise about 75% of the genome sequence. The DNA based transposons make up about another 9% of the total genome size. Another type of transposable element, Helitrons, is represented by several families in maize that together have a copy number of about 20,000. The gene repertoire consists of about 32,000 genes.

In terms of genetic diversity, maize as a species has more than any other that has been analyzed to date (Buckler et al. 2009). A wide collection of inbred lines has been characterized for their relationships and selected lines have been identified that capture a substantial fraction of this variation (Liu et al. 2003; Buckler et al. 2009). This subset of lines has been crossed to B73 and recombinant inbred lines (RILs) produced that can be used to identify the extent of variation for a variety of traits. Additional variation occurs in teosinte species because there was a bottleneck during domestication. However, there is so much variation in maize that there has been no attempt to harness this resource and linkage drag of undesired characteristics complicates the use of novel alleles from teosinte.

The karyotype of maize consists of 10 pairs of chromosomes. There is also a supernumerary B chromosome that is basically inert although it does produce effects on the A chromosomes such as an altered distribution of crossovers in meiosis. It also possesses genetic properties that foster its own perpetuation through nondisjunction at the second pollen mitosis and preferential fertilization of the egg as opposed to the polar nuclei by the sperm carrying two B chromosomes (Roman 1947, 1948). The ability to identify each of the ten chromosome pairs in meiosis was established by Barbara McClintock (McClintock 1930). This advance led to many important findings in cytogenetics by allowing genetic markers to be correlated with the behavior of chromosomes. The features of meiotic chromosomes that are useful identifying characteristics include the arm ratios, the distribution of staining intensity along the length of the chromosomes and the presence of deeply staining heterochromatic "knobs" that are located at specific places on the chromosomes but have considerable variability from one line to another.

A technique to identify each of the ten chromosome pairs in somatic cell preparations was developed (Kato et al. 2004). They used tandemly arrayed repeats in fluorescence *in situ* hybridization (FISH) to provide a distinct pattern of fluorescent probes for each chromosome. The probes include the 28S ribosomal repeats on chromosome 6, the 5S ribosomal RNA repeats on chromosome 2, the CentC centromere satellite sequence, two types of knob repeats, a microsatellite (TAG) that has several clusters on the chromosomes and two types of subtelomeric repeats. The CentC and knob repeats are highly variable in their intensity of labeling and, in the case of the knobs, in the locations in the genome. Nevertheless, these markers are successful in distinguishing the ten pairs of chromosomes in the karyotype in every inbred line tested (Albert et al. 2010). In some cases, an initial confirmation of specific chromosomes is made using single gene probes.

The ability to detect individual genes in maize was also developed (Kato et al. 2006; Lamb et al. 2007). Using this technique it is possible to determine the location of any gene that is about 2–2.5 kb in length to chromosome. These can be used to clarify the patterns for the karyotyping sets mentioned above but an appropriate set can also serve to identify the chromosomes directly. In addition, this technique can serve to locate any gene to position on the chromosomes, to determine the number and sites of insertion of transgenes, to examine the behavior of transposable elements (Yu et al. 2007) and to determine the nature of chromosomal aberrations among other uses.

All plants have experienced cycles of polyploidy formation followed by deletion of genes back toward a diploid state (Freeling 2009; Jiao et al. 2011). The most recent tetraploid event in maize occurred about 4–5 million years ago (Swigona et al. 2004). As a consequence there are many duplicate genes in maize that are remnants of this whole genome-doubling event.

If the ten chromosomes in maize are considered the basic chromosome set then the $2n = 20$ state is considered the diploid. Monoploid (or haploid) plants have been recovered that carry only ten chromosomes (Coe 1959; Kermicle 1969). They have a distinctive phenotype that have less stature than the diploid and a characteristic form but develop reasonably well. Maternal haploids can be induced by using stock 6 (Coe 1959) or derivatives as a male parent that increase the frequency of induction. Rates of haploid induction exceeding 10% can be obtained but the frequency depends on the maternal line in many cases. Paternal haploids can be produced using the *indeterminate gametophyte* mutation (Kermicle 1969). Haploid breeding is used to produce homozygous lines with a unique genetic constitution by producing monoploids from crossing inducer lines to hybrids and then doubling the haploids recovered. This procedure bypasses a procedure of extended inbreeding that would otherwise be needed to produce such materials.

Triploid maize with three sets of each chromosome has been recognized (McClintock 1929; Punyasingh 1947). Their morphology is quite similar to diploids but the male and female fertility is highly compromised. They can occur spontaneously but can also be produced by crosses of diploid and tetraploid lines. These interploidy crosses produce highly defective kernels for the most part but those that develop usually contain triploid embryos that can be grown into plants (Brink and Cooper 1947; Cooper 1951).

Tetraploid maize has been produced in a variety of ways (Randolph 1935, 1942; Kato and Birchler 2006). Heat treatment of early developing ears, colchicine treatment, and nitrous oxide treatment at the time of the early embryonic divisions have succeeded in producing tetraploids. Spontaneous cases are also known and the *elongate* mutation, which produces unreduced gametes, has been used (Rhoades and Dempsey 1966). The stature and biomass of tetraploid maize depends on the status of heterozygosity. Strictly homozygous tetraploids are typically shorter than their diploid progenitors (Kato and Birchler 2006; Riddle et al. 2006, 2010; Riddle and Birchler 2008). However, the tassel is larger and floppy and the kernels are larger as well. Cell size correlates with ploidy level (Rhoades and Dempsey 1966; Yao et al. 2011). Hybrid tetraploids are quite robust and double cross tetraploids that are produced from crossing two different types of single cross hybrids produce even more sturdy specimens. Pentaploid, hexaploid and heptaploid plants have all been reported in maize and the trend is a decline in vigor with increasing ploidy (Rhoades and Dempsey 1966). Octoploid maize plants have been reported but are described as highly defective or lethal in the seedling stage (Randolph 1942; Yao et al. 2011).

Hybrid vigor or heterosis is strong in maize (Birchler et al. 2010) but depends on the inbred lines that are used to produce the hybrids. Heterosis has played an important role in maize production. Different combinations produce differing levels of hybrid vigor—some have a large increase in biomass over the better parent and some have much less (Flint-Garcia et al. 2009). The different characteristics that exhibit heterosis appear to be independent of each other so harnessing heterosis for grain production versus stover mass might focus on combining different sets of inbreds (Flint-Garcia et al. 2009).

Economic Importance

Among the major agricultural producers in the world, the United States has the largest land area planted in maize, produces the largest crop, and has the second highest yield after Canada (Table 1). The US states with the highest production levels are the Midwest farm states of Iowa, Illinois, Nebraska, Minnesota and Indiana (Table 2). Internationally, China is a close second in terms of crop area planted, but has a much lower yield. After the US and China there is a major dropoff in national output and area planted of the maize crop. Among the major producers of corn ethanol, besides the US and China only Canada is also a world leader in maize production yet it only produces a few percent of the US crop size (Table 1). Other large

Table 1 Maize crop area, production and yield by top producers in 2010.

Country	Crop Area (ha)	Production (Mt)	Yield (Mt/ha)
US	32,960,400	316,165,000	9.59
China	32,519,900	177,548,600	5.46
Brazil	12,814,800	56,060,400	4.37
Mexico	7,148,050	23,301,900	3.26
Argentina	2,902,750	22,676,900	7.81
Indonesia	4,143,250	18,364,400	4.43
India	7,180,000	14,060,000	1.96
France	1,571,000	13,975,000	8.90
South Africa	2,742,000	12,815,000	4.67
Ukraine	2,647,600	11,953,000	4.51
Canada	1,202,900	11,714,500	9.74
World	161,821,251	844,358,253	5.22

Source: FAO 2011a.

Table 2 Maize crop area, production and yield by US States in 2010.

State	Grain Crop Area (1000 Acres)	Production (1000 Bushels)	Yield (Bushels/Acre)
Iowa	13,050	2,153,250	165
Illinois	12,400	1,946,800	157
Nebraska	8,850	1,469,100	166
Minnesota	7,300	1,292,100	177
Indiana	5,720	898,040	157
Kansas	4,650	581,250	125
S. Dakota	4,220	569,700	135
Wisconsin	3,100	502,200	162
Missouri	3,000	369,000	123
Michigan	2,100	315,000	150

Source: USDA 2012a.

maize producers, such as Brazil, Argentina, Mexico, India and France, use their crop largely for animal and human food and have focused on other feedstocks such as sugarcane, molasses and sugar beets to produce ethanol (Solomon 2010).

Given the dominance of the US in the maize sector, we will review the economic importance of the crop for the four major usage categories. Maize is the second most important agricultural commodity in the country after indigenous cattle meat, accounting for over $28 billion in value in 2009 (FAO 2011b). In the 2010–2011 growing season 12.45 billion bushels of maize were produced and 13.06 billion bushels were used, while there also was a carryover of 1.71 billion bushels in beginning stocks (USDA 2012b). The largest quantity of the crop was used for livestock and poultry feed and residual at 5.12 billion bushels (39.2%, though in the 1960s livestock and poultry feed accounted for as much as 75% of the total crop usage). Maize accounts for approximately 90% of the total feed grain production in the country (the other feed grains being sorghum, barley and oats). The second largest quantity of maize is dedicated to the production of fuel ethanol plus several farm and nonfarm coproducts, close behind at 4.59 billion bushels (35.1%). A total of 1.98 billion bushels (15.2%) was exported for various uses (mostly to Japan, South Korea, Mexico and Egypt). Maize imports into the US are insignificant. The remaining production, at 1.37 billion bushels, was used for nonbiofuel markets under the Food, Seed & Industrial category (10.5%).

While the main purpose of growing maize is to provide food for humans and farm animals, a clear majority of it is grown for livestock/poultry feed and residual demand, primarily as feed grain (and a small portion as silage). In comparison to other feed grains, maize is slightly higher in energy content while lower in protein and is around 70% starch (NRC 2001). While feed grain accounts for the largest production cost of livestock, it provides excellent nutritional value for stocker cattle, cows, beef heifers, poultry and other animals (Table 3). The cost of feed grain underscores the sensitivity of farm profits to conditions in other (often volatile) markets such as ethanol, though increased maize production puts downward pressure on prices.

Table 3 Maize silage nutrient composition (DMB - dry matter basis).

Category	Immature	Normal	Mature
	< 25% DM	32-38% DM	> 40% DM
Total Digestible Nutrients (TDN) %	65.6	68.8	65.4
Crude Protein (CP) %	9.7	8.8	8.5
Neutral Detergent Fiber (NTF) %	54.1	45.0	44.5
Acid Detergent Fiber (ADF) %	34.1	28.1	27.5
Calcium %	0.29	0.28	0.27
Phosphorus %	0.24	0.26	0.25

Source: NRC (2001).

Maize is also grown for direct human consumption, and is processed (refined) in dry and wet mills to make a large variety of additional human foods, animal feed ingredients, and industrial products. The latter includes: sweeteners (corn syrup, high-fructose corn syrup, crystalline fructose, dextrose and glucose syrups); feed (corn gluten feed, corn gluten meal, corn germ meal and steepwater); cornstarch; corn oil (cooking oil); corn grits, meal, flour and breakfast cereals; beverage alcohol (whiskey, and to a lesser extent beer); and bioproducts (e.g., lysine, and polymers such as polylactic acid and polyhydroxyalkanoates). Fuel ethanol is also an industrial product but because its use has grown so dramatically in the last few decades it is treated separately. The human nutritional value of maize is derived from its high level of carbohydrates, protein, fiber, and B vitamins, and low level of fat and cholesterol (Table 4).

The production of fuel ethanol in the US has been a rapidly growing usage category, having increased 10-fold from 1994 to 2011 (RFA 2012). Approximately 98% of total ethanol output in the US uses maize as feedstock. Fuel ethanol in the US is largely used in E10 (10% ethanol and 90% gasoline) blends, though a modest quantity is used in E85 (85% ethanol and 15% gasoline). The gross amount of maize being used at ethanol mills somewhat overstates its use for manufacturing fuel, however, since a significant portion of maize (~ 30%) results in several coproducts that are nonfuel (Fig. 1). The coproducts include distillers grains (both dried and wet) and solubles, gluten feed and gluten meal for livestock and poultry feed; corn oil, corn starches and syrups; and carbon dioxide (CO_2), which is used to carbonate beverages, manufacture dry ice and to flash freeze meat.

Table 4 Sweet corn nutritional composition.

Category	Nutritional Value per 100 g (3.5 oz)
Energy	360 kJ (86 kcal)
Carbohydrates	19.02 g
Sugars	3.22 g
Dietary Fiber	2.7 g
Fat	1.18 g
Protein	3.22 g
Water	75.96 g
Vitamin B_1	0.2 mg (17% RDA)
Vitamin B_3	1.7 mg (11% RDA)
Vitamin B_6	0.6 mg (46% RDA)
Vitamin B_9	46 mcg (12% RDA)
Vitamin C	6.8 mg (8% RDA)

Source: USDA (2011c).

Use of Maize as a Model Plant for Bioenergy Research

A multitude of considerations drive the development of a biomass production strategy leading to a sustainable, low cost fuel. The government seeks to minimize our dependence on foreign fossil fuels in order to establish independence from the energy policy vagaries associated with political instability in the Middle East and Latin America. The consumer, in turn, is looking for a predictable lowering of energy costs that green technology housed in the United States would provide. Both people and government have a common interest. Reduction in fossil fuel usage should lead to a decrease in CO_2 emissions and moderate the current trend in global warming.

The successful commercialization of a biomass alternative energy program comes neither without price nor careful reflection. Essential to the adoption of any successful biofuel business plan is a strategy that incorporates the development of accountable land use policies. For example, food costs are likely to rise significantly, especially if the land currently used to produce food shifts and becomes dedicated to biomass production. Indeed, approximately four out of ten maize plants are dedicated to bioethanol manufacture in the US with the accompanying result that the price per bushel exceeds $6. Clearly, then, candidate biomass crops, as well as the model species being studied, should be adaptable to growth on marginal land and not on acreage dedicated to food and feed consumption as well as traditional manufactured goods. The choice of this property presents no small challenge. Specifically, these marginal lands should be selected with the intention to keep costs to a minimum with regard to growing inputs and proximity of processing centers. Candidate biomass crops should be capable of being modified so they are better adapted to low cost soil treatment coupled with minimal irrigation needs. Also of importance is the requirement that high efficiency, low cost technologies are available for gene isolation, transformation and the regeneration of genetically modified high biomass crops. The need to produce more high quality dietary food grains that gain in distribution while identifying biomass crops that produce low cost, environmentally friendly fuel becomes especially important given that the Food and Agriculture Organization of the United Nations (FAO) predicts a 25–30% rise in global population topping at nine billion by 2050.

The objective of this section is to present the case for maize both as a vehicle for the production of fuel and to argue its suitability as a model species for study. Specifically, will it be possible to take what is learned from maize and apply it to other cereals, leading to low cost energy recovery platforms. It is clearly arguable that maize provides a robust tool box beginning with the availability of numerous, well characterized cultivars, ease of breeding and gene delivery, evolving and expanding "omics" tools, and numerous maps featuring both genetic and molecular markers that is really second to none (for review see Bennetzen and Hake 2009). Since the 2009 publication of the *Handbook of Maize: Genetics and Genomics*, the B73 genome has been sequenced; multiple physical maps continue to be generated; high throughput technologies are being continually expanded to apply what has been learned from transcriptional and genomic analyses; and the insertion lines associated with transposable elements and T-DNA are growing in coverage and ease of use (Schnable et al. 2009; McMullen et al. 2009; Vermerris 2011). In addition robust regeneration protocols coupled to high frequency DNA transfer are rapidly becoming mature technologies (Sairam et al. 2002, 2003; Wang et al. 2009).

Since maize molecular tools exist in such abundance, it might prove possible to adopt a macro approach and alter the plant's development, so that a majority of its resources become redirected, to significantly increasing its biomass effectively producing a *Miscanthus* mimic. Moreover, the cell wall, its biosynthesis and development, its molecular linkages, and the effect of modified lignin production or its association with cellulose are also obvious targets to modify that resonate in terms of energy harvest microdissection. Certainly, the brown midrib mutants manifest tissue specific changes in the composition of the cell wall with respect to lignin composition and deposition, to say nothing of the bk_2 mutant that effects patterning of cellulose–lignin connections (Vermerris and Boon 2001; Ching et al. 2006; Sindhu et al. 2007; Vermerris 2009, 2011). Yet understanding of the compositional aspects of the cell wall not to mention differential tissue specific expression during development presents only limited avenues of inquiry. The greatest advantages to biomass production and processing are likely to be gained by collaborative approaches. New information relevant to cell wall biosynthesis must be incorporated into research programs whose efforts are designed to increase biomass through alterations in plant development. Such labors are already beginning, albeit in their infancy (Sang 2011).

Clearly, maize as a model organism has tools that make possible the identification, cloning and transfer of single genes. With the ability to add genes, and to modify their expression comes the capacity to change pathways or pathway interactions so as to maximize processing-efficient biomass and thus insure cheaper supplies of biofuel. Another value of maize as an object of study lies in its ability to generate new information that can be applied to other cereals or grass species leading to increases in biocelluosic fuel production. Specifically, cereals, even though they vary significantly in size, morphology, life cycle strategy, and seed set to name but four morphological variants, possess great genome conservation from species to species with respect to the genes themselves and also with regard to their arrangement (Bennetzen 2009). Our objective here is to briefly look at the many issues that maize may address and to see how these may be applied to the production of a fruitful high cellulosic biomass energy industry.

Abiotic and Biotic Stress

Candidate biofuel crop species capable of cultivation on marginal land must be competent for sustained, robust growth with minimum losses to harvest if there is to be no reduction in grain delivery to its traditional markets. Transgenic maize lines meeting these criteria have already been produced that can grow under conditions of drought, high salt, and low temperature (Al-Abed et al. 2007). Maize segregating *DREB/CBF3* can recover 75% of its initial planting when deprived of water for 21 days. The cold tolerant phenotype expressed in these plants could, in principle, also alter traditional planting dates, thus temporally separating maize from its maximum susceptibility to various highly virulent pathogens. Minimizing harvest losses due to either biotic or abiotic stresses are potential places in the value added stream for reducing biomass fuel costs. Finally, the ability to cultivate corn under conditions of high salt directs the promise of bringing into production land currently lying fallow, while giving this land new value by generating a tax base associated with the recovery of a commercial product.

Concomitant with the use of novel lands for maize cultivation is the need to combat and address infections following exposure to new pathogens. It has already been established that corn expresses both quantitative and qualitative response mechanisms to disease challenge; thus the opportunity to manipulate these holds promise for the future (Smith and Hulbert 2005; Wisse et al. 2006). This said, in principle, loss appears to be geography specific, being most severe in regions of the world having poorly developed disease resistance breeding programs. Given this, the possibility exists for importing modified lines into new ranges that express desired disease resistances.

Photosynthesis

One way to increase biomass production is through the modification of photosynthesis. Sorghum and *Miscanthus* can produce as much as 15 to 40 Mg/ha and 45 Mg/ha, respectively, an output that maize cannot currently match (Rooney et al. 2007; Venuto and Kindiger 2008). Although cereals as a group differ widely with respect to biomass production, the genes among the *Poales* regulating its expression are likely conserved. *Miscanthus*, maize, and sorghum use a C_4 photosynthetic mechanism to produce carbohydrate.

At high temperatures, leaf water loss is minimized due to the closure of the stomata. Furthermore, in C_4 plants, CO_2 is concentrated in the mesophyll cells and hence photorespiration is held to a minimum leading to greater production of sugar and carbohydrate as compared to C_3 plants. While the enzymology concerning the fixing of carbon is well understood, the anatomical features of the leaf that partitions Rubisco activity in the sheath from CO_2 sequestration in the mesophyll cells are not well defined. Opportunities are being explored to further improve C_4 photosynthetic capacity by focusing on this partitioning and sequestration. While maize is a C_4 plant, a possible area of study could also focus on modifying or engineering out the inefficiencies of C_3 photosynthate accumulation, leading to identification of new cellulosic biomass sources (Sage and Zhu 2011).

Development

Shifts in biomass production may be regulated by modulating expression of a number of phytohormone pathways among them the gibberellins and the brassinosteroids, which increase vegetative growth (Fleet and Sun 2005; Müssig 2005; Salas Fernandez et al. 2009; Sun 2011). Many of these genes are well conserved and the opportunity to engineer increased plant size, even at the expense of kernel development may be an opportunity for maize's utility, not only as an object of study, but also as a means for increasing cellulosic mass. Delaying the time required to flower can also result in increased biomass. This means that biomass accumulation and availability, with its subsequent translation into alternate fuel is not only affected by the genes regulating the production of cellulose, its component parts, the deposition of lignin but also by genes affecting the entire plant's development. For a full discussion see Chapter 4 in this volume.

Reproductive Strategy

Reproductive strategy also must enter into any consideration of the utility of a high biomass crop as a source of alternate energy. For example, the ability to introgress genes that modify either lignin production or that mediate saccharification is much more easily accomplished in diploids than in polyploids. From this point of view, maize and sorghum are ideal candidates, as each of these species may be readily proliferated by outcrossing or inbreeding. In contrast, the reproductive options open to *Miscanthus* are more complicated, especially when they involve the introgression of new genes. Typically, chromosome number may vary from accession to accession ($2n = 3x = 57$) giving rise to variations in chromosome number and concomitant sterility. In addition, *Miscanthus'* sexual reproductive cycle is restricted to outcrossing. Given these liabilities it must be emphasized that once a desired high biomass *Miscanthus* line is maximized for cellulose processing and degradation, it must be proliferated vegetatively (for a review, see Vermerris 2011).

Mapping Tools and Breeding Populations

The case for maize as the model organism both as an object of study as well as a producer of biomass continues to advance. Sophisticated mapping tools embracing everything from the segregation of traits to whole-genome resequencing continue to be developed. Hence, it has become commonplace to discern those putative genes likely to regulate anything from cellulose biosynthesis, to lignin deposition to say nothing of developmental reprogramming using a variety of different techniques all of which may lead to enhanced cellulosic accumulation.

Ever since Emerson et al. (1935) published the first linkage map, an array of sophisticated mapping tools has been developed that connects a variation in phenotype to the precise location of its underlying cause in the genome. The initial genetic maps that charted mutations in genes controlling anthocyanin deposition, chlorophyll biosynthesis, and plant morphology to chromosome position were complemented by the construction of molecular maps, which were responsive to unique characteristics of maize DNA (Neufer et al. 2009). Restriction fragment length polymorphisms (RFLPs) and simple sequence repeats (SSRs) maps, to name but two, have been constructed. These simple molecular maps have matured to the point where we can now discriminate single nucleotide polymorphisms (SNPs) throughout an entire genome and link these DNA polymorphisms to candidate genes underlying phenotypic variation (Yan et al. 2010). Molecular marker maps that were originally developed in the 1980s have reached a level of articulation that could not have been anticipated, for example, high throughput array-based marker technology is routinely used to generate linkage relationships (Gupta 2008; Xu et al. 2009).

Concomitant with the development of improved higher-throughput mapping technologies, sophisticated mapping populations have been constructed that allow phenotypic variation to be connected to chromosomal location, such as, novel RILs (Burr et al. 1988). The construction of RILs has proven efficacious in identifying quantitative trait loci (QTLs) that affect important agronomic traits (Stuber et al. 1999). In principle, two phenotypically contrasting lines are crossed from which a segregating F_2 is produced. Single seed descent populations are derived through the F_6. These can express extreme trait variations among different lines of descent but the trait remains near homogenous within a given line. The phenotypic variation serves as the basis to identify chromosomal loci through linkage to known polymorphic DNA markers. The details of this technology have been well explained by Vermerris (2011). These initial mapping efforts continue to expand and have led to the development of novel mapping populations such as the nested association mapping (NAM) population (McMullen et al. 2009). The NAM map is distinctive in that nearly every centimorgan of the maize genome has been marked. Using innovative statistical analysis, trait variation can now be associated to genome-wide DNA polymorphisms (Yan et al. 2010). Indeed, it is now possible to map loci controlling distinctive phenotypic variations that lack a visual phenotype, such as putative cell wall mutants that can only be identified as spectral variants (Vermerris 2011).

The utility and success of marker-assisted breeding is summarized by way of example by Xu et al. (1998). The ability then to follow the cosegregation of DNA polymorphisms with genes or QTLs results in placement to chromosomal loci. Then advantage can be taken of the maize physical map to enable positional cloning and the subsequent isolation and analysis of the gene(s) of interest. Most importantly this effort has been complimented by various transcriptional profiling technologies. Xu et al. (2009) has reported 17,000 probe sets that integrate 15,000 transcripts approximating 13,500 genes. Indeed this technology has reached a level of sophistication where the variation in transcript accumulation itself is the trait mapped, such that the linkage of expression QTLs (eQTLs) to chromosomal loci now proceeds with robust proficiency (Stupar and Springer 2006). The current status of mapping in maize and transcriptional analysis is second to none. In principle, any gene candidate modifying some aspect of cell biosynthesis or deposition should be able to be mapped, cloned and its expression assessed.

Transgene Delivery in Maize

Historically, efforts to produce transgenic maize centered on solving three major problems. These included overcoming the seemingly genotype specificity with respect to regeneration competence. In particular, some lines proved regeneration competent while others did not (for a review, see Sairam et al. 2002). Once a cultivar is identified that is amenable to regeneration from a single cell, two issues remained. The first of these was to identify the explants that contained the maximum number of cells that were competent to rapidly yield fertile plants and the second, cells that, were amenable to high frequency DNA transfer.

During the last decade, this problem has been addressed and largely solved by Goldman and Sairam and their coworkers (Sairam et al. 2002, 2003; Al-Abed et al. 2006, 2007). *Tripsacum dactyliodes* var. "Pete" is a mixture of genotypes that were proved amenable to high frequency T-DNA transfer.

This technique was subsequently applied to maize inbreds and hybrids with increasing success. The efficiency of T-DNA transfer increases, as measured by the uniformity of expression of screenable markers along the surface of the explants but also by the number of putatively transformed shoots that can be separated, rooted and transferred to the greenhouse. Finally, the development of the new split-seed explants allowed as many as 58 plants to be separately regenerated from a single seed and grown to fertility thus allowing for the immediate rapid, year round proliferation of the original T_1 transgenic plants. Recently similar success with respect to shoot output has been reported for sorghum (Bushra et al. 2011).

Mutagenesis

The ability to increase the incidence of forward mutation in maize has had a long, distinguished history beginning with Stadler and Spragues' (1936) observation that UV frequency dependent wavelength increased the frequency of forward mutation. Since then, a host of different mutagens have been tested, some causing direct change to the DNA molecule itself; others due to secondary events. These include physical agents such as X- and gamma rays, chemicals such as methyl-methane sulfonate (MMS) and ethyl methane sulfonate (EMS) and certain biologicals that both cause and mark mutations through the physical insertion of either T-DNA or transposable elements.

Indeed, although maize has long been adaptable to forward and reverse genetics, the power of these techniques has made more robust recently by the complete sequencing of the B73 genome (Schnable et al. 2009). Forward genetics relies on the ability to recognize altered phenotypes and to track them to their causative gene. These mutations occur spontaneously in nature or are induced by mutagens and were crucial in developing the first maize genetic map (Emerson et al. 1935). Reverse genetics, by contrast, requires knowledge of a gene sequence and the ability to derive variants with altered function, for example by transposon mutagenesis. Specifically, insertional mutation can lead to the recovery of null or altered expression variants, which can have phenotypic consequences. The nature of the mutant phenotype ultimately clarifies the mechanism(s) of gene action in controlling the trait. These techniques are well developed in maize (Sairam et al. 2005; Lisch and Benetzen 2011). Transposition mutagenesis indeed feeds the reverse genetics strategy. This technology has been successfully used to delineate the function of *An1*, the diversification of C-function during flower development, and more recently meristem fate (Bensen et al. 1995; Mena et al. 1996; Chuck et al. 2007). The utility of this technology has steadily improved and has made possible the use of the Uniform *Mu* maize population to identify genes involved in some aspects of cell wall development (Penning et al. 2009). With this in mind, the impact of the complete sequencing of B73 maize genome on the future impact of comparative genomics is inestimable (Schnable et al. 2009).

Final Thoughts

It is clear that maize today, from the perspective of biomass harvest per hectare, is hardly the only choice as the basis for the cellulosic biofuel industry. Nevertheless, maize development can be altered such that greater investments can be made in cellulosic biomass production. Whether these changes will have commercial application will have to be continually revaluated as more sophisticated "omics" tools become available that can be coupled to new rapid high throughput assays (see Chapters 4 and 5 in this volume). New engineered constructs would have to be modified to reduce maize's investment in flowering, while exposing linkages within the cellulose polymer, to say nothing of how these bind to lignin facilitating more efficient cellulose degradation. Already, mutants that discriminate different maize cell wall variants with no visible phenotypic difference have been identified. The complexity of synthesizing "super" maize will certainly involve the additional modification of genes, modifications to their expression, as well as the fine tuning and articulation of multiple key pathways. The question remains however whether the current tools are sufficient to identify those genes and to manipulate them such that cultivation of a new giant "*Zea*" is possible. Will it be possible to engineer maize and alter its energy allocations such that its modified development produces a massive biomass plant that propagates vegetatively, possibly as a perennial? Alternatively is industry better served by isolating these maize genes and delivering these to *Miscanthus* to produce even greater giants whose walls are subject to ease of digestibility? These answers lie in the future and it is to that end that our labor is directed.

Overall Sustainability of Maize as a Bioenergy Plant

Bioenergy is renewable energy derived from biological sources that can be used for heat, electricity and transportation fuels. In response to rising petroleum prices, the need for national energy security, and fossil fuel-induced climate change, transportation fuels derived from plant materials attract much attention and this industry is growing rapidly. In the United States, biomass derived ethanol (i.e., bioethanol) as a gasoline substitute or additive accounts for the vast majority of biofuel produced domestically.

Bioenergy Products and Conversion Technologies

Maize is by far the major source for ethanol production in the US (USEPA 2010; RFA 2011). The next largest user of maize for ethanol is China, which uses less than 5% the total in the US (RFA 2011a). As was shown in Table 1, these two nations are by far the largest producers of maize. Maize can be used as a bioenergy crop in two ways: the grains can be used to produce starch-based ethanol via dry or wet milling processes, and the corn stover as crop residuals could potentially be converted to lignocellulosic biofuels through biochemical or thermochemical processes (Mosier et al. 2005). Dry and wet milling processes are mature and commercialized technologies for maize-based ethanol production, as shown in Fig. 1. The processes for biofuel derived from corn stover are still largely at the demonstration or pilot scale.

Biofuel Production

The annual ethanol production in the US increased from 6.4 billion liters (1.7 billion gallons) from 54 processing facilities in 2000 to 49.2 billion liters (13.3 billion gallons) from 209 biorefineries as of August 2013 (RFA 2012). Among these 209 biorefineries, over 190 facilities use maize as the feedstock, accounting for 98% of the US annual ethanol production. With additional facilities currently under construction and expansion, total annual production will reach 51.1 billion liters (13.5 billion gallons) within one to two years (RFA 2012). This can replace about 10% of annual US motor gasoline consumption, which was 522 billion liters (137.9 billion gallons) in 2009 (EIA 2011b).

Suitability from an Energy Perspective

The reliance of ethanol production in the US on maize as the feedstock has raised concerns about the long-term environmental impacts and societal implications of corn ethanol. Many studies (Pimentel 1991; Wang et al. 1997; Shapouri et al. 2003; Kim and Dale 2005; Gnansounou and Dauriat 2005; Pimentel and Patzek 2005; Farrell et al. 2006; Tilman et al. 2006; von Blottnitz and Curran 2007; Wang et al. 2007; Davis et al. 2009; Liska et al. 2009) have evaluated the energy balance of corn ethanol from a life-cycle perspective. In those studies, net energy value (NEV)[1] or net energy ratio (NER)[2] are the most commonly used indicators to assess if maize is a

---

[1]NEV is defined as the energy outputs from a biofuel crop minus the fossil energy used to produce the biofuel.
[2]NER is defined as the energy outputs from a biofuel crop divided by the fossil energy used to produce the biofuel.

suitable crop for biofuel production in terms of energy. The results from those studies vary significantly, from a NEV of –9.35 MJ to +8.53 MJ per liter of ethanol (Shapouri et al. 2003). The NEV reported in the unit of MJ/ m^2 ranges from –2.52 to 2.30 (Davis et al. 2009). The NER varies from 0.69 to 2.2 (Gnansounou and Dauriat 2005; Liska et al. 2009). Negative NEVs or NER values less than 1.0 indicate that the corn ethanol is not energy efficient and requires more energy input than the energy produced. Major sources of variation in these studies are in the differences in efficiency term and functional unit, coproduct allocation method, variation in life cycle inventory data, and different system boundaries (Davis et al. 2009; van der Voet et al. 2010).

In most recent studies, positive NEVs were consistently reported with thorough evaluation of recent technological improvements in crop production, ethanol biorefining, and coproduct utilization (Liska et al. 2009). For example, the majority of ethanol biorefineries built after 2004 use the dry milling process powered by natural gas, which results in a smaller energy requirement. With the increase in maize yield per acre, ethanol yield per bushel of maize, fertilizer industry energy efficiency, and ethanol plants energy efficiency, the NEV can increase significantly to 2.04–4.52 MJ/m^2 (Liska et al. 2009).

Suitability from a Climate Change Mitigation Perspective

Greenhouse gas (GHG) emissions are another indicator commonly evaluated in maize biofuel studies, because biofuel development is partially driven by climate change mitigation through GHG reduction. Similar to energy analysis, the results and conclusions from those studies are contradictory. Searchinger et al. (2008) concluded that maize-based ethanol will double GHG emissions over 30 years if including indirect GHG emissions from land use change is counted. However, Liska et al. (2009) estimated that direct GHG emissions from US ethanol biorefineries with improved technologies and efficiencies will be reduced by 48% to 59% (see Fig. 2) as compared to gasoline. Even with inclusion of indirect emissions from land use change, the GHG emission reduction can still meet the regulatory reduction threshold of 20% (Liska et al. 2009; USEPA 2010).

Adler et al. (2007) considered both carbon flux and nitrogen flux over the life cycle of biofuels using the DAYCENT biogeochemical model. They found that displaced fossil fuel was the largest GHG sink followed by soil carbon sequestration. Nitrogen flux, specifically N_2O emissions, contributed significantly to GHG emissions. Ethanol and biodiesel from maize-soybean rotations can reduce GHG emissions by 40% compared with the life cycle of gasoline and diesel fuel (Adler et al. 2007). To evaluate suitability of maize

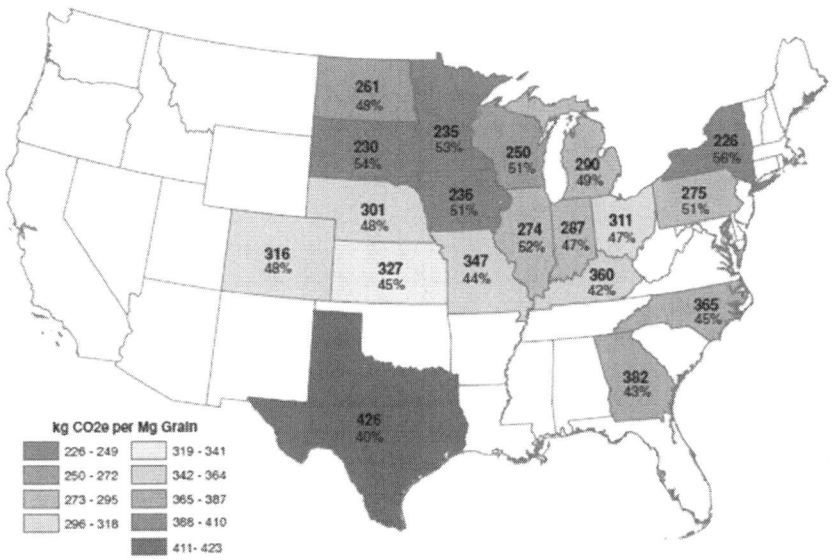

Figure 2 Greenhouse gas (GHG) emissions from corn-based ethanol in the U.S. (Liska et al. 2009).

Color image of this figure appears in the color plate section at the end of the book.

as a bioenergy plant to mitigate GHG emissions, more research is needed to better quantify uncertainties involved in GHG estimation, especially where indirect emissions are concerned.

Suitability from other Environmental Impact Perspectives

When assessing the suitability of maize as a bioenergy crop, it is important to consider not only energy and GHG emissions but also other environment impacts. One major concern is that corn ethanol as a solution to the energy problem may actually pose a threat to water resources (Dominguez-Faus et al. 2009). The water use for biorefinery facilities ranges from 2 to 6 gallons of water per gallon of ethanol produced compared with 1.5 gallon of water per gallon of gasoline for petroleum refining (Pate et al. 2007; NRC 2008). However, this amount is quite small compared with the water needed for crop irrigation. For maize, the feedstock water demand is around 392,000 gallons per acre, which is equal to 980 gallons water per gallon of ethanol with a biofuel yield of 400 gallons per acre (Pate et al. 2007). Clearly, corn biofuel puts additional stress on already unsustainably used water resources. In addition, it can impair the water quality via surface runoff and infiltration to groundwater due to application of fertilizers and pesticides. Nutrients in the runoff will cause eutrophication and have serious effects on aquatic ecosystems, such as the "dead zones" in the Gulf of Mexico and Chesapeake Bay. Soil erosion resulting from agricultural practices will also impact water quality.

Langpap and Wu (2011) investigated the impact of ethanol production on land use and cropping systems environmental quality in the Midwestern US. Results suggest that increasing agricultural commodity prices driven by ethanol production from food-grade maize will result in widespread conversions of noncropland to cropland. Consequently, fertilizer and pesticide application, nitrogen runoff and leaching, and soil wind erosion will all increase. Studies that consider impacts other than energy and GHG emissions consistently show a worse performance of biofuel in terms of eutrophiciation (van der Voet et al. 2010). In other categories such as toxicity and smog formation, corn ethanol is also unfavorable because it has high volatile organic compounds (VOCs) emissions and acetaldehyde concentration during the fuel use phase. VOCs are the precursors of smog and acetaldehyde is a probable carcinogen. It is important to consider all potential environmental impacts to avoid unintended consequences of increased ethanol production from maize.

Suitability from a Societal Perspective

Last but not least, societal implications of maize as a bioenergy plant should be evaluated. The US is a major maize exporter. Increasing maize production for fuel, however, will divert some grain intended for food supply on a global scale. This also drives up the food commodity prices. Indeed, maize prices almost tripled from 2006 to 2008 (Mitchell 2008). There is ongoing debate on food vs. fuel (Pimentel 1991; Mitchell 2008; Escobar et al. 2009). A recent study from The Word Bank (Mitchell 2008) found that the 70 to 75% increase in food commodity prices can be attributed to the increase in biofuels and the related consequences such as low grain stocks and large land use shifts. Other studies (OECD 2008; USEPA 2010), however, concluded that only a less than 10% increase in maize prices is due to the current biofuel support policies alone. Again, different results were reported by studies conducted by different organizations that have different political agendas and used different methods for assessment.

Conclusions

Maize bioenergy production is largely driven by fuel price, climate change mitigation, and political agendas. With the advancement of crop production and biorefinery technologies, the net energy values and life cycle GHG emissions reduction is greatly improving and shows the benefits of maize-based bioenergy. However, it is important to consider other consequences such as water consumption, eutrophication, human health impacts and societal implications. These factors have to be carefully evaluated and negative effects must be mitigated through various approaches such as use of reclaimed water, reduction in nitrogen application rates, use of biosolids as fertilizers, switching from continuous maize production to rotations of maize and soybeans, implementing a ban on conversion of highly erodible land to crop production areas, and using corn stover (waste byproducts) instead of maize grain as a food source.

References

Adler PR, Del Grosso SJ and Parton WJ (2007) Life-cycle assessment of net greenhouse-gas flux for bioenergy cropping systems. Ecol Appl 17: 675–691.

Al-Abed D, Rudrabhatla S, Talla R et al. (2006) Split-seed: A new tool for maize researchers. Planta 223: 1355–1360.

Al-Abed D, Madasamy P, Talla R et al. (2007) Genetic engineering of maize with *Arabidopsis* $DREB_1A/CBF_3$ gene using split-seed explants. Crop Sci 47: 2390–2402.

Albert PS, Gao Z, Danilova TA et al. (2010) Diversity of chromosomal karyotypes in maize and its relatives. Cytogenet Genome Res 129: 6–16.

Baskerville C (1906) Free alcohol in the arts and as fuel. In: Shaw A (ed) The American monthly review of reviews. The Review of Reviews Company, New York, USA, pp 211–214.

Bennetzen JL (2009) Maize genome structure and evolution. In: Bennetzen JL and Hake SC (eds) Handbook of maize: genetics and genomics. Springer, New York, USA, pp 179–200.

Bennetzen JL and Hake SC eds (2009) Handbook of maize: genetics and genomics. Springer, New York, USA.

Bensen RJ, Johal GS, Crane VC et al. (1995) Cloning and characterization of the maize *An1* gene. Plant Cell 7: 75–84.

Bernton H, Kovarik W and Sklar S (2010) The forbidden fuel: a history of power alcohol, new edn. University of Nebraska Press, Lincoln, NE, USA, 328 pp.

Birchler JA, Yao H, Chudalayandi S et al. (2010) Heterosis. Plant Cell 22: 2105–2112.

Brink RA and Cooper DC (1947) The endosperm in seed development. Bot Rev 13: 479–541.

Buckler ES, Holland JB, Bradbury PJ et al. (2009) The genetic architecture of maize flowering time. Science 325: 688–689.

Burr B, Burr F, Thompson KH et al. (1988) Gene mapping with recombinant inbreds in maize. Genetics 118: 519–526.

Carolan MS (2009) Ethanol versus gasoline: The contestation and closure of a socio-technical system in the USA. Soc Stud Sci 39: 421–448.

Ching A, Dhugga KS, Appenzeller L et al. (2006) Brittle stalk 2 encodes a putative glycosyl-phosphatidyl-inostitol anchored protein that affects mechanical strength of maize tissues by altering the composition and structure of secondary cell walls. Planta 224: 1174–1184.

Chuck G, Meeley R, Irish E et al. (2007) The maize tasselseed4 microRNA controls sex determination and meristem cell fate by targeting Tasselseed6/indeterminate spikelet1. Nat Genet 39: 1517–1521.

Coe EH (1959) A line of maize with high haploid frequency. Am Nat 93: 381–382.

Cone KC and Coe EH (2009) Genetic mapping and maps. In: Bennetzen JL, Hake SC (eds) Handbook of maize: genetics and genomics. Springer, New York, USA, pp 507–522.

Cooper DC (1951) Caryopsis development following matings between diploid and tetraploid strains of *Zea mays*. Am J Bot 38: 702–708.

Cummins L (2002) Internal fire: the internal combustion engine, 1673–1900. Carnot Press, New York, USA, 356 pp.

Davis SC, Anderson-Teixeira KJ and DeLucia EH (2009) Life-cycle analysis and the ecology of biofuels. Trends Plant Sci 14: 140–146.

deWet JMJ, Harlan JR, Engle LM et al. (1973) Breeding behavior of maize-*Tripsacum* hybrids. Crop Sci 13: 254–256.

Dimitri C and Effland A (2007) Fueling the automobile: An economic exploration of early adoption of gasoline over ethanol. J Agri Food Ind Organ 5: 1–17.

Doebley JF (2004) The genetics of maize evolution. Annu Rev Genet 38: 37–59.

Doebley JF and Iltis HH (1980) Taxonomy of *Zea* (*Gramineae*). I. A subgeneric classification with key to taxa. Am J Bot 67: 982–993.

Dominguez-Faus R, Powers SE, Burken JG et al. (2009) The water footprint of biofuels: A drink or drive issue? Environ Sci Technol 43: 3005–3010.

EIA (Energy Information Administration) (2011a) Energy Information Administration. various datasets. http: //www.eia.gov.

EIA (U.S. Energy Information Administration) (2011b) Petroleum statistics. http: //www.eia. gov/energyexplained/index.cfm?page=oil_home#tab2 (accessed on Dec 14, 2011).

Emerson RA, Beadle GW and Fraser AC (1935) A summary of linkage studies in maize. Cornell Univ Agric Exp Stn Bull Mem 180.

Escobar JC, Lora ES, Venturini OJ et al. (2009) Biofuels: environment, technology and food security. Renew Sust Energ Rev 13: 1275–1287.

FAO (Food and Agriculture Organization of the United Nations) FAOSTAT (2011a). Agricultural production domain. http: //faostat.fao.org/site/339/default.aspx.

FAO (Food and Agriculture Organization of the United Nations) FAOSTAT (2011b). Crops. http: //faostat.fao.org/?PageID=567#ancor.

Farrell AE, Plevin RJ, Turner BT et al. (2006) Ethanol can contribute to energy and environmental goals. Science 311: 506–508.

Federal Trade Commission (2006) 2006 Report on ethanol market concentration. Washington DC, USA.

Fleet CM and Sun TP (2005) A DELLAcate balance: the role of gibberellin in plant morphogenesis. Curr Opin Plant Biol 8: 77–85.

Flint-Garcia SA, Buckler ES, Tiffin P et al. (2009) Heterosis is prevalent for multiple traits in diverse maize germplasm. PLoS ONE 4: e7433.

Freeling M (2009) Bias in plant gene content following different sorts of duplication: tandem, whole-genome, segmental, or by transposition. Annu Rev Plant Biol 60: 433–453. 2105–2112.

Gnansounou E and Dauriat A (2005) Energy balance of bioethanol: a synthesis. Swiss Federal Institute of Technology & ENERS Energy Concept, Lausanne, Switzerland http: //www. eners.ch/downloads/eners_0510_ebce_paper.pdf, accessed on Dec. 14, 2011.

Gupta PK (2008) Single-molecule DNA sequencing technologies for future genomics research. Trends Biotechnol 26: 602–611.

Herrick RF (1907) Denatured or industrial alcohol. John Wiley, New York, USA.

Jiao Y, Wickett NJ, Ayyampalayam S et al. (2011) Ancestral polyploidy in seed plants and angiosperms. Nature 473: 97–100.

Kato A, Albert PS, Vega JM et al. (2006) Sensitive FISH signal detection in maize using directly labeled probes produced by high concentration DNA polymerase nick translation. Biotechnol Histochem 81: 71–78.

Kato A and Birchler JA (2006) Induction of tetraploid derivatives of maize inbred lines by nitrous oxide gas treatment. J Hered 97: 39–44.

Kato A, Lamb JC and Birchler JA (2004) Chromosome painting in maize using repetitive DNA sequences as probes for somatic chromosome identification. Proc Natl Acad Sci USA 101: 13554–13559.

Kellogg EA (1998) Relationships of cereal crops and other grasses. Proc Natl Acad Sci USA 95: 2005–2010.

Kermicle JL (1969) Androgenesis conditioned by a mutation in maize. Science 166: 1422–1424.

Kiesselbach TA (1980) The structure and reproduction of corn. University of Nebraska Press, Lincoln, NE, USA, 96 pp.

Kim S and Dale BE (2005) Life cycle assessment of various cropping systems utilized for producing biofuels: bioethanol and biodiesel. Biomass Bioenergy 29: 426–439.

Kovarik W (1998) Henry Ford, Charles F Kettering and the fuel of the future. Automot Hist Rev 32: 7–27.

Lamb JC, Danilova T, Bauer MJ et al. (2007) Single-gene detection and karyotyping using small-target fluorescence *in situ* hybridization on maize somatic chromosomes. Genetics 175: 1047–1058.

Langpap C and Wu J (2011) Potential environmental impacts of increased reliance on corn-based bioenergy. Environ Resour Econ 49: 147–171.

Lisch D and Bennetzen JL (2011) Transposable element origins of epigenetic gene regulation. Curr Opin Plant Biol 14: 156–161.

Liska AJ, Yang HS, Bremer VR et al. (2009) Improvements in life cycle energy efficiency and greenhouse gas emissions of corn-ethanol. J Ind Ecol 13: 58–74.

Liu K, Goodman M, Muse S et al. (2003) Genetic structure and diversity among maize inbred lines as inferred from DNA microsatellites. Genetics 165: 2117–2128.

McClintock B (1929) A cytological and genetical study of triploid maize. Genetics 14: 180–227.

McClintock B (1930) A cytological demonstration of the location of an interchange between two nonhomologous chromosomes of *Zea mays*. Proc Natl Acad Sci USA 16: 791–796.

McMullen MD, Kresovich S, Villeda HS et al. (2009) Genetic properties of the maize nested association mapping population. Science 325: 737–740.

Mena M, Ambrose BA, Meeley RB et al. (1996) Diversification of C-function activity in maize flower development. Science 274: 1537–1540.

Mitchell D (2008) A note on rising food prices. Policy Research Working Paper 4682. The World Bank, Washington, DC, USA.

Mosier M, Wyman C, Dale B et al. (2005) Features of promising technologies for pretreatment of lignocellulosic biomass. Bioresour Technol 96: 673–686.

Müssig C (2005) Brassinosteroid-promoted growth. Plant Biol (Stuttg) 7: 110–117.

Neuffer MG, Johal G, Chang MT et al. (2009) Mutagenesis—the key to genetic analysis. In: Bennetzen JL and Hake SC (eds) Handbook of maize: genetics and genomics. Springer, New York, USA, pp 63–84.

NRC (National Research Council) (2001) Nutrient requirements of dairy cattle, 7th rev edn. National Academies Press, Washington DC, USA, 381 pp.

NRC (National Research Council) (2008) Water implications of biofuels production in the United States. National Academies Press, Washington DC, USA, 88 pp.

OECD (Organisation for Economic Co-operation and Development) (2008) Biofuel support policies: an economic assessment. ISBN 978-92-64-04922-2.

Pate R, Hightower M, Cameron C et al. (2007) Overview of energy-water interdependencies and the emerging energy demands on water resources. Report SAND 2007-1349C. Sandia National Laboratories, Los Alamos, NM, USA.

Penning BW, Hunter III CT, Tayengwa R et al. (2009) Genetic resources for maize cell wall biology. Plant Physiol 151: 1703–1728.

Pimentel D (1991) Ethanol fuels: energy security, economics, and the environment. J Agri Environ Ethic 4: 1–13.

Pimentel D and Patzek TW (2005) Ethanol production using corn, switchgrass and wood; biodiesel production using soybean and sunflower. Nat Resour Res 14: 65–76.

Punyasingh K (1947) Chromosome numbers in crosses of diploid, triploid and tetraploid maize. Genetics 32: 541–554.

Randolph LF (1935) Cytogenetics of tetraploid maize. J Agr Res 50: 591–605.

Randolph LF (1942) The influence of heterozygosis on fertility and vigor in autotetraploid maize. Genetics 27: 163.

Rayburn AL, Auger JA, Benzinger EA et al. (1989) Detection of intraspecific DNA content variation in *Zea mays* L. by flow cytometry. J Exp Bot 40: 1179–1183.

RFA (Renewable Fuels Association) (2011) Building bridges to a more sustainable future: 2011 ethanol industry outlook. Washington DC, USA.

RFA (Renewable Fuels Association) (2012) Statistics. http: //www.ethanolrfa.org/pages/ statistics.

Rhoades MM (1950) Meiosis in maize. J Hered 41: 58–67.

Rhoades MM and Dempsey E (1966) Induction of chromosome doubling at meiosis by the elongate gene in maize. Genetics 54: 505–522.

Riddle NC and Birchler JA (2008) Comparative analysis of heterosis in diploid and tetraploid maize. Theor Appl Genet 116: 563–576.

Riddle NC, Jiang H, An L et al. (2010) Gene expression analysis at the intersection of ploidy and hybridity in maize. Theor Appl Genet 120: 341–353.

Riddle NC, Kato A and Birchler JA (2006) Genetic variation for the response to ploidy change in *Zea mays* L. Theor Appl Genet 114: 101–111.

Roman HL (1947) Mitotic nondisjunction in the case of interchanges involving the B-type chromosome in maize. Genetics 32: 391–409.

Roman HL (1948) Directed fertilization in maize. Proc Natl Acad Sci USA 34: 36–42.

Rooney WL, Blumenthal J, Bean B et al. (2007). Designing sorghum as a dedicated bioenergy feedstock. Biofuel Bioprod Biorefin? 1: 147–157.

Ryder OB (1944) Industrial Alcohol. US tariff commission report, War Changes in Industry Series, Washington DC, USA.

Sage RF and Zhu XG (2011) Exploiting the engine of C_4 photosynthesis. J Exp Bot 62: 2989–3000.

Sairam RV, Wilber C, Franklin G et al. (2002) High frequency callus induction and plant regeneration in *Tripsacum dactyloides* (L.). *In vitro* Cell Dev B-Plant 38: 435–440.

Sairam RV, Parani M, Franklin G et al. (2003) Shoot meristem: an ideal explant for *Zea mays* (L.) transformation. Genome 46: 323–329.

Sairam RV, Chennareddy S, Parani M et al. (2005) OPBC symposium: maize 2004 & beyond —plant regeneration, gene discovery and genetic engineering of plants for crop improvement. *In vitro* Cell Dev B-Plant 41: 411–423.

Salas Fernandez MG, Becraft PW, Yin Y et al. (2009) From dwarves to giants? Plant height manipulation for biomass yield. Trends Plant Sci 14: 454–461.

Sang T (2011) Toward the domestication of lignocelluloses energy crops: learning from food domestication. J Integr Plant Biol 53: 96–104.

Schnable PS, Ware D, Fulton RS et al. (2009) The B73 maize genome: complexity, diversity and dynamics. Science 326: 1112–1115.

Searchinger T, Heimlich R, Houghton RA et al. (2008) Use of U.S. croplands for biofuels increases greenhouse gases through emissions from land-use change. Science 319: 1238–1240.

Shapouri H, Duffield JA and Wang M (2003) The energy balance of corn ethanol revisited. Am Soc Agri Engg 46: 959–968.

Sindhu A, Langewisch T, Olek A et al. (2007) Maize *Brittle stalk2* encodes a cobra-like protein expressed in early organ development but required for tissue flexibility at maturity. Plant Physiol 145: 1444–1459.

Smith SM and Hulbert SH (2005) Recombination events generating a novel *Rp1* race specificity. Mol Plant-Microbe Interact 18: 220–228.

Solomon BD, Barnes JR and Halvorsen KE (2007) Grain and cellulosic ethanol: history, economics, and energy policy. Biomass Bioenergy 31: 416–425.

Solomon BD, Barnes JR and Halvorsen KE (2009) From grain to cellulosic ethanol: history, economics and policy. In: Solomon BD and Luzadis VA (eds) Renewable energy from forest resources in the United States. Routledge, London, UK, pp 49–66.

Solomon BD (2010) Biofuels and sustainability. Ann NY Acad Sci 1185: 119–134.

Songstad D, Lakshmanan P, Chen J et al. (2009) Historical perspective of biofuels: learning from the past to rediscover the future. *In vitro* Cell Dev Biol-Plant 45: 189–192.

Stadler LJ and Sprague GF (1936) Genetic effects of ultra-violet radiation in maize: III: genetic effects of nearly mono chronmatic λ2537 and comparison of effects of X-ray and ultraviolet treatment. Proc Natl Acad Sci USA 10: 584–591.

Stuber CW, Polacco M and Senior ML (1999) Synergy of empirical breeding, marker assisted breeding selection and genomics to increase crop yield potential. Crop Sci 39: 1571–1583.

Stupar RM and Springer NM (2006) *Cis*-transcription and variation in maize inbred lines B73 and Mo17 leads to additive patterns in the F_1 hybrid. Genetics 173: 2199–2210.

Sun TP (2011) The molecular mechanism and evolution of the GA-GID1-DELLA signaling module in plants. Curr Biol 21: R338–345.

Swigonova Z, Lai J, Ma J et al. (2004) On the tetraploid origin of the maize genome. Comp Funct Genom 5: 281–284.

Tilman D, Hill J and Lehman C (2006) Carbon-negative biofuels from low-input high-diversity grassland biomass. Science 314: 1598–1600.

Tyner WE (2008) The US ethanol and biofuels boom: its origin, current status, and future prospects. BioScience 58: 646–653.

USDA (U.S. Department of Agriculture) (2011) Nutrient database. http://www.nal.usda.gov/fnic/foodcomp/search/.

USDA (U.S. Department of Agriculture) (2012a) Crop production 2011 summary. http://usda01.library.cornell.edu/usda/current/CropProdSu/CropProdSu-01-12-2012.pdf.

USDA (U.S. Department of Agriculture) (2012b) World agricultural supply and demand estimates. WASDE-512, Washington DC, USA.

USEPA (U.S. Environmental Protection Agency) (2010) Renewable Fuel Standard program (RFS2) regulatory impact analysis. EPA-420-R-10-006, Washington DC, USA.

van der Voet E, Lifset RJ and Luo L (2010) Life-cycle assessment of biofuels, convergence and divergence. Biofuels 1: 435–449.

van Heerwaarden J, Doebley J, Briggs WH et al. (2011) Genetic signals of origin, spread, and introgression in a large sample of maize landraces. Proc Natl Acad Sci USA 108: 1088–1092.

Venuto B and Kindiger B (2008) Forage and biomass feedstock production from hybrid forage sorghum and sorghum-sudangrass. Grassland Sci 54: 189–196.

Vermerris W and Boon JJ (2001) Tissue-specific patterns of lignifications are disturbed in the brownmidrib2 mutant of maize (*Zea mays* L.). J Agri Food Chem 49: 721–728.

Vermerris W (2009) Cell wall biosynthesis genes of maize and their potential for bioenergy production. In: Bennetzen JL and Hake SC (eds) Handbook of maize: genetics and genomics. Springer, New York, USA, pp 741–767.

Vermerris W (2011) Survey of genomics approaches to improve bioenergy traits in maize, sorghum and sugarcane. J Intergr Plant Biol 53: 105–119.

von Blottnitz H and Curran MA (2007) A review of assessments conducted on bio-ethanol as a transportation fuel from a net energy, greenhouse gas, and environmental life cycle perspective. J Clean Prod 15: 607–619.

Wang H, Nussbaum-Wagler T, Li B et al. (2005) The origin of the naked grains of maize. Nature 436: 714–719.

Wang K, Frame B, Ishida Y et al. (2009) Maize transformation. In: Bennetzen JL and Hake SC (eds) Handbook of maize: genetics and genomics. Springer, New York, USA, pp 609–640.

Wang M, Saricks CL and Wu M (1997) Fuel-cycle fossil energy use and greenhouse gas emissions of fuel ethanol produced from U.S. midwest corn. Argonne National Laboratory, Argonne, IL, USA.

Wang M, Wu M and Huo H (2007) Life-cycle energy and greenhouse gas emission impacts of different corn ethanol plant types. Environ Res Lett 2: 024001, 13 p.

Wisser RJ, Balint-Kurti PJ and Nelson RJ (2006) The genetic architecture of disease resistance in maize: a synthesis of published studies. Phytopathology 96: 120–129.

Woodyard C (2012) End of ethanol subsidy will raise the price of gas. USA Today. http: // www.usatoday.com/money/industries/energy/story/2012-01-03/ethanol-subsidy-gas-prices/52355056/1.

Xu Y, Skinner DJ, Wu H et al. (2009) Advances in maize genomics and their value for enhancing genetic gains from breeding. Int J Plant Genom Article ID 957602, 30 pages, doi: 10.1155/2009/957602.

Yan J, Yang Y, Shah T et al. (2010) High-throughput SNP genotyping with the GoldenGate assay in maize. Mol Breed 25: 441–451.

Yao H, Kato A, Mooney B et al. (2011) Phenotypic and gene expression analysis of a ploidy series of maize inbred Oh43. Plant Mol Biol 75: 237–251.

Yergin D (1991) The prize: the epic quest for oil, money & power. Simon and Schuster, New York, USA, 912 pp.

Yu W, Lamb JC, Han F et al. (2007) Cytological visualization of DNA transposons and their transposition pattern in somatic cells of maize. Genetics 175: 31–39.

CHAPTER 2

Basic Biological Research Relevant to Feedstock Conversion

Roman Brunecky,[1,]* Bryon S. Donohoe,[1]
Michael J. Selig,[2] Hui Wei,[1] Michael Resch[1]
and Michael E. Himmel[1]

ABSTRACT

Zea mays L. subspecies is an important feedstock for future bioconversion process of biomass to fuels and chemicals. However, conversion of this lignocellulosic material remains problematic due to the recalcitrant nature of corn stover. Here we examine some of the important considerations in the bioconversion of this feedstock, including a brief overview of the nature recalcitrance phenomenon, examining the anatomical as well as gross chemical factors of recalcitrance. Also examined herein are the major biopolymers and the role each one is thought to play in biomass recalcitrance. We then explore both current and future strategies employed to overcome natural plant cell wall recalcitrance, including an examination of pretreatment technologies and how these strategies inform the choice of enzymatic deconstruction systems. Finally we examine some potential future sources of biomass

[1]National Renewable Energy Laboratory, 15013 Denver West Parkway Golden, CO 80401, USA.
[2]Department of Forest & Landscape, University of Copenhagen, Rolighedsvej 23, 1958 Frederiksberg C, Denmark.
*Corresponding author: roman.brunecky@nrel.gov

deconstructing enzymes, in particular the promise of extremozymes as well as cellulosomal and termite enzyme systems that may be used as future platforms for the conversion of lignocellulosic materials to their constitutative sugars.

Keywords: *Zea mays* L. subspecies, Biofuels, Bioconversion, Thermophile, Termite, Cellulosome, CelA, CelE, Lignin, Cellulose, Xylan

Introduction

When considering industrial scale processes for the conversion of lignocellulosic feedstocks for the production of fuels and chemicals, a number logistical and technical factors must be considered, including feedstock availability, cost, sustainability, transportation, front end processing, and conversion. These considerations can be thought of as the general susceptibility of chosen/available feedstocks to the intended conversion process. The concept of feedstock susceptibility, or recalcitrance, has often been the driving force underpinning the selection of key 2nd generation biomass conversion technologies. The recalcitrance of the feedstock towards current conversion processes involves a number of factors from simple challenges, such as chemical composition, substrate density, particle size, and the initial moisture content of the supplied feedstock; to extremely complex issues associated with the highly integrated nature of the structural carbohydrates that develop within plant cell walls. The latter topic has been studied with growing intensity in recent years and has become deeply invested in a diversity of disciplines, including process and reactor design, enzyme engineering, advanced microscopy, numerical modeling, and studies of the emergent properties of the plant cell wall.

Work in this field over the last thirty years has focused on developing processes to improve the accessibility of enzymes from very specific fungi to plant cell wall carbohydrates from defined energy feedstocks. The most widely studied of these cell wall polymer degrading systems is that secreted by the filamentous fungi, *Trichoderma reesei*. Commercial *T. reesei* preparations are often supplemented with key enzymatic activities from other fungal enzyme systems. The *T. reesei* system, while an efficient cellulase degradation system, is only moderately effective in converting whole lignocellulosic materials to fermentable sugars. Whereas effective at high protein loadings, native *T. resesei* cocktails are known to be inadequate in supplying necessary enzymes, such as β-D-glucosidases for conversion of cellobiose to monomer glucose; as well as key hemicellulose degrading enzymes. The latter are considered necessary to effectively reveal and hydrolyze the complex networks of carbohydrates that reduce access of cellulases to crystalline cellulose.

The current majority of thermal chemical bioconversion technologies have been aimed at providing the most efficient path for the enzyme cocktail to achieve high monomer sugar yields. Unfortunately, the rates of conversion have been moderate at best. The recalcitrant nature of targeted lignocellulosic energy crops to conversion has resulted in the testing of an assortment of pretreatment technologies. In the recent decades, numerous processes have been thoroughly investigated at the lab bench scale; these methods include those performed at acidic, neutral, and alkaline pH; as well as oxidative and steam explosion. The primary contenders at the industrial scale, however, have been dilute acid or base pretreatments at moderate temperatures (< 200°C) (Gould 1984; Kohlmann et al. 1996; Wyman et al. 2005; Larsen et al. 2008; Elander et al. 2009). Gaining an understanding of the fundamental nature of biomass recalcitrance and the mechanisms by which specific technologies prove effective has become increasingly critical as we attempt to further improve the cost effectiveness of these technologies.

In this chapter, we will review the current state of knowledge regarding the recalcitrant nature of plant cell wall materials with respect to the wealth of research specifically for corn stover (e.g., corn stover is the post-harvest residual plant material of *Z. mays*). We will also show how this knowledge applies to feedstock candidates for 2nd generation biofuel crops. We will explore other possibilities for developing enzyme systems that are not based on fungal sources. To date, for example, bacterial systems have not been utilized for industrial bioconversion of biomass, though selections of relevant enzymes from these systems have been characterized. One of the advantages of examining bacterial systems is the large degree of diversity bacterial systems provide. And while it is tempting to speculate, the current enzyme cocktails for bioconversion have been developed for decades in industry, so new novel bacterial systems must be "industrialized" before they can be utilized. Briefly, we will explore the possibilities of using bacterial "extremozymes" to develop novel biomass degrading cocktails, or supplement existing fungal cocktails. Also, from a biomass deconstruction point of view, we will also consider alternative enzyme systems, such as the highly synergistic and tethered cellulosomal system, as well as insect enzyme systems for biomass degradation.

Cell Wall Research

The Recalcitrance Concept, The Opposite of Amenable

The term "recalcitrance" was adopted by the biomass conversion field to portray a concept designed to capture a wide range of phenomena that have collectively confounded scientists and engineers trying to maximize the conversion of plant biomass polymers to fermentable sugars (Himmel et al. 2007). Recalcitrance is clearly a multiscale concept, which covers physical and chemical barriers that extend from plant anatomy, to the cellular architecture, and finally to the chemistry of cell wall polymers. In this section, we will highlight some examples of these barriers and prepare for the greater descriptions given in the following sections. In thinking about the barriers that contribute to recalcitrance, it is useful to envision all of the mechanisms that maize plants have evolved to protect themselves from environmental insults; such as desiccation and assault by microbial pathogens.

An example of a natural physicochemical barrier is that the outermost surface of maize leaves and stems is covered in a layer made of cutin and other waxy compounds that are hydrophobic and water impermeable. This barrier evolved to prevent water loss in the living plant, and in the context of biomass conversion impedes the penetration of pretreatment chemicals or microbial enzymes through those surfaces. Another example at this scale is the trapped air that fills the lumen space of most cells in field-dried, senesced corn stover. Viamajala and coworkers showed how the trapped air can be a significant barrier to bulk transport within corn stover stalks (Viamajala et al. 2006). Until the trapped air is displaced by liquid, usually achieved by soaking in an impregnation step, neither pretreatment chemistries nor cellulolytic enzymes can gain access to the cell wall surfaces from the lumen side (Weiss et al. 2009).

The challenge for enhancing enzyme accessibility is based on the barriers extending from the anatomical to the cell wall ultrastructural level. In conducting biomass pretreatment, generating an enzyme accessible substrate is the goal. One of the most important natural barriers to efficient biomass conversion within plant cell walls is lignin. Lignin is only found in the thickened secondary cell walls, but these cells and their heavy walls contain such a high percentage of cellulose in corn stover that it becomes a critical barrier to overcome to achieve high sugar yields (Donaldson 2001). Lignin is one of the components, along with hemicelluloses, that make up the matrix that surrounds and largely encases cellulose microfibrils in cell walls.

At the macromolecular scale, the main barrier to cellulose depolymerization is the structure of the crystalline microfibril. The energies it takes to decrystallize the microfibril have recently been explored with molecular dynamics (Beckham et al. 2011). There are numerous thermal, chemical, and biological depolymerization systems that work well on purified cellulose, so it seems most appropriate to consider recalcitrance a complex, multiscale phenomenon. The most effective and economical means of overcoming recalcitrance have been the use of thermal chemical pretreatments followed by saccharification using a cocktail of specialized enzymes (Anex et al. 2010).

Anatomical and Tissue-Scale Attributes of Corn Stover as a Feedstock

Corn stover includes all parts of the maize plant, except the roots and the kernels. Anatomically, this includes the following fractions: husks, cobs, leaves, and stalks (Fig. 1). Each of these fractions differs in amenability to conversion and make up different mass fractions of the composite mixture of stover. Because they are the most easily digestible fractions and part of the first harvest from the corn field, some groups have focused exclusively on corn cobs for their biorefinery process models (Jenkins and Alles 2011).

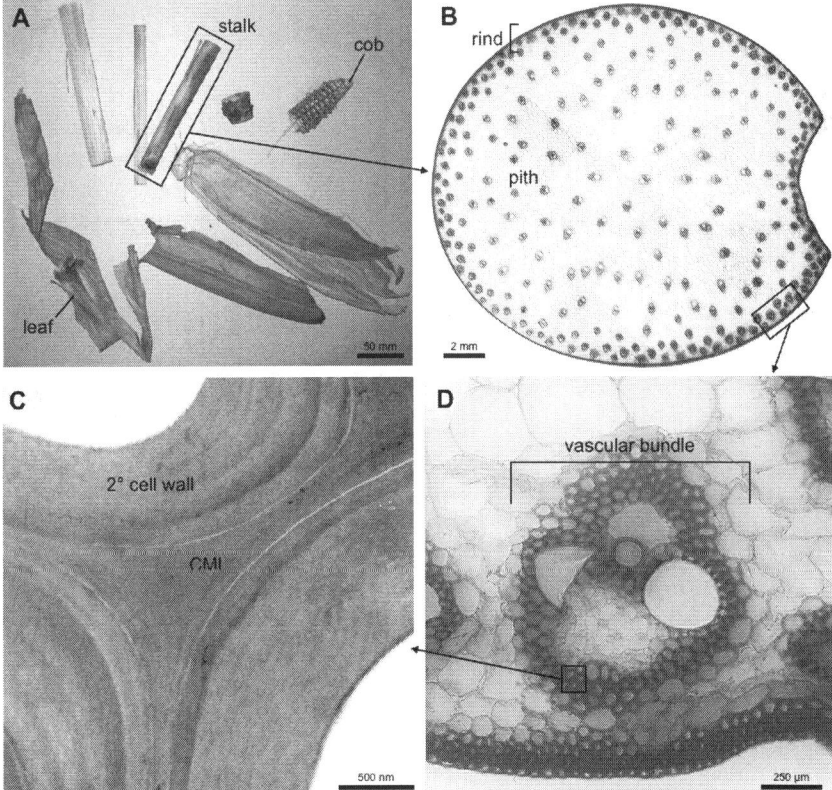

Figure 1 Multi-scale imaging of corn stover structural complexity from tissue types to the cell wall ultrastructure. (A) Photograph of tissue types that comprise corn stover including stalk, cob, leaf, and husk. (B) Corn stover stalk cross-section stained with phloroglucinol to highlight the lignin containing, thick-walled cells in the vascular bundles and epidermis. (C) Transmission electron micrograph of the cell walls between adjacent fiber cells surrounding a vascular bundle. (D) Higher magnification image of a single vascular bundle from the rind of a corn stover stalk.

Color image of this figure appears in the color plate section at the end of the book.

In contrast, much of the reported laboratory research on corn stover conversion focused on the corn stalk and even more specifically, on the rind of the corn stalk. The logic behind this focus is that the stalk comprises ~ 70% of the mass of the corn stover and the rind comprises ~ 70% of the stalk (Zeng et al. 2011). The reason for the high mass content of the rind is that the rind contains a high concentration of vascular bundles. These bundles are dominated by cell types with thick, lignified secondary walls. These xylem and sclerenchyma (also called fiber) tissues are the most recalcitrant tissues in the stalk and; therefore, have been the focus of extensive study. Another component of the rind that contributes to its recalcitrance is the outer epidermis that is coated with a waxy, water impermeable cuticle. One hypothesis is that pretreatment and enzymatic digestion strategies can be developed that can completely deconstruct rind fiber cells. Those processes would then be able to treat any other part of the plant effectively.

The corn stover biomass that is delivered to the biorefinery gate has been homogenized to some extent by grinding and milling. Some amount of particle size reduction is necessary for densifying the biomass for transport from the field to the biorefinery. There is still some controversy about how much milling is economically feasible. The canonical view is that some milling and size reduction is going to be required for transport and transfer issue, but that mechanical size reduction smaller than a particle size of about ¼" is not cost effective (Yang and Wyman 2008). At this size, a great deal of tissue and cellular structure remains intact and the barriers are still a concern.

The Fiber Reinforced Composite Concept of Plant Cell Wall Architecture

Whereas there is some variability in anatomical fractions, all of the tissue types are composed of cells with cell walls made of the same basic components dominated by cellulose, hemicelluloses, and often lignin (Jung and Casler 2006, 2006). In thinking about the architecture of plant cell walls, it is important to consider the time-line and developmental context in which the cell wall layers were deposited. A cell wall originates in a cell within one of the growing/dividing or meristem regions of the plant. As a cell grows to a defined size, it prepares to divide and construct a new wall to partition it into two halves. This initial cell wall, called the phragmoplast, later becomes the middle lamella region of mature biomass cell walls (Fig. 2A). The cell wall continues to develop with cellulose and pectin synthesis into a complete primary cell wall. The primary cell wall exists in cells while the cells are still expanding and elongating. Many cell types, including the parenchyma cells in the corn stalk pith, will only develop a primary cell wall.

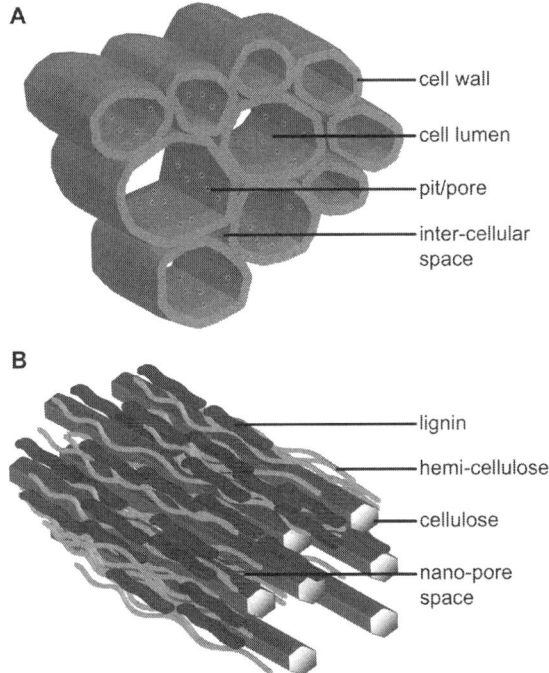

Figure 2 (A) Representation of plant ultrastructural elements affecting enzyme digestion. (B) Representation of chemical factors affecting enzyme digestibility of biomass.

However, fiber cells provide a structural role for the stalk. These cells eventually cease elongating and begin to reinforce their primary cell wall with additional cell wall layers, known as secondary cell walls. The secondary cell wall is built up in layers using the primary cell wall as a foundation and may become up to ten times as thick as the original primary cell wall (Fig. 1D). Once the secondary cell wall is formed, it may be lignified by the sequential deposition and polymerization of lignin monomers into a complex three-dimensional matrix that fills in gaps among the cellulose microfibrils (Boudet 2000; Monties 2005). The cell types that end up with thickened, lignified secondary cell walls conduct water (xylem) through the stalk and leaves or in structurally supporting the water carrying vessels (e.g., fibers) (Joseleau and Ruel 1997).

The macromolecular architecture of the plant cell wall is envisioned as a fiber reinforced composite material, much like fiber glass or steel reinforced concrete. The cellulose microfibrils are the fiber component that provides tensile strength and rigidity. The hemicellulose/lignin composite matrix provides compressive strength and elasticity (Fig. 2B). Deconstructing biomass, therefore, often involves undoing the sequential processes of cell wall assembly.

The matrix components of the cell wall are the most labile and are significantly reduced by pretreatments. Acid and hot water pretreatments hydrolyze and solubilize xylan. Base pretreatments remove lignin. By removing the hemicellulose/lignin matrix (Fig. 2B), the cell wall can become porous and delaminated (Weiss et al. 2009). Once the cell wall is 'opened up,' it becomes accessible to the enzymes that depolymerize the carbohydrate polymers (Jeoh et al. 2007).

Particle Size Reduction

Size reduction is mechanical process of cutting, shredding, or milling that can be incorporated into harvest or post-harvest steps. There is no doubt that particle size reduction is an important step in any bioconversion process. However, there is still some debate as to how much size reduction is economical and at what stage in the harvesting, preprocessing, densification, storage and biorefinery process the size reduction should take place. Also, the final size and form that the biomass is reduced to ultimately represents the best balance of cost and increased performances. On the positive side, particle size reduction increases exposed surface area and generates a material that is more homogenous with improved feeding and mixing qualities (Fig. 3). The drawback is the cost and the complication of an additional process step. Additionally, particle size is an indicator of external surface area, but not necessarily an indicator of accessibility within the biomass particle volume, which is more important for enzyme activity. In the most recent state of technology techno-economic analyses, feedstock price is one of the main cost drivers for the biorefinery, costing approximately \$60 per dry ton (Anex et al. 2010). This price assumes that the biomass is being chopped to a particle size of ~ 5 cm. This moderate amount of size reduction could be done concurrently with the harvest. More extensive size reduction or densification will add significantly to the cost of the delivered feedstock, but those costs may be partially offset by the reduced cost of transporting a more energy dense material (Daniel Ryan 2008). There are still a number of questions and variables that complicate the question of the impact of particle size reduction. For example, the energy consumption of milling varies greatly with the type of biomass and with its moisture content (Cadoche and López 1989). In a study that compared corn stover pretreatment using steam explosion, the authors found that the steam explosion used only 60% of the energy of milling to generate the same increase in surface area. In a recent review, Vidal and colleagues (Vidal et al. 2011) presented a summary of the maximal particle size for a variety of pretreatments. Maximal particle size is defined as the size below which no further particle size reduction results in an increase in pretreatment. Steam explosion has a larger maximal particle sizes (> 10 mm), dilute acid and

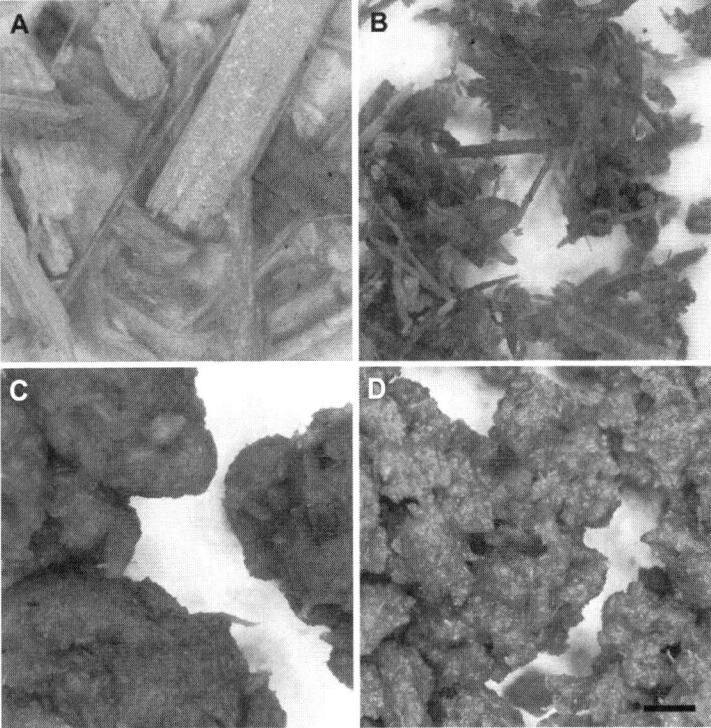

Figure 3 Stereo micrographs of dilute-acid pretreated (2% H2SO4, 160°C, 5 min) corn stover. (A) Control sample preprocessed by knife milling to ¼ inch. (B-D) Pretreated samples from a zipperclave (B), steam gun (C), and horizontal screw feed reactor (D) display varying degrees of particle size reduction and clumping. Bar = 1 mm.

Color image of this figure appears in the color plate section at the end of the book.

base (ammonia fiber expansion (AFEX) lime and dilute alkali) treatments had smaller maximal particle sizes (< 3 mm), and the liquid hot water pretreatments had maximal particle sizes were intermediate of these two pretreatment groups.

Chemistry of the Major Cell Wall Polymers in Senesced Corn Stover

During the life of the plant, the heterogeneous complex of the plant cell wall serves a number of functions which are often thought to include structural stability, maintenance of hydration, and defense against insects and pathogenic micro-organisms. Considering that much of the biological world attempting to invade these carbohydrate-rich materials has utilized numerous, highly adaptable, and complex enzymatic systems to extract desired carbohydrates, it can be expected that the evolution of the plant materials themselves would have logically moved in the direction of being highly resistant to degradation by such systems. With this in mind, we can easily see how the notion that the abundant diversity of degradation systems in nature should make it possible to completely dismantle lignocellulosics by purely enzymatic means is naïve at best.

Most lignocellulosic feedstocks that have been recently considered with respect to the aforementioned "sugar platform" have fallen into a fairly narrow compositional domain. This is typically characterized by the dominance of mature secondary cell-wall material, resulting in a lower abundance of complex pectins and other hemicelluloses. These two polymers have been more recently referred to as "cross-linking glycans," although nomenclature use appears to be a rather preferential debate (Gorshkova et al. 2010; Scheller and Ulvskov 2010). The bulk of this material consists of a highly intermeshed matrix of crystalline cellulose, lignin and hemicelluloses (or cross-linking glycans). According to Gorshkova and coworkers (Gorshkova et al. 2010), "in all cell wall types, the backbone consists of cellulose microfibrils and the differences are observed first of all in the composition of matrix polysaccharides, in particular major cross-linking glycans."

In the case for corn stover, the majority of the cell wall mass is secondary cell wall, and is predominantly characterized by an abundance of long chain, xylan backbone decorated to varying degrees by acetyl, arabinofuranosyl, 4-O-methylglucuronsyl side groups. In addition, these crops normally contain a range of silica contents from a few percent up to 30% of the total lignocellulosic mass; these levels are typically highest among grass species (Currie and Perry 2007).

The primary target product of most conversion processes is the fermentable monomer, glucose. Glucose is, to a large extent, tied up as crystalline cellulose characterized by hydrogen bonded chains of β-1,4 linked glucosyl units. Other available glucose units are also known to be present, to much lesser extents, as xyloglucans, glucomananans, and mixed linkage glucans, such as β-1,3 glucans or mixed linkage β-(1,3); β-(1,4) glucans, depending on the plant species. Natural cellulose itself has almost entirely been reported to exist in crystalline form, characterized primarily by a triclinic conformation, Iα, and a monoclinic form, Iβ, with the latter being the predominant cellulose in higher plants (Matthews et al. 2006; Scheller and Ulvskov 2010). Crystalline cellulose microfibrils, consisting of packed glucan chains ranging from 36 in plants to more than 2,000 per microfibril in some algae, are thought to regularly contain alternating regions of crystallinity and disorder (Matthews et al. 2006). These regions of disorder have been observed with advanced imaging and referred to as "dislocations" by some researchers. The disordered regions have been the subject of increasing interest with respect to the understanding of microfibril formation and function within the plant cell wall (Thygesen et al. 2007). In general, it is thought that cellulose microfibrils are held together loosely by a network of various cross-linking glycans (hemicelluloses) using hydrogen bonding and carbohydrate-carbohydrate interactions (Gorshkova et al. 2010).

The 2nd most abundant target sugar in energy crops, such as corn stover, is the xylan backbone of hemicellulose, which can be converted to xylose monomers and used as a fermentation feedstock. Somewhat less abundant are the complex xylooligomers, which have been proposed as a resource for animal feed (Wyman et al. 2005; Larsen et al. 2008). Xylose sugars are typically present in corn stover in the form of the highly substituted, medium-chain β-(1,4) linked arabinoglucuronoxylan. This polymer does not form crystalline structures and is typically interlaced throughout the cell wall carbohydrate matrix. Xylose can also be present in other hemicellulose forms, such as xyloglucans; however, the xylan backbone is the most predominant in secondary cell wall rich materials. Substitutions on individual xylose units include acetyl, arabinofuranosyl, and methyl glucuronosyl groups that vary in placement and abundance widely depending on the plant material. Additionally, arabinofuranosyl units often cross-link the xylan backbones to lignins in the plant cell wall via feruloyl ester units, which are thought to be ether bonded to the lignin (Jeffries 1990; Scheller and Ulvskov 2010). In corn stover specifically, chemically isolated xylans from the material have been shown to be primarily decorated with 2-*O*- and 3-*O*-monoacetyl, [MeGlcA-α-(1,2)] [3-OAc], and arabinofuranosyl units (Naran et al. 2009).

The third major component of the plant cell wall, lignin, is present as a complex heteropolymer whose development within the matrix has been highly debated and to date not fully understood. In nature, lignins represent one of the most abundant classes of polymer on earth, which are most commonly made up of β-*O*-4 linked phenylpropanoids, specifically para-hydroxyyphenol (H), guaiacyl (G) and syringal (S) depending on the degree of methoxylation. These compounds are primarily derived from coniferyl, synapyl, and p-coumaryl alcohols, respectively. According to Lewis and coworkers, grasses are often complicated by the presence of hydroxycinnamic acids (Lewis and Yamamoto 1990). The primary historical debate is related to the polymerization process, with one view stating that the phenoxy radical coupling resulting in lignin polymerization is a completely random process, while the other view claims that the formation of lignins is under full control by biological means. This dilemma has spawned a significant amount of true scientific debate over the decades (Lewis and Yamamoto 1990; Norman G 1999; Boerjan et al. 2003; Chen and Sarkanen 2010). In the case of corn stover, lignin contents generally range from 10 to 20% of the total dry mass, although this content often depends on the anatomical fractions harvested.

Another major component of plant cell walls, the pectins, are most abundant in the primary cell walls of lignocellulosic plant materials, but also exist within the middle lamella separating adjacent plant cells (Donohoe et al. 2008). They are considered to be "highly heterogeneous polysaccharides" that are the least resistant to extraction from the cell wall matrix (Scheller and Ulvskov 2010). They consist primarily of substituted β-D-galactosyluronic acidic polysaccharides, including polygalacturonan, rhamnoglacturonan I, and rhamnogalacturonan II (Gorshkova et al. 2010). In corn stover, where mass is typically dominated by secondary cell walls post harvest, pectin contents are rather low (usually less than 5% of the total dry weight) and are generally not considered a substantial source of fermentable sugars; although to date there has been little study of these polysaccharides in the context of 2nd generation energy crops.

Finally, the study of silica and the functional roles it may play in plants is an undervalued topic among plant biologists, we feel. This cell wall component has, to date, been almost completely ignored as a factor with respect to biomass recalcitrance in the biofuels arena. Accumulation of silica in plant cells occurs from the uptake of silicic acid from the soil, followed by a "controlled polymerization" in destination cells. While some aspects of silica uptake mechanisms are known, the potential consequence of silica association within the plant cell wall is poorly understood. Whereas only limited knowledge exists to date, there is growing evidence suggesting that silicas play important roles in relieving biological, structural, and toxicological stresses in natural environments (Currie and Perry 2007; Epstein 2009). In the context of this chapter, corn stover is a member of the grass family Poaceae, which is considered to be one of the most silica rich plant families, with secondary cell wall silica contents ranging from 5 to 15% of dry mature biomass (Currie and Perry 2007).

Chemical Factors Contributing to Lignocellulosic Recalcitrance with Respect Corn Stover

In a recent article, Himmel and co-workers (2007) strongly reiterated and strengthened the importance of the phrase "biomass recalcitrance," thus expanding upon the traditionally short explanation describing the difficultly to which lignocellulosic structural carbohydrate can be degraded by biological systems. In their writings, these authors included a number of anatomical factors for plant epidermal tissues, these are "the arrangement and density of the vascular bundles" and "sclerenchymatous" tissues, lignifications within the plant cell wall, and "the structural heterogeneity and complexity of cell-wall constituents, such as microfibrils and matrix polymers." Additional difficulties are associated with the hydrolysis of insoluble substrates, including naturally occurring inhibitory compounds that can be extracted from plant cell walls, the crystalline nature of cellulose in plant cell walls, the presence of "amorphous cellulose" and hemicellulose directly surrounding cellulose microfibrils, and mass transport limitations generally created by the "complex heterogenous nature" of plant cell wall materials (Himmel et al. 2007). It is clear that the classical viewpoint of recalcitrance, which states that recalcitrance is primarily associated with the "physical presence of lignin and hemicellulose and the form of cellulose, with crystalline cellulose being more recalcitrant than the amorphous form," is overly simple (Kohlmann et al. 1996).

Today, there remains considerable enthusiasm for resolving the problem of biomass recalcitrance; however, decades of work have provided new insight which have redefined our approaches the problem. For example, Yang and Wyman (2004) have recently reported that lignin removal positively improves cellulose conversions by fungal enzymatic systems. They also state that removal of lignin from the substrates is not necessary to achieve acceptable conversion levels for commercialization. It has been more historically noted that a direct relationship between hemicellulose removal and enzymatic cellulose conversions exists. The basic message in this field has been that for enzymatic systems to be effective, the substrate must become more accessible to the enzymes (Fig. 4). This apparent requirement for thermal chemical pretreatments, then, addresses this need to physically "open up" cell wall matrices, often referred to as "cell wall delamination." One might ask, what else can we do to enhance the effectiveness of thermal chemical biomass pretreatments?

Recent coordinated assessments of the effectiveness of leading pretreatment chemistries on corn stover have clearly confirmed the notion that there are many specific paths forward for improving the accessibility and conversion of complex carbohydrates in lignocelluloses, depending upon the feedstock type and process outcomes desired (Mosier et al. 2005; Wyman et al. 2005). Dilute sulfuric acid, hot water flow through, hydrothermal and sulfite steam explosion, AFEX, ammonia recycle percolation, and lime pretreatments were all shown to cost effectively prepare corn stover for enzymatic production of fermentable sugars, while imparting differential changes to the substrate matrix (Mosier et al. 2005; Wyman et al. 2005). Acidic and hydrothermal pretreatments are highly effective at hydrolyzing and solubilizing hemicellulosic constituents of the cell wall to varying degrees, ranging from complete reduction to monomer sugars to solubilization of complex hetero-polysaccharide oligomers. This pretreatment chemistry is also known to result in the mobilization and redistribution of cell wall lignins as spherical bodies lodged throughout the biomass solids. Alternatively, ammonia and lime alkali pretreatments are effective at lignin solubilization and depolymerization, but tend to leave the hemicellulosic backbone sugars intact.

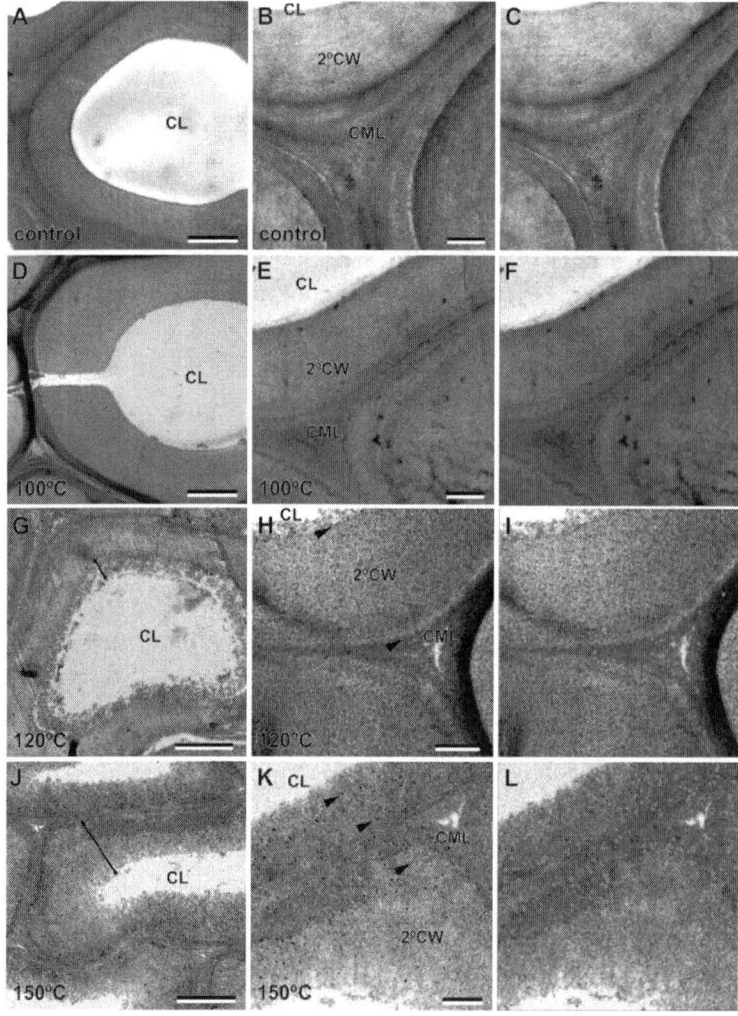

Figure 4 Anti-Cel7A immuno-EM micrographs of corn stover stalk sclerenchyma cells. Control samples before pretreatment (A-C) and samples pretreated at 100°C (D-F), 120°C (G-I), 150°C (J-L) in 2% H2SO4 for 20 min, then digested for 5 days at 50°C with 15 mg/g Spezyme CP. Samples were labeled with anti-Cel7A antibody, and detected with 15 nm gold (arrowheads (H, K), orange circles (I, L)) conjugated secondary antibody. After 100°C (D-F) pretreatment, the cell walls appear little changed and only rarely display any antibody on the lumen surface of the cell wall. Following 120°C (G-I) pretreatment, enzymes are able to penetrate the cell wall (F, arrows), however, the enzymes have only partially penetrated the walls (E, barbell). After 15°C (J-L) pretreatment the penetration of enzyme is throughout the depth of the cell wall (G, barbell, and F, arrowheads). CL, cell lumen; 2°CW, secondary cell wall; CML, compound middle lamella. Scale bars = 1 μm (A, D, G, J), 500 nm (B, E, H, K). Modified and reprinted with permission from Donohoe et al. 2008.

Color image of this figure appears in the color plate section at the end of the book.

Lignin

Numerous studies over recent decades have consistently shown delignification to effectively improve depolymerization costs by commercial cellulase enzymes, primarily cellulases systems produced by *Trichoderma reesei* (Gharpuray et al. 1983; Gould 1984; Converse et al. 1990; Schell et al. 1992; Lu et al. 2002; Yang and Wyman 2004; Hendriks and Zeeman 2009). The necessary degree of lignin removal as been much debated, however. This issue likely stems from traditional experimental design, which until recently operated in a typical engineering style, black-box manner; where biomass was treated as a block of polymers. The reality of what happens to the lignin matrix during pretreatment was not well understood. Additionally, it should be noted that effective conversion of the delignified substrate by commercial cellulase preparations does not necessarily mean that delignification alone improves access to cellulose, because such enzyme preparations typically have hemicellulose degrading enzymes which also improve enzyme access to substrate.

Recently, work at the National Renewable Energy Laboratory (NREL) has shown that during pretreatment, lignins have the tendency to aggregate when taken above their glass transition temperature (~ 120 to $160°C$), which permits their emergence and migration from, and even redeposition of lignin and xylan onto treated cell walls and surfaces (Selig et al. 2007; Donohoe et al. 2008; Brunecky et al. 2009). This concept was suggested in theory nearly two decades ago by (Fukagawa et al. 1991). Selig and co-workers showed that even this lignin, redeposited on pure cellulose surfaces, can impede enzymatic hydrolysis. Furthermore, Jeoh and co-workers (Ishizawa et al. 2009; Jeoh et al. 2007) have shown that delignification of pretreated, low-xylan containing corn stover can improve hydrolysis by purified cellulases; whereas both sets of data support the potential for lignin to impede cellulose hydrolysis, they do not prove that the primary surface affected is cellulose. Digestion of alkaline peroxide delignified corn stover with purified cellulase and xylanase cocktails has shown that lignin removal alone does not directly improve cellulase performance. However, delignification does improve conversions of the residual xylan fractions by purified xylanase mixtures, which, in our study, was a combination of a $β$-(1,4) endoxylanase and $β$-D-xylosidase. Only when the mixtures of purified xylanase and cellulases are introduced into the hydrolysis mixture together, does the direct relationship between lignin removal and cellulose conversion emerge, indicating that most of the impediment lignin represents for cellulase access to cellulose is indirect (i.e., lignin limits access to hemicelluloses by hemicellulases and hemicelluloses impede access by cellulases to cellulose) (Selig et al. 2009). One may speculate that the primary benefit from high temperature hydrothermal and acid treatments is that the lignin fraction, while not removed, is mobilized and potentially distinguished from hemicellulose sufficiently to permit access by hemicellulases.

Interestingly, even though partial removal of lignin seems beneficial to enzymatic conversion and thus, process economics; there has been growing evidence that removing lignins completely may modify plant cell walls. Jeoh and co-workers suggested that, "near complete removal of xylan and lignin may cause aggregation of the cellulose microfibrils resulting in decreased cellulase accessibility to cellulose" (Ishizawa et al. 2009). In this case, the notion that complete removal of cell wall matrixing polymers would cause structural collapse seems intuitive only in hind sight.

Hemicellulose

If we can tie the benefits of lignin removal to improving enzyme access to hemicelluloses, then a good argument could be made that hemicellulose removal is also essential for optimal cellulose conversion. Numerous reports in the pretreatment literature have shown a direct correlation between cellulose conversions and xylan and/or hemicellulose removal. In addition, other studies utilizing purified cellulose and hemicellulose hydrolyzing enzymes have shown that even when the xylan component in lignocellulose remains intact following pretreatment, the enzymatic hydrolysis of this fraction dramatically improves cellulose hydrolysis by cellulases; this has also been repeatedly shown to be a linear relationship (Selig et al. 2008a,b, 2009). With respect to xylan, the primary reason for the impediment is thought to be steric hindrance, but there is currently no conclusive evidence that this is the true mechanism, leaving other plausible mechanisms as potential explanations. We can conclude now that the removal of hemicellulose, xylan in particular in the case of corn stover, is critical if glucose sugar yields are to be maximized.

As we have already stated, the xylan backbone in lignocellulose, and particularly in corn stover, is typically decorated with a number of different constituent side groups; in corn stover these are predominantly acetyl, methyl glucuronosyl, and arabinofuranosyl residues; these side groups can be effectively cleaved from the xylan backbone using enzymes such as acetyl xylan esterase, α-glucuronidase, and α-L-arabinofuransidase enzymes, respectively (Naran et al. 2009). The impact on xylan hydrolysis of such side groups was first pointed out by Biely and co-workers who concluded in their study on acetyl xylan esterases that, "enhancement by acetyl xylan esterase of the extent hydrolysis of glycosidic bonds of acetyl xylan indicated that the role such activity plays will require consideration whenever acetyl xylan is a substrate" (Biely et al. 1986).

In addition, there have been numerous reports regarding the potential for arabinofuransyl side groups to impede hydrolysis of the xylan backbone in arabinoxylans (Sørensen et al. 2003). In corn stover, we have previously shown that remaining acetyl side groups, and additionally feruloyl esters linked to arabinofuranosyl side chains, have a direct impact on the hydrolysis of residual xylan present after chemical pretreatment (Selig et al. 2008b; Selig et al. 2009). In addition, work performed at NREL showed that the impact of arabinofuranosyl and methyl glucuronic acid side groups in xylans isolated from native corn stover appears to be less significant than that of the acetyl groups; although this may simply be a matter of the quantity of substituent versus the degree of recalcitrance imparted to the xylan backbone. A study at NREL found that hydrolysis of all three side chains resulted in a nearly complete xylan hydrolysis by a pairing of endoxylanase and β-D-xylosidase enzymes, compared to about 60% conversion when debranching activities were not present. Whereas typical *T. reesei* cellulase preparations do contain sufficient levels of acetyl xylan esterase, enhanced expression of this activity could be beneficial in some process scenarios, particularly milder, neutral pH pretreatments which leave significant portions of acetylated xylan in place. Additionally, commercially available hemicellulase preparations have been shown to be effective in providing adequate supplementation of debranching enzyme activities; for example Novozymes Ultraflo L has been effectively shown to dramatically enhance the hydrolysis of arabinoxylans (Sørensen et al. 2007).

With secondary cell wall material comprising the bulk of the mass of lignocellulosic energy crops, most of the research regarding hemicellulose and recalcitrance has been focused on the xylan backbone and little work to date has focused on the impact of other less abundant hemicelluloses; particularly those associated with primary cell walls, such xyloglucans, glucomannans, and mixed linkage glucans, β-D-(1,3-1,4) glucan (Scheller and Ulvskov 2010). All of these carbohydrate are known to exist in low to moderate levels in the primary cell walls of grasses and often only in trace amounts in the secondary walls. Such carbohydrate chains certainly have the potential to loosely associate with target carbohydrates, such as cellulose and glucuronoarabinoxylans (Geoffrey B 2009). Whereas limited work has been reported on the impacts of these hemicelluloses on recalcitrance, Benko and co-workers have reported minor benefits with respect to glucose release from pretreated grass species, including corn stover, associated with the addition of a xyloglucanase obtained from *T. reesei* to simple cellulase mixtures; it should be noted here that the minor nature of these improvements does make it difficult to discern whether or not the benefit is additive or synergistic (Benkő et al. 2008). Internal work at NREL has also show, synergistic enhancements in the hydrolysis of variously pretreated corn stover samples with the addition of enzymes capable of hydrolyzing β-D-(1,3) linked glucan.

Crystalline Cellulose

The crystalline nature of cellulose in plant cell walls has always been regarded as a key contributor to lignocellulose recalcitrance; even though cellulose degrading enzyme systems have evolved to digest the crystalline form of cellulose; albeit at slow rates (Fan et al. 1981; Kohlmann et al. 1996; Himmel et al. 2007). Jeoh and co-workers have also reported significant increases in 24 h conversion extents; as well as the binding of a cellobiohydrolase (Cel7A) to amorphous cellulose (following Schroeder et al. 1986) compared to the native and highly crystalline form (Schroeder et al. 1986; Jeoh et al. 2007). Furthermore, Chundawat and co-workers attributed the "amorphous-like" nature of cellulose III produced from commercially available cellulose I (Avicel) to the significant increases in general cellulase synergy they observed on ammonia treated cellulose III compared to the cellulose I; interestingly, these authors also observed higher conversion rates and binding capacities for purified cellulases on cellulose I versus cellulose III (Chundawat et al. 2011). These studies show that some treatments (especially alkaline) known to disrupt the hydrogen bonding networks in cellulose, resulting in non native isoforms, can impact the action of enzymes evolved to act on native celluloses. Furthermore, if we consider the thesis that native cellulose is interspersed with regions of disorder, one could postulate that the acidic thermal chemical pretreatment process results in a modified cellulose, relaxed upon cool down, which contains more extensive regions of crystallinity, interspersed with regions of concentrated disordered and perhaps highly twisted cellulose (native cellulose has been postulated to have a twist (Matthews et al. 2010). If the disordered cellulose is also amorphous-like, this would explain the enhanced cellulase action observed for dilute acid pretreated cellulose.

The notion of non-productive binding of enzymes, both cellulolytic and hemicellulolytic, is also important to understand. Carbohydrate binding domains are commonly part of lignocellulose degrading enzymes and whereas they are often highly targeted, they can also bind to many other structures non specifically. Some proteins also have exposed regions of hydrophobicity which may also be susceptible to interaction with biomass. One example of this is that residual lignins in pretreated biomass have been shown to bind to cellulases and other proteins due to their hydrophobic nature (Yang and Wyman 2006; Gao et al. 2011). Yang and co-workers showed that actively blocking binding sites/regions on lignins by introducing non-functional proteins, such as bovine serum albumin prior to cellulase addition, improved conversion rates and extents (Yang and Wyman 2006). The concept of non-productive binding/loss of enzymes to lignin is also supported by the observation that some surfactants, when added to the hydrolysis mixture, improve enzymatic conversion rates (Qing et al. 2010).

Also important is the notion of the non-productive enzyme interactions with process generated xylose and xylooligodextrins in the hydrolysis mixture. Qing and co-workers showed that the complex xylooligomers, much more so than monomer xylose and insoluble xylan, are strongly inhibitory towards hydrolysis of pretreated lignocelluloses using commercial cellulase complexes (Qing et al. 2010). Even more importantly, these authors noted that prehydrolysis of xylan and xylooligomers prior to cellulase addition was more effective than simultaneous hydrolysis of these carbohydrates, suggesting that the hemicellulosic carbohydrates may irreversibly bind to and essentially inactivate the cellulases. These data clearly raise questions about substrate-enzyme interactions; one could, for instance, speculate that xylose and xylooligomers pose a significant threat to effective cellulose hydrolysis by interacting with the cellulose surface; as well as with cellulases and other key enzymes. Soluble hemicelluloses also reduce the titer of free-water present in the hydrolysis mixture; however, the impact of this effect is not currently known.

Related to the aforementioned examples of non-productive substrate-enzyme binding, Crowley and co-workers recently presented an intriguing set of numerical modeling experiments which may accelerate the deeper understanding of the nature of such interactions. They have recently shown that the free energy of the removal of cellooligomers from the cellulose surface is slightly negative for cellobiose and slightly positive for cellotetraose, suggesting that these oligomers can diffuse away relatively easily from the cellulose surface. The higher DP chains require considerable energy to enter the aqueous phase, suggesting that cellulases play a major role in stabilizing the non crystalline form of these cellodextrins (Payne et al. 2011).

Musings about Recalcitrance and Enzyme-Substrate Dynamics

Much of what has been presented so far has been from the perspective of the chemical structure and physical association of components within the cell wall complex and the mostly steric hindrances they impose on enzymes. However, it would be prudent to consider the chemical nature of the different cell wall components, how they interact chemically with each other, and more importantly, how they interact with the enzyme systems degrading them. Plants have evolved elegant solutions for defense; however, discovering these solutions remains challenging. Ultimately, a deeper understanding of the interaction between cell wall components and degrading enzymes may be required to enable the nascent biofuels industry; however, the practical problem now is to determine how deep this understanding must be.

Finally, when considering industrial scale operation at high solids concentrations, another level of complexity with respect the availability of water and the concentration of inhibitory compounds is encountered. This can include traditional end-product inhibitors, such as glucose, xylose, and cellobiose; as well as the concentration of complex compounds, such as phenolics resultant from the pretreatment process (Tejirian and Xu 2011). A recent talk at the 33rd symposium on Biotechnology for Fuels and Chemicals discussed that beyond traditional end-product inhibition, the concentration of even simple water soluble species, such as glucose, xylose, preservative sugar alcohols, and ethanol has the potential to drastically effect the aqueous environment during hydrolysis and significantly impair enzymatic action. The inhibition, in this case, was tied to depressions in water activity imparted by the presence of soluble species and then also to the impacts this has on the distribution and potential availability of water within the hydrolysis environment. While this work still has uncertainties with regard to its meaning, it does highlight the sensitivity of current enzyme systems to the environment where conversion is taking place. Combined with the dynamics imparted upon enzymes by hydrophobic and non-hydrophobic forces associated with components of a heterogeneous substrate (both native and pretreated lignocellulosics), it seems appropriate to suggest that the current understanding of the molecular nature of biomass recalcitrance, while guided by decades of research, is still insufficient. As the emerging lignocellulose based biofuels industry moves forward using current technologies, it seems that focusing on the dynamics of the enzyme-substrate-environment will have the largest payback and ultimately enable commercial processing solutions.

Technology Platforms for the Deconstruction of Biomass Cell Walls

The end goal of biomass deconstruction is the depolymerization of structural carbohydrate polymers to sugars which will then serve as an intermediate for fuel production. There are many strategies, usually used in combination, for deconstructing plant cell walls, including: mechanical, thermal, chemical, and biological technologies.

Mechanical processing is always employed as an intermediate step between biomass harvesting and conversion. Size reduction by milling and grinding increases reactive surface area, improves heat and mass transfer, and enables efficient mixing. However, mechanical grinding is costly (Stephen et al. 2010). The most critical balance that will have to be reached early on is between the cost of milling and the benefit of densification for transportation. It is impractical to transport corn stover from the field to the biorefinery without some form of densification (Richard 2010).

Corn stover can be converted to fuel intermediates using heat alone. In a pyrolysis process, biomass is heated to 500–550°C at a heating rate of up to 10,000°C/sec in an anoxic environment (Shafizadeh 1982). Pyrolysis treatment depolymerizes carbohydrates through dehydration reactions to generate products such as levoglucosan. The main products of pyrolysis are pyrolysis oils that require substantial stabilization and upgrading before use. Solid biochar can be combusted as a heat source or returned to the field for soil augmentation. The cellular and cell wall structure of biomass particles has an impact on product evolution even at these high temperatures and high heating rates (Haas et al. 2009; Bahng et al. 2011). Vaporized products produced during volatilization have to rapidly escape the biomass particle. It has been shown that when the primary products don't escape, they are trapped in gas-filled pockets that may be the site of secondary reactions and subsequent tar formation (Nimlos et al. 2011).

Even if thermal treatment is not used alone as in pyrolysis, some elevated temperature, typically in the range of 150°C–200°C, is nearly always a part of pretreatment strategies (Elander et al. 2009). This range of temperatures has been shown be sufficient to significantly increase xylan hydrolysis kinetics and to melt lignin, causing substantial relocalization of lignin and xylan throughout the biomass particle (Donohoe et al. 2008; Brunecky et al. 2009). Additionally, expression of certain cellulolytic enzymes within the plant itself during growth may make it more amenable to subsequent pretreatments and digestion (Brunecky et al. 2011). The leading thermochemical pretreatment strategies are acidic, alkaline (basic), and ionic liquids. Today, commonly used acidic pretreatments include dilute sulfuric acid and hot water pretreatments; these treatments are especially successful in increasing enzyme accessibility to substrates by hydrolyzing the hemicellulose component of the cell wall matrix and relocalizing lignin (Donohoe et al. 2008; Brunecky et al. 2009). Common alkaline pretreatments include soaking biomass in aqueous ammonia and lime treatment; these treatments are especially effective at removing significant amounts of lignin (Kumar et al. 2009). Another pretreatment strategy that doesn't appear to remove cell wall matrix components, but in fact does cause extensive depolymerization and relocalization, is ammonia fiber expansion (AFEX) (Chundawat et al. 2011). In contrast, ionic liquids have the ability to completely solubilize cellulose and have recently been proposed for biomass pretreatment (Singh et al. 2009). Pretreatments are needed to prepare the biomass for enzymatic saccharification.

Biological deconstruction is carried out by cellulose degrading fungi and bacteria or their isolated enzymes. In fact, most work on saccharification has been done with isolated enzymes with substantial effort going into determining the best, minimal enzyme cocktail for a given pretreated feedstock. More recently, there has some interest in how the microbes themselves contribute to the deconstruction process. The concept of consolidated bioprocessing (CBP) involves using a single organism, an engineered bacteria or yeast, for both the biomass deconstruction step and the fermentation of sugars to fuels (Lynd et al. 2005; van Zyl et al. 2007).

The next wave of biofuels research will focus on third generation, or advanced Biofuels, that may or may not use sugars as the intermediate (Cheng and Timilsina 2011). Advanced biofuels will strive for a depolymerized, deoxygenated product with a higher heat capacity than ethanol and the ability to be "dropped into" the existing petroleum fuel-processing infrastructure.

Enzymological Investigation—Novel Sources of Enzymes from Bacteria and Insects

Currently, commercial cocktails for biomass conversion are derived from fungal sources, typically from various strains of *T. reesei*. To date, the various individual enzymes from these cocktails have been studied extensively and their secretomes compared (Herpoel-Gimbert et al. 2008; Chundawat et al. 2011). It is becoming better understood that the most important enzymes that constitute these broths include specific cellulases and xylanases; as well as critical accessory enzymes that hydrolyze the wide variety of chemical linkages (see above). And while the generalities of glycoside hydrolase biochemistry are becoming clearer, recently discovered biomass oxidizing enzymes (GH family 61) demonstrates that new paradigms in nature may still be found (Harris et al. 2010). Therefore, although plant cell wall degrading products are commercially available today, the likelihood of improving these systems with enzyme engineering is high. In this section, we will examine several potential enzyme systems from extremophiles; as well as bacterial and insect sources of enzymes in the context of corn stover conversion.

Extremophiles and Extremozymes: An Opportunity for Developing More Pretreatment-compatible Enzymes

In the past decade, significant progress has been made in the pretreatment of biomass for its conversion to simple sugars. However, although currently required, these pretreatment technologies pose hurdles for downstream enzymatic hydrolysis due to the chemical residues (e.g., from acids, bases, solvents, or salts) used in the pretreatment process and the inhibitory substances generated from the biomass itself, such as furfural, hydroxymethylfurfural, and soluble phenolics (Engel et al. 2010). To solve this problem, post-pretreatment neutralization steps; as well as temperature reduction, are often required before further processing can begin. Unavoidably, these neutralization steps produce a salty solution which compromises the efficiency of subsequent enzymatic hydrolysis. To meet these challenges in a more cost-efficient way, the search has begun for thermo-, acid-, alkaline- and halo-tolerant enzymes, which could be used directly for enzymatic hydrolysis during or following pretreatment. These enzymes allow the lignocellulosic hydrolysis to be conducted under higher temperature, low or high pH, or high-salt conditions, and thus have the advantage of reducing the cost in neutralizing and washing the pretreated biomass solids.

The term "extremophiles" refers to microorganisms that have adapted to survive in one or more environmentally extreme conditions, such as at high temperatures, high acidity or alkalinity, high salinity, or high pressures (Podar and Reysenbach 2006). With regard to biomass conversion to simple sugars, we will focus primarily on thermophiles, acidophiles, alkalophiles, and halophiles, which also produce unique enzymes/biocatalysts that function under extreme conditions. These extreme conditions are comparable to those prevailing in some of the biomass pretreatment processes as described below. Accordingly, extremozymes refer to the enzymes that are produced by extremophiles and can function under extreme conditions. Various extremozymes have been reported, including amylases, cellulases, xylanases, proteases, pectinases, keratinases, lipases, esterases, catalases, and peroxidases (Gomes and Steiner 2004). While pectinases, esterases and peroxidases can be used as accessory enzymes in biomass conversion to simple sugars, we will focus on the cellulases and xylanases, the extremozymes that directly attack the major components of the plant cell wall.

To survive in these extreme ecological niches, extremophiles produce both intracellular and extracellular enzymes. Extremophiles generally employ specific mechanisms to control their intracellular environment, such as proton pumps to maintain neutral cytosolic pH in the case of acidophiles and alkalophiles (Gomes and Steiner 2004). As a result, the intracellular enzymes may not be adapted to the environment that their hosts are exposed to. In contrast, the secreted (extracellular) enzymes must function under extreme conditions using mechanisms not fully understood. In this regard, the secreted extremozymes are likely to be more desirable than their intracellular counterparts in meeting the challenge to be used in harsh industrial processes. Prompted by demands for enzymes that can function under various experimental and industrial processing conditions, a variety of thermo- acid-, alkaline- and halo-philic enzymes have been isolated and characterized in the past decade. We will briefly review these systems, as well as the bacterial cellulosomal systems which will provide important clues for the functioning of certain key biological ecosystems, as well as termite enzyme systems.

Thermophilic Cell Wall Degrading Enzymes

Thermophilic enzymes are of key interest for biofuels production for many reasons; in addition to thermostability, they also typically exhibit improved pH tolerances. Thermostable enzymes are often highly specific and thus have considerable potential for bioconversion processes (Haki and Rakshit 2003). One obvious advantage of utilizing thermophilic enzymes industrially is the advantage that few contaminating organisms can survive at the elevated temperature conditions the enzymes require to function effectively. Another advantage noted in starch industry which utilizes thermophilic amylases, are improved substrate and product solubility at elevated temperatures. However in the context of lignocellulosic bio-conversion these solubility advantages may not be as great due to the overall recalcitrance of biomass at even at elevated temperatures.

Examples of cellulases from the Caldi clade are among the most thermophilic cellulose degrading enzymes known to date. One example, the multifunctional the *Caldicellulosiruptor bescii* CelA, has a temperature optimum of 85 to 95°C; and a pH optimum between pH 5.5 and 6 (Zverlov et al. 1998). Another multifunctional enzyme from *C. saccharolyticus*, CelE, has a broad pH stability (pH 4–11); as well as thermal stability up to 75°C (Gibbs et al. 2000). *Bacillus lichenformis* has interesting enzymes belonging to GH families 9 and 12. These enzymes work optimally between 60–70°C and show good activity; as well as synergy, on synthetic substrates (Liu et al. 2004). *Pyrococcus furiosus* produces a thermophilic cellulase, EglA, which demonstrates a temperature optimum of 100°C at pH 6.0, with a half life of 1.6 h. At 95°C, its half life is 40 h (Cady et al. 2001). Graham and coworkers recently reported a cellulase that demonstrates maximal activity at 109°C, with a half-life of 5 h at 100°C. This enzyme resists denaturation in strong detergents, high-salt concentrations, and ionic liquids (Graham et al. 2011). Thus, many classical and novel sources for extremophiles have been demonstrated; however, very little systematic evaluation of these enzymes on relevant lignocellulosic substrates has been performed.

Thermophilic xylanases are also common in the literature; an early example was the thermostable enzyme XynA from *Thermatoga neapolitana*, with a reported pH and temperature optima at pH 5.5 and 102°C. This endoxylanase was stable at 90°C and retained 50% activity when incubated for 2 h at 100°C (Zverlov et al. 1996). Another more recent example from *T. maratima* is XynB which exhibits activity against xylan substrates. The temperature and pH optima (87°C and pH 6.5) make this an important enzyme as well (Reeves et al. 2000). Typically, thermostable xylanases are found in most thermophilic organisms that also express cellulases and other accessory enzymes required for conversion of lignocellulosic materials.

Even though there are no commercial thermophilic enzyme cocktails available for lignocellulosic biomass conversion today, novel sources of thermophilic enzymes continue to be found and characterized. Recently Kataeva and coworkers, have characterized a novel organism similar to *C. saccharolyticus* (Kataeva et al. 2009). However, a far greater challenge for both thermophilic cellulases and xylanases is the proper evaluation and characterization of these enzymes for their suitability for industrial bioconversion processes.

Practical Considerations Regarding Thermophilic Enzymes

Thermophilic enzymes are currently not used in biomass conversion processes, with the exception of the starch utilization industries which employ thermophilic amylases (Haki and Rakshit 2003). A major consideration with regard to the use of thermophilic enzymes individually or in cocktails is one of process cost, which is embodied in both the expense of producing industrially relevant quantities of these enzymes and in obtaining high yields of active enzymes compared to current fungal enzyme production systems. Other considerations are the energy costs associated in maintaining large reaction vessels at an elevated temperature for long periods of time and the requirement many extremophiles have for high volumetric inocula. Of course, where possible, these enzymes would be produced in suitable recombinant hosts which may be grown at reduced cost.

Another consideration for process use of thermophilic enzymes is that whereas the published temperature activity optimum of any given enzyme may be indeed be very high, the actual half life of the enzyme may be short at that temperature, as noted in some of the examples above. For most of the aforementioned enzymes, the temperature optimum was determined over a relatively short period of time, typically on the order of minutes on a synthetic substrate; in several cases, the half life at any given temperature was not reported. The actual process stability, or robustness, of the enzyme at a given temperature and pH is also critical consideration. It is clear that an enzymatic half life on the order of several days is a likely requirement for any industrial conversion process. Current enzymatic saccharification technologies require as much as 5 days to achieve acceptable biomass conversion yields (Kazi, Fortman et al. 2010). Thus, even if thermophilic enzyme systems doubled the current rate of an enzymatic saccharification, they would still have to remain stable for 2 to 3 days to reach suitable conversion levels. One example of this concern noted by Lochner and coworkers is that when testing enzyme mixtures obtained from *C. obsidiansis* and *C. bescii*, they found that whereas the enzyme mixtures showed optimal activity at 85°C, pre-incubation of these enzymes at these temperatures for 1.5 hours either abolished or significantly reduced enzyme activities (Lochner et al. 2011). Thus, the temperature and activity optima for any thermophilic enzyme cocktail must be carefully evaluated in the context of long term stability expected for use conditions.

Another consideration is that almost all of the enzymes noted above were tested on some form of synthetic substrate, in lieu of lignocellulosic biomass. It is well known that performance on synthetic substrates, while easy to perform in the laboratory, often does not translate well to enzyme performance on lignocellulosic materials, such as corn stover. Particle size of substrate is also another important parameter to consider when evaluating enzymes for scale up. Conversions performed on finely ground biomass will appear much more promising than when using realistic materials (usually 0.5 to 4 inches in length for corn stover).

Acid-tolerant Cell Wall-degrading Enzymes

Acidophiles grow optimally between pH 1 and 5. Acidophilic xylanases are members of GH families 10 and 11. *T. reesei* xylanases have an optimal pH of 3.5 (Beg et al. 2001) and *Aspergillus niger* xylanases have an optimal pH of 3.0 (Krengel and Dijkstra 1996). Three other xylanases have been reported with an optimal pH of 2.0 and stability over a broad pH range, including *A. kawachii* (Fushinobu et al. 1998), Penicillium spAO (Fushinobu et al. 1998), and the yeast, *Cryptococcus* sp. S-2 (Iefuji et al. 1996).

Acid-tolerant xylanases have increasingly been used for pulp bleaching by the pulp and paper industry. When compared with alkali-tolerant xylanases (Damiano et al. 2003) and haloalkaline-tolerant xylanases (Gupta et al. 2000), acid-tolerant xylanases have an advantage when utilized in biobleaching processes—acid-tolerant xylanases minimize the need for pH adjustment prior to the ozone or chlorine dioxide bleaching stages that are conducted at moderately acidic pHs (Tenkanen et al. 1997). The xylanases that have been used or tested in pulp bleaching include the *A. kawachii* xylanases, which are active at pH 2.0 to 2.5 (Crous et al. 1995; Kohlmann et al. 1996; Tenkanen et al. 1997). Another acid-tolerant xylanase from *Scytalidium acidophilum* (optimal pH of 3 and temperature of 50°C) was also used for prior-to-bleaching pretreatment of hardwood kraft pulps (Kuligowski 2005).

The prospects of utilizing acid-tolerant xylanases in the enzymatic hydrolysis of biomass pretreated by thermochemical techniques, especially dilute-acid pretreatment, looks promising; considering that the overlap in acidity between the dilute sulfuric acid pretreatment (Schell et al. 2003), usually 0.5 to 1.5% H_2SO_4 (w/w) with a theoretical pH range of pH 0.8 to1.2, and the aforementioned extreme acid-tolerant xylanases, which have optima around pH 2.0 and can retain 75% activity at pH 1.0 (Iefuji et al. 1996; Fushinobu et al. 1998; Kimura et al. 2000). [Note: this pH can be partially neutralized by biomass during pretreatment, the final post-pretreatment pH can be as low as pH 1.1 to 2.1]. Such a close match in pH requirements between pretreatment and enzymes activity can minimize the steps and cost related to biomass residue neutralization after pretreatment. As current commercial cellulases, such as GC220 Cellulase (Genencor, Rochester, NY) are typically used at pH 4.8 (Selig et al. 2008), future searches for cellulases from these sources that function effectively at more acidic pHs, or enzyme engineering approaches, may lead to the development of extreme acid-tolerant xylanase-cellulase cocktails, which could be used at pretreatment or near-pretreatment pHs.

Alkaline-tolerant Cell Wall-degrading Enzymes

Alkalophiles are extremophilic microorganisms that grow optimally at pH 9.0 and above. Alkaline-tolerant xylanases produced by alkalaphiles have been studied extensively because of their potential application in numerous industrial processes, including pulp and paper, textile, food, feed and bio-fuel industries. For the pulping industry, which is technically related to biomass processsing for biofuels, the value of alkaline-tolerant xylanases is that their use in biological bleaching can replace or reduce the use of the traditional, environmentally damaging chemical processes, such as chlorine. Alkalophiles and the alkaline-tolerant enzymes, including xylanases, have been extensively reviewed recently in the literature (Horikoshi 1999; Gomes and Steiner 2004; Wiegel and Kevbrin 2004; Fujinami and Fujisawa 2010; Sarethy et al. 2011). To compile and compare results reported recently, Table 1 describes characteristics of some selected alkaline-tolerant xylanases (Chang et al. 2004; Sharma and Bajaj 2005; Wang et al. 2010). Recent progress in the field of note is the production of alkali-tolerant xylanase using renewable agricultural wastes. For example, *Penicillum citrinum* cultured on wheat bran produces a xylanase with an optimal pH of 8.5 (Dutta et al. 2007). More recently, using agricultural biomass waste residues such as wheat bran, rice bran and sawdust as carbon sources, a newly isolated alkaline-tolerant *Penicillium* sp. SS1 produced a xylanase for which the maximal activity was observed to be pH 8. This enzyme retained 70 to 80% activity at pH 9 to 10 (Bajaj et al. 2011).

Table 1 Production of extremophilic enzymes by alkaliphiles.

Alkaliphiles	Alkaliphilic enzymes	Topt/oC	pHopt	Alkaline-tolerance	References
Bacillus firmus	xyn10A, xyn11A	70°C	pH 5–9	n/a	Chang et al. 2004
Bacillus pumilus	XynBYG	50°C,	pH of 8.0–9.0	pH 9.0, 120 min	Wang et al. 2010
Streptomyces sp.	endocellulase	50°C	pH 8	n/a	Solingen et al. 2001
Streptomyces sp. CD3[1]	xylanase	50°C	pH 8	pH 10.0	Sharma and Bajaj 2005
Thermomonospora (actinomycete)	endocellulase	50°C	pH 5	pH 10.0	George et al. 2001

Note 1. The activity of xylanase from Streptomyces sp CD3 was found to be enhanced by Fe^{3+}, Ca^{2+} and Zn^{2+} ions, but inhibited by Hg^{2+} ions.

In addition to alkaline-tolerant xylanases, alkalophilic microorganisms also produce alkaline-stable cellulases. For example, an alkalothermophilic Actinomycete secretes a cellulase that was stable at pH 7.0 to 10.0, while retaining 100% activity at 50°C for 72 h and showing a half-life of 3 h at 70°C (George et al. 2001) (Table 1). Furthermore, a novel *Streptomyces* species isolated from east African soda lakes produced an endocellulase with an optimal pH of 8.0 (Solingen et al. 2001). In addition, 48 psychrotolerant bacteria were recently isolated from soils in the Qinghai-Tibet Plateau; they survived in pHs from 6.5 to 10.5 and grew optimally at pHs from 9.0 to 9.5; many of these strains produced extracellular cellulases (Zhang et al. 2007) and can be considered future sources of additional enzymes.

Halo-tolerant Cell Wall-degrading Enzymes

Halophilic microorganisms can be classified into the following categories (Kushner 1985; Ventosa et al. 2008): slight halophiles, grow best in media containing 1 to 3% NaCl (ca. 0.2 to 0.5 M); moderate halophiles, grow best at 3 to 15% NaCl (ca. 0.5 to 2.5 M); and extreme halophiles, grow best at 15 to 25% NaCl (ca. 2.5 to 4.2 M) and are able to grow even at saturated salt concentrations (~ 30%; 5.2 M). To put this in perspective, the common LB broth for *E. coli* culture contains 1% NaCl, and the seawater generally contains 3.5% NaCl, belonging to slight and moderate halophilic range, respectively. Most microorganisms found in high salt environments belong to moderate halophilic bacteria, and moderate halophilic bacteria and archaea. Two extremely halotolerant endoxylanases have been isolated and characterized from moderately halophilic Proteobacteria strain CL8 (Wejse et al. 2003) (see Table 2) and the haloalkaline xylanase from a marine *Bacillus pumilus* strain, GESF-1 (Menon et al. 2010). Intrinsically, halo-tolerant enzymes are compatible with thermochemically pretreated biomass, due to

Table 2. Production of extremophilic enzymes by halophiles and haloalkaliphiles.

Halophiles	Halo-tolerant enzymes	Topt°C	pHopt	Halo-tolerance	Activity enhanced by	inhibited by	References
Bacillus aquimaris[1]	cellulase	45°C	pH 11	20% solvents	Ionic liquids		Trivedi et al. 2011
Bacillus pumilus[1]		40°C	pH 8	2.5 M NaCl	Ca^{2+}, Mn^{2+}, Mg^{2+}, Na^+	Fe^{3+}, Cu^{2+}, Cd^{2+}, Zn^{2+}	Menon et al. 2010
Bacillus flexus[1]	cellulase	45°C	pH 10	2.5 M NaCl	Cd^{2+} and Li^+	Cr^{2+}, Co^{2+}, Zn^{2+}, EDTA	Trivedi et al. 2011
Bacillus subtilis[1]	xylanase	55°C	pH 9	n/a	Fe^{2+}, Ca^{2+}, and Mg^{2+}	Hg^{2+}, EDTA	Annamalai et al. 2009
Enterobacter sp. MTCC 5112[1]	xylanase	50 °C	pH 9	n/a	Co^{2+}, Zn^{2+}, Fe^{2+}, Cu^{2+}, Mg^{2+}, Ca^{2+}	Hg^{2+}, EDTA	Khandeparkar and Bhose 2006
Halomonas sp.	endoglucanase Cel8H	45°C	pH5	5 M NaCl	Fe^{2+}		Huang et al. 2010
*Halophilic bacterium*CL8	Xylanase 1	60	6	5 M NaCl	n/a	n/a	Wesjse et al. 2003
	Xylanase 2	65	6	5 M NaCl	n/a	n/a	
Halorhabdus utahensis	β-xylanase	55, 70	NA	4.5 NaCl	n/a	n/a	Waino et al. 2003
	β-xylosidase	65	NA	5 M NaCl	n/a	n/a	
Martelella mediterranea	endoglucanase Cel5D	60°C	pH5	n/a	Cu^{2+}, Fe^{2+}, Mn^{2+}, Zn^{2+}		Dong et al. 2010
Salinivibrio sp. NTU-05	cellulase	35 °C	pH 7.5	4 M NaCl	K^+, Mg^{2+}, Na^+	Hg^{2+}	Wang et al. 2009
Staphylococcus sp. SG-13[1]	xylanase	50 °C	pH 7.5 & 9.2	n/a	Fe^{2+}, Ni^{2+}, Cu^{2+}	Co^{2+}, Hg^{2+}, Pb^{2+}	Gupta et al. 2000
Thermoanaerobacter tengcongensis[2]	β-1,4-endoglucanase	75 °C	pH 6–6.5	3 M NaCl[3]	Ca^{2+}, K^+, Mg^{2+}, Mn^{2+}	Zn^{2+}, Ni^{2+}	Liang et al. 2011

Note 1.   These listed strains tolerant to both alkaline and high salinity (haloalkalitolerant): *Bacillus aquimaris*, *Bacillus flexus*, *Bacillus pumilus*, *Bacillus subtilis*, and *Staphylococcus* sp. SG-13

Note 2.   This halotolerant enzymes (β-1,4-endoglucanase that tolerant up to 3 M NaCl, Table) were derived from nonhalophilic microorganism

Note 3.   It is also tolerant to to 1 M ionic liquors

their ability to tolerate salty environments with low water activity, making them some of the most promising enzymes biomass conversion.

Cellulosomal Synergetic Enzyme Systems

Many microbes have developed systems to hydrolyze cell wall polysaccharides into sugar monomers and oligosaccharides. A unique approach used by the gram-positive anaerobic bacteria *Clostridium thermocellum*, utilizes large (> 1 Mda) extracellular multi-enzyme systems, known as cellulosomes (Bayer et al. 2004), for plant cell wall sugar digestion. *C. thermocellum* strain ATCC 27405 was originally isolated from manure and later described by McBee over 50 years ago (McBee 1948). ATCC 27405 has been studied extensively since that time (Demain et al. 2005; Fontes and Gilbert 2010). *C. thermocellum* is of great interest to industrial biofuel production, due to its ability to deconstruct biomass and generate Biofuels in the CBP process.

Cellulosomes contain a diverse array of cellulases, hemicellulases and other plant cell wall deconstructing enzymes. These multi-component structures are assembled onto large, non-catalytic proteins, known as scaffoldins. Primary scafoldins in *C. thermocellum*, CipA, can bind up to nine enzymes via the type I cohesin domains (Fig. 5). Enzymes strongly bind to the type I scafoldin through their non-catalytic module, known as the type I dockerin domain (Kruus et al. 1995; Pages et al. 1996). Scafoldins also contain a carbohydrate binding domain (CBM) and type II dockerin domain (Bayer et al. 2004). The type II dockerin binds secondary scafoldins (OlpB, SdbA, Orf2p) which can bind up to seven primary scafoldins (Raman et al. 2009). Secondary scafoldins adhere to the cell surface layer (S-layer) using S-layer homology (SLH) domains. Thus, the cellulosomal architecture can facilitate dramatically diverse assemblies of enzymes and

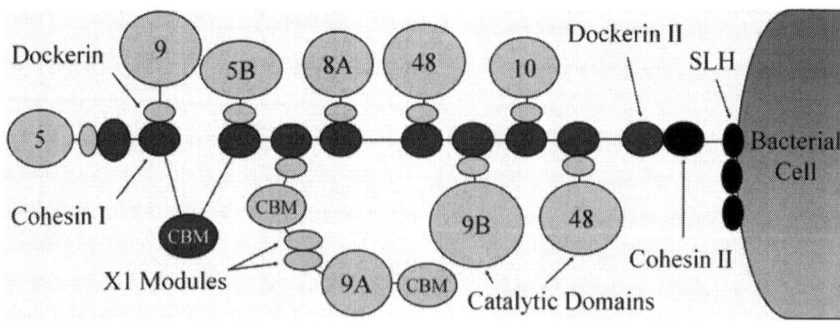

Figure 5 A schematic of the cellulosome illustrating some of the possible cohesion dockerin interacting species.

Color image of this figure appears in the color plate section at the end of the book.

CBMs with molecular weights from 1.5 MDa (CipA fully loaded) up to tens of MDa (Seven loaded CipA bound to OlpB). This organization facilitates cell adhesion to the biomass substrate and a dynamic assembly of enzyme specificities.

Many labs have reported the most abundant enzymes associated with the cellulosome (Zverlov et al. 2005; Raman et al. 2009). They have found a mix of cellulases, hemicellulases, and scafoldin proteins which have the potential for degrading the diversity of plant cell wall carbohydrates. The close proximity of CBMs, and glycosyl hydrolases, which have multiple binding preferences and hydrolytic specificities, may help the system by synergistically degrading any available glycosidic bond.

It has been shown that the growth of *C. thermocellum* on different carbon sources influences the expression and secretion of enzymes extracellularly (Raman et al. 2009). Cellulosome enzymes also contain signal peptides, which suggest that the assembly of enzymes into cellulosomes happens extracellular (Beguin and Aubert 1994). From this work, we assume that the level of enzyme and scaffold protein secretion from the cells dictates the composition of the cellulosomes, resulting in different enzymatic activities. Of course, such a system is necessarily polydisperse with respect to enzyme type and number per scaffoldin domain. The question of how the composition of the large molecular weight cellulosomes influences the overall enzymatic activity remains unanswered.

Another recent and related field of study has been that of creating "designer" cellulosome scaffolds for biomass conversion from the Bayer lab (Heyman, Barak et al. 2007). Indeed, various types and degrees of synergy utilizing engineered cellulosomal systems have been reported, including a 30% improvement in overall enzyme synergism utilizing a scaffold linked exoglucanase and a free endoglucanase (Moraïs, Heyman et al. 2010). A more recent study explored the mechanisms of designer cellulosome synergism by coupling a xylanases with an endoglucanase (Moraïs, Barak et al. 2010). While this approach is relatively new, such scaffold engineering may prove to be a viable method to improve overall enzyme cocktail performance.

Promising Extremozymes Compatible with Current Pretreatment Technologies

There are several promising candidates among thermophilic enzyme systems from the Caldi clade, notably CelA from *C. bescii* which has been grown on both Avicel and switchgrass (Dam P. et al. 2011). This organism has been studied in recent years as a possible bioconversion platform, because it can also ferment sugars to ethanol. CelA from *C. bescii* has been reported to have high thermostability; as well as a half life compatible with industrial processes (Zverlov et al. 1998). From genomic analysis of *C. bescii*, there are several other promising xylanases and cellulases produced; as well as some accessory enzymes, thus a *C. bescii* based enzyme cocktail may prove desirable for high temperature biomass deconstruction.

At least one halotolerant endoglucanase and two halotolerant xylanases, described in Table 2, can tolerate 5 M NaCl (30%), which is close to the saturating concentration of NaCl. Using these enzymes in the saccharification of thermochemical pretreated biomass can not only reduce the requirement in desalting the pretreated biomass, but also permit the enzymatic hydrolysis to be conducted at high solids biomass loading; considering that halotolerant enzymes maintain their stability at least partially because of their ability to withstand very low water activity. Together, these two aspects will increase the cost efficiency of pretreatment and subsequent saccharification.

Furthermore, some halotolerant enzymes are enhanced by some metal ions. As indicated in Table 2, the stimulation of enzyme activity in the presence of Fe^{2+} had been reported for numerous halotolerant cellulases from *Halomonas* sp. and *Martelella mediterranea* (Dong et al. 2010; Huang et al. 2010); and xylanases from *B. subtilis* (Annamalai et al. 2009), *Enterobacter* sp. MTCC 5112 (Khandeparkar and Bhosle 2006), *Staphylococcus* sp. SG 13 (Gupta et al. 2000), and *Thermomyces lanuginosus* (Cesar and Mra 1996). In another report, the stimulation of enzyme activity by Fe^{3+} had also been reported for alkaline-tolerant xylanase (Table 1) (Sharma and Bajaj 2005). These features distinguish the above enzymes as specially suitable to be used in dilute-acid/iron co-catalyzed pretreatments of biomass (Nguyen and Tucker 2002; Wei et al. 2011).

As some extremophiles live in complex harsh environments, they have adapted to multiple stresses simultaneously (referred as polyextremophile), such as thermoacidophiles and haloalkalophiles). Not surprisingly, some extremozymes can function under multiply adverse conditions. For example, in Table 2, a recent marine isolate of bacterium *B. aquimaris* showed a relative growth yield of 86% on acetone, 71% on methanol, and 52% on benzene (Trivedi et al. 2011). Accordingly, in the presence of 20% (v/v) solvents, the cellulase it secreted showed increased or substantial activities in benzene (122% of activity in the absence of solvent), methanol (85%), acetone (75%), and toluene (73%). This cellulase had an optimum activity at pH 11 and 45°C, and was stable even at pH 12 and 75°C (see Table 2). Remarkably, a pre-incubation of this enzyme in ionic liquids (1-ethyl-3-methylimidazolium methanesulfonate or 1-ethyl-3-methylimidazolium bromide) increased its activity to approximately 150% (Trivedi et al. 2011). In combination, this enzyme is thermophilic, alkaline-, halo- and solvent-tolerant, making it suitable for application in the saccharification of biomass pretreated by various pretreatments including those treated by ionic liquids. For example, the b-D-(1,4)-endoglucanase from *Thermoanaerobacter tengcongensis* (Liang et al. 2011) (see Table 2) can tolerate up to 3 M NaCl. It can retain 55 to 65% of its activity in a 1 M ionic liquor of 1-butyl-3-methylimidazolium chloride or 1-allyl-3-methylimidazolium chloride, making it another promising candidate to be used in the saccharification of ionic liquor pretreated biomass.

The Termite: A Bioreactor for Saccharification of Biomass

Termites have generated broad interest in recent years because of their high efficiency in degrading lignocelluloses and their potential to be used as a reservoir for novel biomass degrading enzymes. Termites belong to the order Isoptera and can be classified into seven families, which can be further classified into about 200 genera and about 2,300 species (Engel and K. Krishna 2004). According to their level of evolution, their behavior and anatomy, five families of termites (Mastotermitidae, Kalotermitidae, Termopsidae, Hodotermitidae, and Rhinotermitidae) referred as the lower termites, while the remaining two (Serritermitidae and Termitidae) are referred as the higher termites. The lower termite and higher termite can degrade 74 to 99% and 91 to 97%, respectively, of the lignocellulosic material they intake within about 24 h (Breznak and Brune 1994; Sun and Scharf 2010; Itakura et al. 1995).

In recent years, the progress in three areas has contributed to the partial demystification of the termite's high efficiency in breaking down biomass. Primarily, metagenomic sequencing studies reveal that about 200 microbial species inhabit in termite hindgut and directly aid the termite in the digestion of biomass (Warnecke et al. 2007). Secondly, a study of the dampwood termite, *Zootermopsis angusticollis*, demonstrated the oxidation of propyl side-chain (depolymerization), demethylation of the ring methoxyl group, as well as general ring hydroxylation of lignin after its passage through the termite gut (Geib et al. 2008). A more recent study identified some lignin degrading/modifying gene candidates in the gut of the termite *Reticulitermes flavipes* (Coy et al. 2010). Using a dissolved oxygen microelectrode to monitor oxygen profiles in intestinal microhabitats of two termites, *Coptotermes formosanus* and *R. flavipes*, it was demonstrated that lignin modification and disruption are initiated in the foregut, and continues in the midgut of termites (Ke et al. 2011). Together, these studies provide chemical, biological and physical evidences for the efficient attack and disruption of lignin by termite. Thirdly, in addition to the role of microorganisms and protozoa in degrading lignocelluloses in the termite gut (Warnecke et al. 2007), meta-transcriptomic, physiological, genetic, and enzymatic studies have also been conducted to investigate the involvement of enzymes produced by host termites in the lignocellulose degradation process (Scharf and Tartar 2008; Tartar et al. 2009; Walker and Wilson 1991). The results indicated that the host itself produces enzymes that work in synergy with the enzymes produced by those symbiotic microorganisms in degrading the lignocelluloses (Scharf et al. 2011).

In addition to above mechanisms for termite degradation of lignocelluloses, a key mechanism is the particle size reduction of lignocellulose. This size reduction is attributed to the fine grinding action of the insect's mandibles (i.e., the lower jaw is a hard, biomineralized structure with considerable mechanical strength) and muscular gizzard (located in foregut). This comminution increases biomass surface area and may decrease the crystallinity of cellulose, thus rendering the polysaccharides more susceptible to enzymatic hydrolysis (Walker and Wilson 1991). The particle size reduction performed by the termites is quite remarkable and reduces particles to a size of 8 to 15 mm, an extremely effective size reduction unmatched by current, viable pretreatment technologies (Schmidt 1956). Additional factors contributing to the degrading efficiency of termites includes the alkaline conditions in the gut that favors the solubilization lignin and breaking of ester linkages in hemicellulose, as well as the anatomical or physiological features in the gut that increase the retention time of lignocelluloses in thus its exposure to digestive enzymes (Breznak and Brune 1994). These characteristics sum to make the "termite bioreactor" extremely efficient in particle size reduction and alkaline pretreatment, both of these characteristics greatly facilitate subsequent enzyme hydrolysis. There have been a few proposals to design and develop a termite-inspired "biomechano-reactor" that combines fine grinding with unique a set of enzymes from termite guts. Such bio-mechano-reactor concepts are not only supported by the termite studies themselves, but also supported by chemical and enzymatic pretreatment studies of biomass in literature. In the 1980s, an attrition bioreactor was designed and tested; it consisted of a jacketed stainless-steel vessel with shaft, stirrer, and milling media, combined the mechanical action of wet milling with cellulose enzymatic hydrolysis and had good performance (Ryu and Lee 1983; Jones and Lee 1988). Another study confirmed that when ball milling was applied simultaneously to enzymatic hydrolysis of delignified, steam-exploded *Douglas-fir* wood chips, it achieved up to ca. 100% conversion rate Mais et al. (2002). Unfortunately, while ball milling is extremely effective in increasing hydrolysis rates of all enzymes, presumably by increasing the reactive surface area of the biomass and reducing cellulose crystallinity, it is very expensive from an energy cost point of view, requiring approximately 500 to 800 Wh/kg for milling wood chips into fibers (Zhu et al. 2010). Because mechanical particle size reduction consumes significant amounts of mechanical energy this makes mechanical reduction schemes for reactors very unattractive from a process efficiency standpoint (Yang and Wyman 2008; Stephen et al. 2010).

Future Prospects for Termite based Enzymes

Despite the recent progress in understanding the mechanisms in termites for degrading woody lignocellulose, several remaining questions should be addressed in order to implement the concept of a termite based biomechano-reactor. The first question is that among the ~200 microbial species that were revealed by the metagenomic sequencing of the microbial community in the termite hindgut (Warnecke et al. 2007), how many are the essential species and what are the major microbial enzymes that involved in the breakdown of woody lignocelluloses? An answer to these questions will help generate minimal synthetic microbial consortia with a defined list of microbial enzymes for degrading lignocellulose. Such a list of microbial enzymes can be used, together with the recently identified host termite-secreted, lignocellulose-degrading enzymes (Scharf et al. 2011), to optimize cocktail enzymes for a termite based enzyme cocktail.

The second question is how these major microbial and host enzymes are distributed in termite gut, i.e., the need to construct an *in situ* high-resolution spatial mapping of the individual major enzymes. Such spatial mapping of major enzymes can reveal not only the interactions among the enzymes/proteins, but also the interactions between the enzymes and the lignocellulosic substrates. Progress in this field will be useful to design a biomechano-reactor that better mimics the termite's machinery in degrading the lignocellulose.

And the final question is to evaluate these enzyme systems on biomass that is not milled to microscopic levels, but rather biomass that is treated by a viable pretreatment process to evaluate the real world efficiency of these systems. Given the termite gut environment that these enzymes operate in, corn stover treated by one of the alkaline pretreatments would seem to be a logical choice to test in conjunction with termite enzymatic systems.

Genomic Approaches to Explore the Diversity, Multiplicity and Function Mechanisms of Extremozymes

To date, there are at least 1,000 completed bacterial and archaeal genomes (Wu et al. 2009), among which a portion of the genomes are extremophiles, including marine microorganisms. As enzyme multiplicity is likely to be a common characteristic for cellulases and xylanases that exist in extremophiles (Wong et al. 1988), the availability of these genomes provide an opportunity for increasing the efficiency of characterizing the targeted extremozymes and identifying promising biocatalysts for polysaccharide degradation at each individual genome level. For example, recently the first complete genome sequence for a member in halotolerant bacterial genus

Glaciecola was published (Klippel et al. 2011). Their in-depth annotation analysis identified numerous glycoside hydrolases, glycosyl transferases, and carbohydrate esterases.

Furthermore, at the polygenomic level, the availability of a increasing number of extremophile genomes; as well as improved annotation techniques, make it possible to conduct comparative genomic analyses, which help reveal novel acid-, alkaline- and halotolerant lignocellulolytic enzymes of commercial interest that can be used at a large scale (Li et al. 2011). Recently, Anderson and coworkers comparatively analyzed five recently completed haloarchaeal genomes with five previously sequenced ones (Anderson et al. 2011). They found that the soil/sediment-derived halophiles seem to have greater machinery and capacity for polysaccharide degradation, cell wall modification, and siderophore synthesis. When we consider another recent report that under iron-depleted conditions, bacteria produce siderophores to bind iron, which are then actively taken up by bacterial cells (Gulick 2009), we can speculate that iron (and probably some other metal ions as well) are among the essential chemicals involved in the utilization of polysaccharides existing in these habitats. This in turn may provide an explanation why a considerable portion of halotolerant enzymes described in Table 2 had an increased enzyme activity in the presence of some metal ions. With the recent advances in genome sequencing, gene expression libraries, protein engineering, and high-throughput enzyme functional screening (Lorenz and Eck 1988; Short and Keller 2001; Short 2002; Lorenz and Eck 2004; Voget et al. 2006; Duck et al. 2010; Kennedy et al. 2011; Li et al. 2011), it can be expected that the rate of discovery for novel extremozymes will be increased dramatically in the coming years.

Future Prospects for Extremozymes

To date, considerable effort has been applied to the study of lignocellulolytic extremophiles and extremozymes, especially the discovery of novel species, as well as the characterization, cloning, and protein engineering of novel enzymes (Hough and Danson 1999; DeLong 2000; Jenney and Adams 2008; Miller and Blum 2010). Understanding the mechanisms related to acid- and alkaline-adaptation of enzymes has also been the subject of considerable work (Shirai et al. 2001; Liu et al. 2002; Collins et al. 2005; Shibuya et al. 2005; Siddiqui and Thomas 2008; Liu et al. 2009). However, relatively modest progress has been made in understanding how the halo-tolerant extremozymes function in high salt environments and thus, additional studies are warranted. Further studies are also needed to evaluate the current collection of known extremozymes on real world substrates and under process relevant conditions. Additionally, more efficient engineering of thermal chemical pretreatment-compatible extremozymes using rational structural design is also warranted. To enable this work, a deeper understanding of the function of these enzymes at the molecular scale is needed. We believe this work will enable a new generation of tailored enzyme cocktails which will outperform current industrial standard formulations for biomass conversion.

Acknowledgment

This work was funded by the DOE EERE Office of the Biomass Program; by the DOE OSC Biological and Environmental Research (BER), Bioenergy Research Center (BESC); and by the DOE OSC Basic Energy Sciences (BES), Energy Frontier Research Center (C3Bio).

References

Anderson I, Scheuner C, Gorker M et al. (2011) Novel Insights into the Diversity of Catabolic Metabolism from Ten Haloarchaeal Genomes. PLoS ONE 6: e20237.

Anex RP, Kazi FK, Joshua A et al. (2010) Techno-economic comparison of process technologies for biochemical ethanol production from corn stover. Fuel 89: S20–S28.

Annamalai N, Thavasi R, Jayalakshm S et al. (2009) Thermostable and alkaline tolerant xylanase production by Bacillus subtilis isolated form marine environment. Ind J of Biotech 8: 291–297.

Bahng MK, Donohoe BS, Nimlos MR et al. (2011) Application of an Fourier Transform-Infrared Imaging Tool for Measuring Temperature or Reaction Profiles in Pyrolyzed Wood. Energy & Fuels 25: 370–378.

Bajaj BK, Sharma M, Sharma S et al. (2011) Alkalistable endo-b-1,4-xylanase production from a newly isolated alkalitolerant *Penicillium* sp. SS1 using agro-residues. Biotech 3.

Bayer EA, Belaich JP, Shoham Y et al. (2004) The cellulosomes: multienzyme machines for degradation of plant cell wall polysaccharides. Annu Rev Microbiol 58: 521–554.

Beckham GT, Matthews JF, Peters B et al. (2011) Molecular-Level Origins of Biomass Recalcitrance: Decrystallization Free Energies for Four Common Cellulose Polymorphs. J of Phys Chem B 115: 4118–4127.

Beg QK, Kapoor M, Hoondal GS et al. (2001) Microbial xylanases and their industrial applications: a review. Appl Microbiol Biotechnol 56: 326–338.

Beguin P and Aubert JP (1994) The biological degradation of cellulose. FEMS Microbiol Rev 13: 25–58.

Benkő Z, Siika-aho M, Vilkan L et al. (2008) Evaluation of the role of xyloglucanase in the enzymatic hydrolysis of lignocellulosic substrates. Enzyme and Microbial Technology 43: 109–114.

Biely P, MacKenzie CR, Puls J et al. (1986) Cooperativity of Esterases and Xylanases in the Enzymatic Degradation of Acetyl Xylan. Nat Biotech 4: 731–733.

Boerjan W, Ralph J and Baucher M (2003) Lignin Biosynthesis. Ann Rev Plant Biology 54: 519–546.

Boudet AM (2000) Lignins and lignification: Selected issues. Plant Physiol Biochem 38: 81–96.

Breznak JA and Brune A (1994) Role of microorganisms in the digestion of lignocellulose by termites. Ann Rev of Ento 39: 453–487.

Brunecky R, Selig M, Vinzant TB et al. (2011) In planta expression of A. cellulolyticus Cel5A endocellulase reduces cell wall recalcitrance in tobacco and maize. Biotech. for Biofuels 4: 1.

Brunecky R, Vinzant TB, Porter SE et al. (2009) Redistribution of xylan in maize cell walls during dilute acid pretreatment. Biotech and Bioeng 102: 1537–1543.

Cady SG, Bauer MW, Callen W et al. (2001) β-Endoglucanase from Pyrococcus furiosus. Methods in Enzymology. R. M. K. Michael W.W. Adams, Academic Press 330: 346–354.

Cesar T and Mra V (1996) Purification and properties of the xylanase produced by Thermomyces lanuginosus. Enz and Microb Techn 19: 289–296.

Chang P, Tsai WS, Sai CL et al. (2004) Cloning and characterization of two thermostable xylanases from an alkaliphilic Bacillus firmus. Biochem Biophys Res Communs 319: 1017–1025.

Chen YR and Sarkanen S (2010) Macromolecular replication during lignin biosynthesis. Phytochem 71: 453–462.

Cheng JJ and Timilsina GR (2011) Status and barriers of advanced biofuel technologies: A review. Renew Ener 36: 3541–3549.

Chundawat SPS, Bellesia G, Uppugundla N et al. (2011) Restructuring the Crystalline Cellulose Hydrogen Bond Network Enhances Its Depolymerization Rate. J Amer Chem Soc 133: 11163–11174.

Chundawat SPS, Donohoe BS, da Costa Sousa L et al. (2011) Multi-scale visualization and characterization of lignocellulosic plant cell wall deconstruction during thermochemical pretreatment. Energy & Environ Sci 4: 973–984.

Chundawat SPS, Lipton MS, Purvine SO et al. (2011) Proteomics-based Compositional Analysis of Complex Cellulase Hemicellulase Mixtures. J of Proteome Res 10: 4365–4372.

Collins T, Gerday C, Feller G et al. (2005) Xylanases, xylanase families and extremophilic xylanases. FEMS Microbiol Rev 29: 3–23.

Converse A. Ooshima H, Burns DS et al. (1990) Kinetics of enzymatic hydrolysis of lignoce losic materials based on surface area of cellulose accessible to enzyme and enzyme adsorption on lignin and cellulose. Appl Biochem and Biotech 24–25: 67–73.

Coy M, Salem T, Denton JS et al. (2010) Phenol-oxidizing laccases from the termite gut. Insect Biochem. Molec Biol 40: 723–732.

Crous JM, Pretorius IS, van Zyl WH et al. (1995) Cloning and expression of an Aspergillus kawachii endo-1, 4-xylanase gene in Saccharomyces cerevisiae. Curr Genet 28: 467–473.

Currie HA and Perry CC (2007) Silica in Plants: Biological, Biochemical and Chemical Studies. Annals Bot 100: 1383–1389.

Dam P, Kataeva I, Yang SJ et al. (2011) Insights into plant biomass conversion from the genome of the anaerobic thermophilic bacterium Caldicellulosiruptor bescii DSM 6725. Nucleic Acids Research 39: 3240–3254.

Damiano V, Bocchini D, Gomes E et al. (2003) Application of crude xylanase from Bacillus licheniformis 77-2 to the bleaching of eucalyptus Kraft pulp. World J of Microbiol Biotech 19: 139–144.

Demain AL, Newcomb M, Wu JHD et al. (2005) Cellulase, clostridia, and ethanol. Microbiol Mol Biol Rev 69: 124–154.

Donaldson LA (2001) Lignification and lignin topochemistry—an ultrastructural view. Phytochem 57: 859–873.

Dong J, Hong Y, Shou Z et al. (2010) Molecular cloning, purification, and characterization of a novel, acidic, pH-stable endoglucanase from Martelella mediterranea. J Microbiol 48: 393–398.

Donohoe BS, Decker SR, Tucker MP et al. (2008) Visualizing Lignin Coalescence and Migration Through Maize Cell Walls Following Thermochemical Pretreatment. Biotech and Bioeng 101: 913–925.

Duck NB, Koziel MG, Carozzi N et al. (2010) Integrated system for high throughput capture of genetic diversity, US Patent Number 7,689,366 B2.

Dutta T, Sengupta R, Sahoo R et al. (2007) A novel cellulase free alkaliphilic xylanase from alkali tolerant Penicillium citrinum: production, purification and characterization. Letters in Appl Microbiol 44: 206–211.

Elander RT, Dale BE, Hotzapple M et al. (2009) Summary of findings from the Biomass Refining Consortium for Applied Fundamentals and Innovation (CAFI): corn stover pretreatment. Cellulose 16: 649–659.

Engel MS and Krishna K (2004) Family-group names for termites (Isoptera). American Museum Novitates 1–9.

Engel P, Mladenov R, Wulfhorst H et al. (2010) Point by point analysis: how ionic liquid affects the enzymatic hydrolysis of native and modified cellulose. Green Chem.

Epstein E (2009) Silicon: its manifold roles in plants. Ann of Appl Biol 55: 155–160.

Fan LT, Lee YH, Beardmore DR et al. (1981) The influence of major structural features of cellulose on rate of enzymatic hydrolysis. Biotech and Bioeng 23: 419–424.

Fincher GB (2009) Exploring the evolution of (1,3;1,4)-β-d-glucans in plant cell walls: comparative genomics can help! Curr Opin Plant Biol 12: 140–147.

Fontes CM and Gilbert HJ (2010) Cellulosomes: highly efficient nanomachines designed to deconstruct plant cell wall complex carbohydrates. Ann Rev Biochem 79: 655–681.

Fujinami S and Fujisawa M (2010) Industrial applications of alkaliphiles and their enzymes -past, present and future. Environ Technol 31(8-9): 845–856.

Fukagawa N, Meshitsuka G, Ishizu A et al. (1991) A Two-Dimensional NMR Study of Birch Milled Wood Lignin. J of Wood Chem and Tech 11: 373–396.

Fushinobu S, Ito K, Konno M et al. (1998) Crystallographic and mutational analyses of an extremely acidophilic and acid-stable xylanase: biased distribution of acidic residues and importance of Asp37 for catalysis at low pH. Protein Engin 11: 1121.

Gao D, Chundawat SPS, Uppugundla N et al. (2011) Binding characteristics of Trichoderma reesei cellulases on untreated, ammonia fiber expansion (AFEX), and dilute-acid pretreated lignocellulosic biomass. Biotech and Bioeng 108: 1788–1800.

Geib SM, Filley TR, Hatcher PG et al. (2008) Lignin degradation in wood-feeding insects. Proc Natil Acad of Sciences 105: 12932–12937.

George SP, Ahmad A, Rao MB et al. (2001) Studies on carboxymethyl cellulase produced by an alkalothermophilic actinomycete. Bioresource Techn 77: 171–175.

Gharpuray MM, Lee Y-H, Fan LT et al. (1983) Structural modification of lignocellulosics by pretreatments to enhance enzymatic hydrolysis. Biotech and Bioeng 25: 157–172.

Gibbs MD, Reeves RA, Farmington GK et al. (2000) Multidomain and Multifunctional Glycosyl Hydrolases from the Extreme Thermophile Caldicellulosiruptor Isolate Tok7B.1. Current Microbiology 40: 333–340.

Gilbert HJ (2010) The Biochemistry and Structural Biology of Plant Cell Wall Deconstruction. Plant Physiol 153: 444–455.

Gomes J and Steiner W (2004) The biocatalytic potential of extremophiles and extremozymes. Food Tech, and Biotech 42: 223–235.

Gorshkova T, Mikshina P, Gurjanov MP et al. (2010) Formation of plant cell wall supramolecular structure. Biochem (Moscow) 75: 159–172.

Gould JM (1984) Alkaline peroxide delignification of agricultural residues to enhance enzymatic saccharification. Biotech and Bioengin 26: 46–52.

Graham JE, Clark ME, Nadler DC et al. (2011) Identification and characterization of a multidomain hyperthermophilic cellulase from an archaeal enrichment. Nat Commun 2: 375.

Gulick AM (2009) Ironing out a new siderophore synthesis strategy. Nat Chemical Biol 5: 143–144.

Gupta S, Bhushan B and Hoondal GS (2000) Isolation, purification and characterization of xylanase from *Staphylococcus* sp. SG 13 and its application in biobleaching of kraft pulp. J of Appl Microbiol 88: 325–334.

Haas TJ, Nimlos MR, Donohoe BS et al. (2009) Real-Time and Post-reaction Microscopic Structural Analysis of Biomass Undergoing Pyrolysis. Energy & Fuels 23: 3810–3817.

Haki GD and Rakshit SK (2003) Developments in industrially important thermostable enzymes: a review. Bioresource Techn 89: 17–34.

Harris PV, Welner D, McFarland KC et al. (2010) Stimulation of Lignocellulosic Biomass Hydrolysis by Proteins of Glycoside Hydrolase Family 61: Structure and Function of a Large, Enigmatic Family. Biochem 49: 3305–3316.

Hendriks ATWM and Zeeman G (2009) Pretreatments to enhance the digestibility of lignocellulosic biomass. Bioresource Tech 100: 10–18.

Herpoel-Gimbert I, Margeot A, Dolla A et al. (2008) Comparative secretome analyses of two Trichoderma reesei RUT-C30 and CL847 hypersecretory strains. Biotech for Biofuels 1: 18.

Heyman A, Barak Y, Caspi J et al. (2007) Multiple display of catalytic modules on a protein scaffold: Nano-fabrication of enzyme particles. J Biotech 131: 433–439.

Himmel ME, Ding SY, Johnson DK et al. (2007) Biomass recalcitrance: Engineering plants and enzymes for biofuels production. Science 315: 804–807.

Horikoshi K (1999) Alkaliphiles: some applications of their products for biotechnology. Microbiol. Molec Biol Rev 63: 735.

Hough DW and Danson MJ (1999) Extremozymes. Curr Opin Chem Biol 3(1): 39–46.

Huang X, Shao Z, Hong Y et al. (2010) Cel8H, a novel endoglucanase from the halophilic bacterium *Halomonas* sp. S66–4: molecular cloning, heterogonous expression, and biochemical characterization. J of Microbiol 48: 318–324.

Iefuji H, Chino M, Kato M et al. (1996) Acid xylanase from yeast *Cryptococcus* sp. S-2: purification, characterization, cloning, and sequencing. Biosci Biotech Biochem 60: 1331–1338.

Ishizawa C, Jeoh T, Adney WWS et al. (2009) Can delignification decrease cellulose digestibility in acid pretreated corn stover? Cellulose 16: 677–686.

Itakura S, Ueshima K, Tanaka H et al. (1995) Degradation of wood components by subterranean termite, Coptotermes formosanus Shiraki. J Japan Wood Res Soc 41: 580–586.

Jarvis H, Haas TJ, Donohoe S et al. (2011) Elucidation of biomass pyrolysis using a laminar entrained flow reactor and char particle imaging. Energy & Fuels 25: 324–336.

Jenkins R and Alles C (2011) Field to fuel: developing sustainable biorefineries. Ecolog Applic 21: 1096–1104.

Jenney FE Jr and Adams MW (2008) The impact of extremophiles on structural genomics (and vice versa). Extremophiles 12: 39–50.

Jeoh T, Ishizawa CI, Davis MF et al. (2007) Cellulase digestibility of pretreated biomass is limited by cellulose accessibility. Biotech Bioengin 98: 112–122.

Joseleau JP and Ruel K (1997) Study of lignification by noninvasive techniques in growing maize internodes—an investigation by Fourier transform infrared cross-polarization magic angle spinning C-13-nuclear magnetic resonance spectroscopy and immunocytochemical transmission electron microscopy. Plant Physiol 114: 1123–1133.

Jung HG and Casler MD (2006a) Maize stem tissues: Cell wall concentration and composition during development. Crop Sci 46: 1793–1800.

Jung HG and Casler MD (2006b) Maize stem tissues: Impact of development on cell wall degradability. Crop Sci 46: 1801–1809.

Kataeva IA, Yang S-J, Dam P et al. (2009) Genome Sequence of the Anaerobic, Thermophilic, and Cellulolytic Bacterium *Anaerocellum thermophilum* DSM 6725. J Bacteriol 191(11): 3760–3761.

Kazi FK, Fortman JA, Anex RP et al. (2010) Techno-economic comparison of process technologies for biochemical ethanol production from corn stover. Fuel 89, Supplement 1(0): S20–S28.

Ke J, Laskar DD, Singh D et al. (2011) *In situ* lignocellulosic unlocking mechanism for carbohydrate hydrolysis in termites: crucial lignin modification. Biotech Biofuels 4: 17.

Ke J, Sun JZ, Nguyen HD et al. (2010) *In situ* oxygen profiling and lignin modification in guts of wood feeding termites. Insect Sci 17: 277–290.

Kennedy J, O'Leary ND, Kiran GS et al. (2011) Functional metagenomic strategies for the discovery of novel enzymes and biosurfactants with biotechnological applications from marine ecosystems. J Appl Microbiol 111: 787–799.

Khandeparkar R and Bhosle NB (2006) Purification and characterization of thermoalkalophilic xylanase isolated from the *Enterobacter* sp. MTCC 5112. Res in Microbiol 157: 315–325.

Kimura T, Ito J, Kawano A et al. (2000) Purification, characterization, and molecular cloning of acidophilic xylanase from *Penicillium* sp. 40. Biosci Biotech and Biochem 64: 1230–1237.

Klippel B, Lochner A, Bruce DC et al. (2011) Complete genome sequence of the marine, cellulose and xylan degrading bacterium *Glaciecola* sp. 4H-3-7+ YE-5. J of Bacteriol 193: 4547–4548.

Kohlmann KL, Westgate P, Velayudhan A et al. (1996) Enzyme conversion of lignocellulosic plant materials for resource recovery in a controlled ecological life support system. Adv Space Res 18(1-2): 251–265.

Krengel U and Dijkstra BW (1996) Three-dimensional structure of Endo-1,4-beta-xylanase I from *Aspergillus niger*: molecular basis for its low pH optimum. J Mol Biol 263: 70–78.

Kruus K, Lua AC, Demain AL et al. (1995) The anchorage function of CipA (CelL), a scaffolding protein of the *Clostridium thermocellum* cellulosome. Proc Natl Acad Sci US A 92: 9254–9258.

Kuligowski C, Brochier B, Petit-Conil M et al. (2005) Development and optimization of biotechnology use in the manufacture of bleached chemical pulps. TAPPI Engeering, Pulping and Environmental Conference, Proceeedings, Philadelphia, USA.

Kumar R, Mago G, Balan V et al. (2009) Physical and chemical characterizations of corn stover and poplar solids resulting from leading pretreatment technologies. Bioresource Tech 100: 3948–3962.

Kushner D (1985) The Halobacteriaceae. The Bacteria Vol. 8: Archaebacteria CR Woese and RS Wolfe (Eds.). New York, Academic Press 171–214.

Lewis NG and Yamamoto E (1990) Lignin: Occurrence, Biogenesis and Biodegradation. Ann. Rev Plant Physiol and Plant Molec Biol 41: 455–496.

Li LL, Taghavi S, McCorkle SM et al. (2011) Bioprospecting metagenomics of decaying wood: mining for new glycoside hydrolases. Biotechn Biofuels 4: 23.

Liang C, Xue Y, Fioroni M et al. (2011) Cloning and characterization of a thermostable and halo-tolerant endoglucanase from Thermoanaerobacter tengcongensis MB4. Appl Microbiol Biotech 89: 1–12.

Liu T, Wang B, Chen H et al. (2009) Rational pH-Engineering of the thermostable zylanase based on computational model, Process Biochem 44: 912–915.

Liu X, Qu Y, Liu Y et al. (2002) Studies on the key amino acid residues responsible for the alkali-tolerance of the xylanase by site-directed or random mutagenesis. J Molec Cataly B: Enzymatic 18(4-6): 307–313.

Liu Y, Zhang J, Liu Q et al. (2004) Molecular Cloning of Novel Cellulase Genes cel9a and cel12A from Bacillus licheniformis GXN151 and Synergism of Their Encoded Polypeptides. Current Microbiol 49: 234–238.

Lochner A, Giannone RJ, Rodriguez M et al. (2011) Label-free quantitative proteomics distinguish the secreted cellulolytic systems of *Caldicellulosiruptor bescii* and *Caldicellulosiruptor obsidiansis*. Appl Environ Microbiol 77: 4042–4054.

Lorenz P and Eck J (1988) Metagenomics and industrial applications. Genomics 2: 231–239.

Lorenz P and Eck J (2004) Screening for novel industrial biocatalysts. Engin Life Sci 4: 501–504.

Lynd LR, Zyl WH, McBride JE et al. (2005) Consolidated bioprocessing of cellulosic biomass: an update. Curr Opin Biotechnol 16: 577–583.

Mais U, Esteghlalian AR, Saddler JN et al. (2002) Enhancing the enzymatic hydrolysis of cellulosic materials using simultaneous ball milling. Appl Biochem and Biotech 98: 815–832.

Matthews JF, Skopec CE, Mason PE et al. (2006) Computer simulation studies of microcrystalline cellulose Iβ. Carbohydrate Res 341: 138–152.

Menon G, Mody K, Kishri J et al. (2010) Isolation, purification, and characterization of haloalkaline xylanase from a marine Bacillus pumilus strain, GESF-1. Biotech Bioprocess Engineering 15: 998–1005.

Miller PS and Blum PH (2010) Extremophile-inspired strategies for enzymatic biomass saccharification. Environ Technol 31(8-9): 1005–1015.

Monties B (2005) Biological variability of lignin. Cellulose Chem and Tech 39(5-6): 341–367.

Moraïs S, Heyman A, Barak Y et al. (2010) Enhanced cellulose degradation by nano-complexed enzymes: Synergism between a scaffold-linked exoglucanase and a free endoglucanase. J Biotech 147(3-4): 205–211.

Mosier N, Wyman C, Dale B et al. (2005) Features of promising technologies for pretreatment of lignocellulosic biomass. Bioresource Tech 96: 673–686.

Naran R, Black S, Decker SR et al. (2009) Extraction and characterization of native heteroxylans from delignified corn stover and aspen. Cellulose 16: 661–675.

Nguyen Q and Tucker M (2002) Dilute acid/metal salt hydrolysis of lignocellulosics, US Patent 6423145.

Norman G, L (1999) A 20th century roller coaster ride: a short account of lignification. Curr Opin in Plant Biol 2: 153–162.

Pages, S, Belaich A, Tardif C et al. (1996) Interaction between the endoglucanase CelA and the scaffolding protein CipC of the *Clostridium cellulolyticum* cellulosome. J Bacteriol 178: 2279–2286.

Payne CM, Himmel ME, Crawley MF et al. (2011) Decrystallization of Oligosaccharides from the Cellulose Iβ Surface with Molecular Simulation. J Physic Chem Letters 2: 1546–1550.

Podar M and Reysenbach AL (2006) New opportunities revealed by biotechnological explorations of extremophiles. Curr Opin Biotech 17: 250–255.

Qing Q, Yang B, Wyman CE et al. (2010) Impact of surfactants on pretreatment of corn stover. Bioresource Techn 101: 5941–5951.

Qing Q, Yang B and Wyman CE (2011) Xylooligomers are strong inhibitors of cellulose hydrolysis by enzymes. Bioresource Techn 101: 9624–9630.

Raman B, Pan C, Hurst GB et al. (2009) Impact of pretreated Switchgrass and biomass carbohydrates on *Clostridium thermocellum* ATCC 27405 cellulosome composition: a quantitative proteomic analysis. PLoS One 4: e5271.

Reeves RA, Gibbs MD, Morris DD et al. (2000) Sequencing and Expression of Additional Xylanase Genes from the Hyperthermophile *Thermotoga maritima* FjSS3B. 1. Appl Environ Microbiol 66: 1532–1537.

Richard TL (2010) Challenges in Scaling Up Biofuels Infrastructure. Science 329: 793–796.

Ryabova O, Vršanská M, Kaneko S et al. (2009) A novel family of hemicellulolytic α-glucuronidase. FEBS Let 583: 1457–1462.

Sarethy IP, Saxena Y, Kapoor A et al. (2011) Alkaliphilic bacteria: applications in industrial biotechnology. J Indus Microbiol Biotech 38: 1–22.

Scharf ME, Karl ZJ, Sethi A et al. (2011) Multiple Levels of Synergistic Collaboration in Termite Lignocellulose Digestion. PLoS ONE 6: e21709.

Scharf ME. and Tartar A (2008) Termite digestomes as sources for novel lignocellulases. Biofuels, Bioproducts and Biorefining 2: 540–552.

Schell D, Farmer J, Newman M et al. (2003) Dilute-sulfuric acid pretreatment of corn stover in pilot-scale reactor. Appl Biochem and Biotech 105: 69–85.

Scheller HV and Ulvskov P (2010) Hemicelluloses. Ann Rev Plant Biol 61: 263–289.

Schmidt H (1956) Studien an darmbewohnenden Flagellaten der Termiten Parasit Res 17: 269–275.

Schroeder LR, Gentile VM and Atlla RH (1986) Nondegradative Preparation of Amorphous Cellulose. J Wood Chem and Tech 6: 1–14.

Selig M, Knoshaug E, Decker SR et al. (2008) Heterologous Expression of *Aspergillus niger* β-D-Xylosidase (XlnD): Characterization on Lignocellulosic Substrates. Appl Biochem and Biotech 146: 57–68.

Selig M, Adney W, Himmel ME et al. (2009a) The impact of cell wall acetylation on corn stover hydrolysis by cellulolytic and xylanolytic enzymes. Cellulose 16: 711–722.

Selig M, Vinzant T, Himmel ME et al. (2009b) The Effect of Lignin Removal by Alkaline Peroxide Pretreatment on the Susceptibility of Corn Stover to Purified Cellulolytic and Xylanolytic Enzymes. Appl Biochem and Biotech 155: 94–103.

Selig M, Weiss N and Ji Y (2008) Enzymatic saccharification of lignocellulosic biomass. Laboratory Analytical Procedure (LAP), NERL/TP-510-42629, National Renewable Energy Laboratory Golden, CO.

Selig MJ, Knoshaug EP, Andney WS et al. (2008) Synergistic enhancement of cellobiohydrolase performance on pretreated corn stover by addition of xylanase and esterase activities. Bioresource Tech 99: 4997–5005.

Selig MJ, Viamajala S, Decker SR et al. (2007) Deposition of Lignin Droplets Produced During Dilute Acid Pretreatment of Maize Stems Retards Enzymatic Hydrolysis of Cellulose. Biotech Prog 23: 1333–1339.

Shafizadeh F (1982) Introduction to Pyrolysis of Biomass. J Analyt Appl Pyrolysis 3: 283–305.

Sharma P and Bajaj BK (2005) Production and partial characterization of alkali-tolerant xylanase from an alkalophilic *Streptomyces* sp. CD3. J Sci Indus Res 64: 688.

Shibuya H, Kaneko S and Hayashi K (2005) A single amino acid substitution ehances the catalytic activity of family 11 xylanase at alkaline pH. Biosci Biotech Biochemis 69: 1492–1497.

Shirai T, Ishida H, Noda J et al. (2001) Crystal structure of alkaline cellulase K: insight into the alkaline adaptation of an industrial enzyme1. JMB 310: 1079–1087.

Short JM (2003) Sequence based screening, Application Number US 6455254.

Short JM and Keller M (2000) High throughput screening for novel enzymes, Application Number EP 1009858 A1.

Siddiqui KS and Thomas T (2008) Protein adaptation in extremophiles, Nova Biomed.

Singh S, Simmons BA, Vogel KP et al. (2009) Visualization of biomass solubilization and cellulose regeneration during ionic liquid pretreatment of switchgrass. Biotech Bioengine 104: 68–75.

Solingen P, Meijer D, KLeij WA et al. (2001) Cloning and expression of an endocellulase gene from a novel streptomycete isolated from an East African soda lake. Extremophiles 5: 333–341.

Sørensen HR, Meyer AS and Pedersen S (2003) Enzymatic hydrolysis of water-soluble wheat arabinoxylan. 1. Synergy between α-L-arabinofuranosidases, endo-1,4-β-xylanases, and β-xylosidase activities. Biotech Bioengine 81: 726–731.

Sørensen HR, Pedersen S, Meyer AS et al. (2007) Synergistic enzyme mechanisms and effects of sequential enzyme additions on degradation of water insoluble wheat arabinoxylan. Enzy. Microbial Tech 40: 908–918.

Stephen JD, Mabee WE and Sadler JN (2010) Biomass logistics as a determinant of second-generation biofuel facility scale, location and technology selection. Biofuels Bioprod & Biorefining-Biofpr 4: 503–518.

Sun JZ and Scharf ME (2010) Exploring and integrating cellulolytic systems of insects to advance biofuel technology. Insect Sci 17: 163–165.

Tartar A, Wheeler MM, Zhou X et al. (2009) Parallel metatranscriptome analyses of host and symbiont gene expression in the gut of the termite Reticulitermes flavipes. Biotech Biofuels 2: 1–19.

Tejirian A and Xu F (2011) Inhibition of enzymatic cellulolysis by phenolic compounds. Enzyme Microbial 48: 239–247.

Tenkanen M, Viikari L and Buchert J (1997) Use of acid-tolerant xylanase for bleaching of kraft pulps. Biotech Tech 11: 935–938.

Trivedi N, Gupta V, Kumar M et al. (2011) An alkali-halotolerant cellulase from Bacillus flexus isolated from green seaweed *Ulva lactuca*. Carbo. Polymers 83: 891–897.

Trivedi N, Gupta V, Kumar M et al. (2011) Solvent tolerant marine bacterium *Bacillus aquimaris* secreting organic solvent stable alkaline cellulase. Chemosphere 83: 706–712.

van Zyl WH, Lynd LR, den Haan R et al. (2007) Consolidated bioprocessing for bioethanol production using Saccharomyces cereviside. Biofuels 108: 205–235.

Ventosa A, Mellado E, Sanchez-Porro C et al. (2008) Halophilic and Halotolerant Micro-Organisms from Soils. In: Microbiology of. Extreme Soils. P Dion and CS Nautiyal (eds) Springeer-Verlag Berlin/Heidelberg, pp 87–115.

Viamajala S, Selig MJ, Vizant TB et al. (2006) Catalyst Transport in Corn Stover Internodes: I: Elucidating Transport Mechanisms Using Direct Blue-I. Appl Biochem Biotechnol 130: 509–527.

Voelker SL, Lachenbruch B, Meinzer FC et al. (2010) Antisense down-regulation of 4CL expression alters lignification, tree growth and saccharification potential of field-grown poplar. Plant Physiol 154: 874–886.

Voget S, Steele H and Steit WR (2006) Characterization of a metagenome-derived halotolerant cellulase. J Biotech 126: 26–36.

Wainø M and Ingvorsen K (2003) Production of ß-xylanase and ß-xylosidase by the extremely halophilic archaeon Halorhabdus utahensis. Extremophiles 7: 87–93.

Wang CY, Hsieh YR, Ng CC et al. (2009) Purification and characterization of a novel halostable cellulase from *Salinivibrio* sp. strain NTU-05. Enzyme Microbial Tech 44: 373–379.

Wang J, Zhang WW, Liu J et al. (2010) An alkali-tolerant xylanase produced by the newly isolated alkaliphilic *Bacillus pumilus* from paper mill effluent. Mol Biol Rep 37: 3297–3302.

Warnecke F, Luginbuhl P, Ivanova N et al. (2007) Metagenomic and functional analysis of hindgut microbiota of a wood-feeding higher termite. Nature 450: 560–565.

Wei H, Donohoe BS et al. (2011) Elucidating the role of ferrous ion co-catalyst in enhancing dilute acid pretreatment of lignocellulosic biomass. Biotech Biofuels (submitted).

Weiss ND, Nagle NJ, Tucker MP et al. (2009) High Xylose Yields from Dilute Acid Pretreatment of Corn Stover Under Process-Relevant Conditions. Appl Biochem Biotech 155: 418–428.

Wejse PL, Ingvorsen K, Mortensenet KK et al. (2003) Purification and characterisation of two extremely halotolerant xylanases from a novel halophilic bacterium. Extremophiles 7: 423–431.

Wiegel J and Kevbrin VV (2004) Alkalithermophiles. Biochem. Soc Trans 32 (Pt 2): 193–198.

Wong K, Tan L, Saddler JN et al. (1988) Multiplicity of beta-1, 4-xylanase in microorganisms: functions and applications. Microbiol Molec Biol Rev 52: 305.

Wu D, Hugenholtz P, Mavromatis K et al. (2009) A phylogeny-driven genomic encyclopaedia of Bacteria and Archaea. Nature 462: 1056–1060.

Wyman CE, Dale BE, Elander RT et al. (2005) Coordinated development of leading biomass pretreatment technologies. Bioresource Tech 96: 1959–1966.

Yang B and Wyman C (2008) Pretreatment: the key to unlocking low-cost cellulosic ethanol. Biofuels, Bioprod Biorefin 2: 26–40.

Yang B and Wyman CE (2004) Effect of xylan and lignin removal by batch and flowthrough pretreatment on the enzymatic digestibility of corn stover cellulose. Biotech Bioengine 86: 88–98.

Yang B and Wyman CE (2006) BSA treatment to enhance enzymatic hydrolysis of cellulose in lignin containing substrates. Biotech Bioengine 94: 611–617.

Zhang G, Ma X, Niu F et al. (2007) Diversity and distribution of alkaliphilic psychrotolerant bacteria in the Qinghai–Tibet Plateau permafrost region. Extremophiles 11: 415–424.

Zverlov V, Mahr S, Riedel K et al. (1998) Properties and gene structure of a bifunctional cellulolytic enzyme (CelA) from the extreme thermophile *Anaerocellum thermophilum* with separate glycosyl hydrolase family 9 and 48 catalytic domains. Microbiol 144: 457–465.

Zverlov V, Piotukh K, Dakhova O et al. (1996) The multidomain xylanase A of the hyperthermophilic bacterium *Thermotoga neapolitana* is extremely thermoresistant. Appl Microbiol Biotechn 45: 245–247.

Zverlov VV, Kellermann J and Schwarz WH (2005) Functional subgenomics of Clostridium thermocellum cellulosomal genes: identification of the major catalytic components in the extracellular complex and detection of three new enzymes. Proteomics 5: 3646–3653.

CHAPTER 3

Special Requirements
Agricultural and Industrial Infrastructure

Luis E. Rincón,[a] *Carlos A. Cardona*[b],* and
Carlos E. Orrego[c]

ABSTRACT

Maize (*Zea mays* L.), a native tropical crop from America, has served as basic food for humanity for centuries. Structurally a typical maize seed has four main components: the pericarp, the germ or embryo, the tip cap and the endosperm. The elements contained in these parts include mainly vegetable oil, starches and sugars minerals and proteins, which yield different added value products for maize. Indeed, today maize is directly present in the diet of more than 200 million people as a breakfast cereal or dairy product. In this chapter, the agricultural features of maize planting, as well as different aspects of the industrial milling process. The first section comprises general characteristics of maize planting, harvesting and storage. The second section considers the two types of maize milling processes: wet and dry, as well as a brief description of the products derived. The final section includes

Instituto de Biotecnología y Agroindustria, Departamento de Ingeniería Química, Universidad Nacional de Colombia sede Manizales, Cra. 27 No. 64-60, Manizales, Colombia.
[a]Email: lerinconp@unal.edu.co
[b]Email:ccardonaal@unal.edu.co
[c]Email: ppba_man@unal.edu.co
*Corresponding author

two case-studies oriented to biodiesel and ethanol production from maize, establishing and describing their basic infrastructure, yields and production costs.

Keywords: Maize plantation, Maize Tillage, Maize Irrigation Dry milling, Wet Milling, Maize Ethanol, Biodiesel

Introduction

Maize (*Zea mays* L.*)*, a native tropical crop from America, has served as basic food for humanity for centuries. It is one of the oldest domesticated plants and its origins are dated more than 7,000 years ago from Central Mexico. There, the Mesoamerican natives adapted this crop from a wild grass to a productive food source (Abbassian 2006). Currently, maize is well adapted to different climates with a wide range of maturities from 70 days to 210 days (Belfield and Brown 2008). Maize can be produced in both hemispheres; the northern hemisphere from April through November and the southern from September through May. Maize is the world's third largest cultivated crop after wheat and rice, and the first, most harvested cereal crop worldwide. In 2009, the reported world production was 818 million tons (see Table 1) (FAOSTAT 2012). Moreover, the United States is currently the world's largest, producer, consumer, and exporter of maize (Abbassian 2006).

Maize includes a number of different varieties that incorporate flour maize, flint maize, dent maize, sweet maize, pop maize, waxy maize, and amylomaize, in addition to grades and classes related to a set of its physical descriptions or qualities. These include the minimum test-weight, feeding values, maximum limits of damaged kernels, starch and gluten composition, as well as foreign material content (Singh et al. 2011). For instance, in the United States, maize is classified according to color. Maize is divided into yellow, white and mixed and graded from 1 to 5 (Abbassian 2006). However, a more general classification for maize is defined by its amylopectin content: i) waxy maize, which contains 100% of amylopectin; ii) high amylose maize

Table 1 The Top 5 World Cereal Production in 2009 per Area Harvested and Total production.

Cereal	Harvested area (Ha)	Production (Tons)
Maize	158.628.747	818.823.434
Wheat	225.000.000	685.614.399
Rice, paddy	158.300.068	685.240.469
Barley	54.059.705	152.125.329
Sorghum	39.969.624	56.098.260

Source: (FAOSTAT 2012)

(amylose content between 40–70%); and iii) sugary maize, containing lower starch and higher sucrose levels (Singh et al. 2006).

Structurally a typical maize seed has four main components (see Fig. 1), and from each, value-added products are derived. These features include: The seed coat or pericarp, the germ or embryo, the tip cap, and the endosperm. The pericarp, the outer part of the seed, is composed of cellulose, hemicelluloses, lignin and various waxes. These cellular layers protect the kernel from diseases, insects and moisture losses. The germ or embryo contains a miniature plant composed of the plumule (leaves), radicle (roots) and sculletum. The embryo is rich in oil (25% of its weight) and contains the genetic information required both for shoot development during germination and for development of the sexually mature reproducing plant. The genetic material is required to develop roots and shoots during germination. The tip cap is the point into where the kernel is attached to the cob. It provides the major entry point for food and water into the kernel. The endosperm is formally the seed's food storage tissue comprising 80% of the kernel. It is composed primarily of starches and sugars, minerals and proteins in proportions that vary according to the variety (Espinoza 2001; Eckhoff 2004).

Today, maize is directly present in the diet of more than 200 million people as a breakfast cereal or dairy product (du Plessis 2003). This consumption rate is growing annually. The Food and Agriculture Organization (FAO) of the United Nations estimates that human and animal consumption demand for maize will be increased by nearly 300 million tons by 2030. This (Fig. 1) does not include the demand for industrial applications (O'Gara 2007). Indeed, maize plays a major role as a vital food in different parts of the world, particularly in Africa, Asia as well as Central and South America. This is why maize is included among the leading commodities that make up the bulk of international food aid (Abbassian 2006). However, despite this high demand for maize for human food and animal feed, not all planted maize is used towards this end. During the 20 year period beginning with 1980, industrial use of maize was below 20% according to FAOSTAT. By

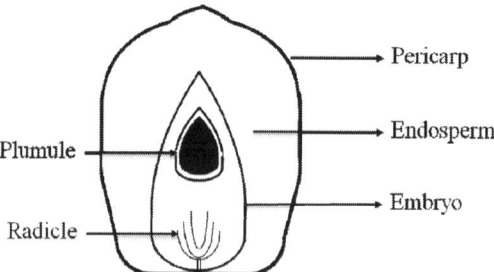

Figure 1 Simplified components of maize seed.

2007 however, only 110 of the 789 million of tons produced were employed for human consumption signaling a profound change in market patterns (FAOSTAT 2011). Thus, although maize is used and traded as a food crop, today, it is also an important food staple and industrial feedstock. Currently, conservative estimates indicate that industrial use of maize could be between 15%–30% of the total production, as its use increases due to its value to new bio-based initiatives, to renewable energy production and to a growing sweeteners industry (Davis 2001).

By way of introduction, industrially maize is extensively employed for a number of purposes that include feedstock for the production of organic chemicals, dietary complements, sweeteners (syrup), biodegradable polymers, fibers and, in a rising proportion, for ethanol production (O'Gara 2007). These choices are dictated by the availability of value-added molecules that are synthesized in various parts of the plant. For example, the oil found in the germ may be processed to produce different salad oils and similar products. In addition, the starch found in the endosperm is a polymer composed of a large number of repeating glucose units that can be processed and used for food, pharmaceutics, cosmetics, paints products, adhesives in the paper industry, and as filler in pharmaceuticals. Enzymatic starch degradation produces sorbic and lactic acids (Halm et al. 2004; Pirgozliev et al. 2008). Starch can also be used in a huge assortment of chemical derivatives such as biodegradable bioplastics substituting for petroleum based plastics (van Soest et al. 1996; Chaudhary et al. 2008). In addition, maize starch can be fermented and used by the ethanol and brewing industries.

This list in no way exhausts the application of maize products in the world market. In this connection, the silage industry is worldwide in its scope as is maize's impact on the production of sweeteners. Indeed, maize used industrially for different purposes including ethanol and sweetener production, may potentially generate about 3.2 USD per each 1 USD expensed on-farm economic activity (Gardisser 2001).

Maize Agriculture

The successful agricultural production of maize depends upon the interaction among different determinant parameters such as plant population, soil tillage, fertilization, harvesting, storage, and others (du Plessis 2003). A full growing season can range between 130 and 150 days from planting to harvesting (O'Keeffe and Byrne 2010). Maize growth is an activity favored by warm daytime conditions, mild nights and well-drained soils. The plant grows well without irrigation or little agricultural care in tropical, coastal land table areas (O'Keeffe and Byrne 2010). However, maize's full potential can only be reached when it is grown without any stress under optimal conditions (Espinoza 2001). While Mother Nature and plant genetics define the limits of plant development, many important factors, which can be controlled, are related to good agricultural practices. In this sense, the addition of agrochemicals, coupled with tillage conditions, sowing time, irrigation, and planting management are as important as the selection of suitable soils and climates to assure successful maize planting. In this section, the main aspects related to agricultural practices during maize planting, growing, harvesting, and storage are briefly described.

Land Selection

Land selection is the first factor that should be considered before planting maize, in order to assure high product yields. Maize can be planted from 50°N to 40°S, and from sea level up to 4,000 m altitude (Belfield and Brown 2008). This crop grows optimally between 18°C and 32°C. It is not recommended to expose maize to temperatures above 35°C. Maize crops require 500 mm to 1200 mm of rainfall, although it can grow with yields reduced to only 40%–60% with 300 mm (Belfield and Brown 2008).

While maize can be cultivated using a wide range of soil types, well-drained loamy soils with pH 5–8 are preferred. Sandy soils with pH < 5 can produce aluminum toxicity problems for growing. Plant vigor is better on deep, well drained soils (Ross et al. 2001). Indeed, maize is moderately sensitive to salinity, in which case nutrient uptake and dry matter production is reduced. Maize is not a drought tolerant crop (Belfield and Brown 2008).

However, the most important consideration for land selection is drainage. It is necessary to drain water in depressions or low spots in less than 24 hours (Wright et al. 2002). Poor drainage hampers field operations from field preparation through the harvest. Insufficient drainage limits the effectiveness of irrigation, and may reduce yields and production potential (Ross et al. 2001). Proper drainage can preserve natural soil productivity by reducing field rutting that requires additional tillage (Tacker et al. 2001). This task of smoothing and forming field surface can be completed before bedding. In this way, surface drainage of the field is improved.

Maize Tillage

Maize seeds need soil conditions that guarantee an environment that is warm, moist, and well aerated, providing enough contact between the seed and the soil for rapid germination (Roozeboom et al. 2007). Sufficient soil cover and a firm contact must be guaranteed for optimum establishment

(O'Gara 2007) and driven by appropriate tillage practices. Tillage is the mechanical manipulation of the soil profile to prepare the field for planting (Soil Science Glossary Terms Committee 2008). The purposes of tillage are to loosen the soil, incorporate chemicals such as fertilizers and nutrients, control weeds, provide for soil and water conservation as well as for seedbed preparation. All of the above functions can be performed together depending on the availability of equipment. Tillage systems can be classified as no-till, reduced-till, mulch-till, ridge-till, and strip-till (Roozeboom et al. 2007). Some aspects of above tillage systems are described in Table 2.

The extent of tillage will varyingly impact the soil. Tilling can invert the surface soil, bury plant residues and stands in contrast to no till. Indeed, modern management systems use less tillage and leave more residues on the surface (Robinson 2005). Conservation tillage has advantages over full tillage because it minimize adverse impacts on water quality, conserves soil moisture, reduces soil erosion and reduces labor and energy requirements (Roozeboom et al. 2007). Specifically, maize can be cultivated under a no-till system where crop residues remain on the field as stubble. Stubble can provide a good microclimate and a good rainfall filtration system for the next round of planting. Other potential benefits of mulching embrace

Table 2 Tillage systems used in maize cultivation.

Tillage system	Feature
No-Till	This system leaves soil untreated from planting to harvesting. In addition, it helps to reduce production costs. This kind of cultivation requires at least 30% of the soil to be covered by plant residues after planting to achieve effective water erosion reduction
Stubble-mulch tillage	In this type of cultivation, soil is disturbed before planting, but without burying or destroying it. Chisel ploughs, discs, spring-tooth implements or V-type blades are used. Weeds are controlled either chemically or mechanically
Reduced tillage	This type of tillage leaves 15%–30% of soil covered with stubble. Weeds are controlled either chemically or mechanically
Conventional tillage	Conventional tillage systems use a combination of cultivation and herbicides to control weeds. Once the weeds are killed following cultivation, the crop is planted into a clean seedbed. This option requires less than 15% of the soil surface covered with stubble. Conventional tillage requires plough action or other intensive range of cultivations
Conservation tillage	In this tillage, crops are sown with minimum soil disturbance and herbicides replace cultivation as the primary means of weed control. This option requires at least 30%–60% of the soil surface to be covered with residues; in order to ensure moisture conservation.

Source: (du Plessis 2003 and O'Gara 2009)

reduced crusting of soil surface, reduced soil temperature, reduced surface evaporation, reduced emergence of weeds, reduced soil erosion, and reduced sandblasting damage to seedlings (Belfield and Brown 2008). However, independently of the type of tillage system selected, hardpans and compacted layers from over-cultivation must also be retired as these conditions block moisture penetration and prevent root growth. When these damaged layers are detected, they must be broken, in order to assure better drainage, conservation, and water utilization (du Plessis 2003; Belfield and Brown 2008). Over-cultivation consumes energy and leads to a decline in soil structure and depletion of organic matter (O'Gara 2007).

A variety of different machinery can be used during maize tillage. An overview of the equipment used in maize tillage, as well as their main features can be found in Table 3. The next section explains three more important tasks to be accomplished during tillage and includes seedbed preparation, weed control, and fertilizing.

Seedbed preparation

An important step of tillage is seedbed preparation. Plowing is employed to ensure profitable maize production and to minimize soil erosion due to wind and water (Roozeboom et al. 2007). Maize must be planted carefully with spatial accuracy, in order to achieve the best germination and emergence. Thus, if soil is too wet or dry or seeds are planted too deeply, germination will either be delayed or fail. An adequate seedbed, which guarantees a good seed-to-soil contact requires a depth of 5 cm–7 cm of fine firm soil and no weeds. Good practice dictates planting seeds in moist soils at an even depth of 2 cm–5 cm (Belfield and Brown 2008). Furthermore, maize should be planted on raised rows or beds, especially on relatively flat fields. This arrangement reduces the effect that cold, wet soil may have on planting and early crop development (Tacker et al. 2001). An ideal seedbed should have the following elements: a) weed control, b) moisture conservation, c) tilth preservation or improvement, d) water quality protection, e) erosion control and f) suitability for planting with available equipment (Roozeboom et al. 2007). Maize can be planted in either a bed or a flat. However, the bed system is the most used. Beds normally warm more quickly earlier in the planting season and provide drainage following heavy rainfall; the top of the bed can be knocked down during a dry spring, to improve soil moisture (Ross et al. 2001).

Table 3 Main machinery used in tillage.

Machinery	Feature	Used in
Ripper hipper	Rippers are used to break up compacted soil in order to improve drainage and aeration. Rippers are used when deep cultivation is necessary and turning of the soil is undesirable. Otherwise, when soils are tilled annually to the same depth, plough-sole can be developed, preventing infiltration and root development.	Soils should be quite dry to be ripped; in moist soils ripping does not shatter the subsoil, smearing and sealing the soil beside the ripper-tine.
Mouldboard plough	The Mouldboard plough is one of the basic implements of tillage, and is used to turn sod up to 300 mm depth. It is particularly useful on heavier well-structured soils. Turning soil helps to bury weeds and unwanted plant residues deep within the soil.	Mouldboard ploughs are not recommended on sandy soils, where their granular structure might promote erosion with this implement.
Disc ploughs	This slicing device can break undisturbed soil, turning it to bury surface weeds and trash. Regular use of disc ploughs reduces soil aggregates to small particles and produces a compacted layer or plough pan. Disc ploughs are used to remove soil surface by inverting it. In addition, this cultivator may be employed for weed control and seedbed preparation where it is necessary to roll the soil or to use press-wheels to enhance seedling emergence.	Better penetration is obtained under dry, hard conditions with lower wear than the Mouldboard plough. It is useful on hard, dry soils where loss of structure is not too critical. Disc ploughs are not recommended for sandy soils.
Chisel ploughs	This equipment is used to loosen soil to a limited depth of 250 mm. Best results are obtained if soils are relatively dry, because moist soils can be smeared and the soil surface sealed.	When conditions are too dry, big clods can be formed, restricting plant development.
Rotary Tiller	This device aerates the soil and provides a fine seedbed.	This implement is used to prepare seedbeds on moist clay soils; however, it is not recommended for sandy soils where it can destroy the integrity of the stratum.
Tined cultivators	Tined cultivators include a variety of hoeing implements used in seedbed preparation. They are used for controlling young weeds and breaking surface cuts.	These implements are more effective in moist soils, and not effective in clay soil.
Harrows	Harrows include different maize farming implements. A tined harrow is primarily used to level the seedbed once it is in good condition while a disk harrow is designed to break surface crusts; also, harrows can be used to break clods to obtain a fine seedbed.	Any soil type, however if they are used regularly, they will break down and pulverize the soil structure.

Source: (du Plessis 2003; Lines-Kelly 2004; Coulter 2005)

Weed and pest control

Weed control is another important task during maize planting. Weeds compete for light, water, and nutrients. Lack of any of these essentials can dramatically reduce maize yields (O'Gara 2007). Weeds must be destroyed before they deplete soil moisture, in order to permit full development of the maize grain (Roozeboom et al. 2007). This said, a maize crop is more susceptible to weed competition at early stages of growth and remains so until the seedling reaches 0.8 m height, approximately 8 weeks after planting. Common weed species, which may potentially affect maize include: (i) broadleaf species including pigweeds, velvetleaf, cocklebur, kochia, smartweed, and puncturevine, (ii) common grass weeds including shattercane, large crabgrass, foxtails, field sandbur, fall panicum, barnyard grass, and prairie cupgrass (O'Keeffe and Byrne 2010), and (iii) perennial weed species such as johnsongrass, field bindweed, and bur ragweed (Wright et al. 2002; Regehr et al. 2007).

Conventional tillage systems control weeds using mechanical cultivation, herbicides or a combination of both. Once the weeds are eliminated, the crop is planted into a clean seedbed (O'Gara 2007). Mechanical weed control includes either false seedbed or stale seedbed or pre-plant weed control strategies. The false seedbed approach involves preparing a seedbed to enhance weed germination, followed, for example, by hoeing, and the preparation of a new seedbed with a resulting reduction in weed emergence. A stale seedbed approach is similar except that weed seedlings are killed to avoid bringing new weed seeds up to the soil surface where they have a better chance of germinating.

Mechanical control is necessary during the first six weeks after planting, but weeds that emerge later will not cause yield reduction (Coulter et al. 2005). Herbicides offer more flexibility for weed control and these are used to manage weeds not controlled by the other practices. Each kind of chemical herbicide has specific restrictions and application procedures. Application rates will depend on soil texture, organic content and targeted weed species (Wright et al. 2002). However, the final choice of a herbicide combination will depend on the weed spectrum (O'Gara 2007). Maintaining weed control during whole growing stage is essential for harvesting and preventing contamination of final grain with bacterium (O'Keeffe and Byrne 2010). Weed reduction can be enhanced with good production practices such as crop rotation, timely planting, soil fertility management and cultivation (Wright et al. 2002).

Insects and fungi are also a threat to maize, causing yield and economic losses. Insect species can cause crop damages either by direct feeding or by serving as a vector for delivery of a plant virus such as maize dwarf mosaic. Control requires having a knowledge of insect and microbial biology, as well as management options available (Wright et al. 2002). Maize seed should be treated with insecticides to prevent infestation. Cutworms attack maize early in the season. However, the most damaging and prevalent insect is the maize earworm. This invasion occurs in the spring, but becomes more problematic in the fall (Motes et al. 2001). Fungicide protects maize against seed rot and damping-off. The crop can also be attacked by leaf blights, leaf rust, smut, bacterial wilt, and maize dwarf mosaic virus, to name but a few (Motes et al. 2001).

Fertilizing

In order to obtain high maize yields, high soil fertility is required. Hence, the nitrogen, phosphorus, and potassium contents of the fertilizer are dependent on factors such as cropping and fertilizer history, age of cultivation, fallow conditions, and yield targets (O'Keeffe and Byrne 2010). Each soil is depleted in part as a function of its species planting history. If these nutrients are not replaced, the soil fertility declines. In consideration of these losses, good agricultural practices require the establishment of a soil nutrition program. Following germination, maize takes up small amounts of nutrients. Nutrient uptake rapidly increases after four weeks. More than 90% of potassium uptake occurs between four and seven weeks after planting. At this stage of maturity, less than half of the final, above ground dry matter has been formed. Nitrogen uptake increases rapidly with 55% occurring in the short window from seven weeks after planting until the end of silking. Nitrogen uptake is virtually complete two weeks after flowering. Phosphorus uptake is complete four weeks after flowering. Banding fertilizer at sowing ensures that the crop can access nutrients from the very early stages of root development (O'Keeffe and Byrne 2010). Additionally, irrigation has a determinant effect over the transport of chemical fertilizers helping to incorporate and activate them, while enhancing crop germination. This task requires an irrigation system which guarantees at least an average water level of 6.0 mm to 10.0 mm immediately after application (O'Gara 2007).

Maize Planting

Agronomic factors such as planting space, depth and seed dropping rate are critical to achieve and maximize maize crop yields. Thus, seeds should be spaced uniformly within the row by releasing them at predictable intervals rates so as to eliminate spacing variability (Ross et al. 2001). In theory, the highest maize yields can be achieved by limiting plant population between 53,000 to 66,000 plants/ha, planting rates of 15 to 20 kg/ha with 2 plants/hill on 70 cm of row spacing and 50 cm between hills (Belfield and Brown 2008).

However, these optimum values may change according to soil conditions, agricultural practices, and available irrigation systems.

Planting space

In maize plantations, the row spacing range is 75 cm–110 cm, although width may be ultimately determined by factors such as the availability of planters, tractors, harvesters or other farming equipment. Currently, a variety of different row spacing sizes and arrangements exist, and each has its own advantages and disadvantages. Narrow rows (75 cm) can potentially produce slightly higher yields than broad spaced plants. This requires good in-crop rainfall or irrigation, high soil fertility and a saturated plant population (O'Keeffe and Byrne 2010). Reduced row width can enhance weed control and reduce intra-row plant competition. Nevertheless, narrow row maize production has higher costs due to modification in planting and harvesting equipment (Roozeboom et al. 2007). Row spacing less than 75 cm can be too narrow to provide enough surface drainage and nutrient uptake (Ross et al. 2001). Rows greater than 100 cm may result in comparatively lower yields, due to a comparative lower solar light access and potentially water losses (Ross et al. 2001).

With relation to row arrangement, twin-row (double row) is a configuration that would allow advantages over narrow rows without requiring additional harvest equipment. Nowadays, this row arrangement has been found that shows and inconsistent performance in relation to agro-ecological conditions (Roozeboom et al. 2007). Inter-row planting, can be also employed, but it needs to be undertaken early to prevent damage to the roots and the crop. Once this configuration is chosen, post-emergence herbicides are usually necessary for extended weed control. This pattern in dry land requires skip row planting, in order to minimize yield variability. In this arrangement, rows of maize are skipped and not planted, but plants per hectare remain the same as if every row were planted. Using a single skip pattern, the number of seeds dropped is doubled per planted row. Therefore, the final stand will be similar to that for a field with every row planted (Roozeboom et al. 2007). In this way, single or double skip on 100 cm rows, can be successful in helping to conserve soil moisture for grain fill (O'Keeffe and Byrne 2010).

Planting depth

Maize roots will develop either directly from the kernel and be temporary or arise from nodes above the seed and remain permanent. Optimal planting depth for maize crop will vary according to soil conditions affecting the speed of germination and emergence. Although an adequate planting depth for maize approximates 4.5 cm–5 cm, this value can change using wet soils, where seeds might be planted at 3.8 cm. A shallow planting increases the risk for poor establishment of the nodal roots, especially in fluffy soils challenged by heavy rains and results in shallower seed displacement (Coulter et al. 2005). Planting too superficially may force the crown to develop at soil surface and thus inhibit the development of the secondary and brace root systems, limiting nutrient uptake.

Planting deeper than 9 cm, might cause emergence problems as well. For example, a sandy soil holds less water than a finely textured one. Here, it is necessary to plant 5 cm–7 cm deep, to prevent the seed from drying. Another factor that affects maize emergence is temperature. At 10°C–13°C it takes 18–21 days, while at 16°C–19°C the seedling emerges in 8–10 days. Temperatures below 10°C result in the complete absence of germination. Soils are colder at greater depths, slowing germination and increasing the seed's vulnerability to disease or insect attack (Roozeboom et al. 2007).

Seeding rate and plant population

The seeding rate is the speed of seed planting, while plant population is the number of plants that finally survives (Coulter et al. 2005). The seeding rate of maize is determined according to an optimal number of seeds planted per hectare that itself is regulated by the number/kilogram, germination percentage, row spacing, planting date, and soil conditions (Coulter et al. 2005; O'Gara 2007). In contrast, plant populations vary according to moisture conditions, seed and product costs, land fertility and management practices (Coulter et al. 2005). Thus, different maize varieties and management conditions require specific plant populations. A population of 40,000–85,000 plants per hectare is recommended for fully irrigated maize acreage (O'Gara 2007). For dry lands, 40,000–60,000 plants per hectare is common (Ross et al. 2001; Roozeboom et al. 2007). Not all seeds planted will germinate and a seeding rate established should account for these losses (O'Gara 2007). Generally, the final or harvest plant population is about 85% of planting rate, although some varieties of maize can compensate this drop in yield by producing larger ears (Roozeboom et al. 2007). For instance, 5%–10% may be lost by insects, disease or other pests (Wright et al. 2002). Otherwise, excessive plant populations, indeed as few as planting 10–15 more seeds per Ha may reduce yield because of inadequate watering or rain fall (Wright et al. 2002).

Variety Selection

Depending upon the kinds of environmental challenges, different maize varieties should be planted in order to face different potential situations such as: cool temperatures, lower germination rates or frost damage. Adequate selection among early, normal and late varieties must be made to accommodate different considerations. In this way drops in productivity or even total losses are avoided (Motes et al. 2001). Moreover, the decision as to which hybrid is chosen is driven by additional factors such as yield, maturity, stay green, lodging, shuck cover, ear placement, disease and insect resistance. Moreover, a combination of plant breeding techniques and new technologies are able to increase the number and quality of new maize hybrids, leading to species potentially more resistant to weed and insects (Ross et al. 2001).

Irrigation Methods

Maize has a better response to irrigation than other crops. Maize varieties take anywhere from 100 to 120 days to reach physiological maturity using 100 days of irrigation. Underestimating water requirements will stress the crop and reduce yield, while overestimating water demand results in nutrient leaching, water logging, and water waste (O'Gara 2007). Although reasonable yields occur without irrigation in years with good rainfall patterns, yields can easily drop and the resulting drought stress can contribute to charcoal rot, aflatoxin contamination, and ultimately crop failure (Tacker et al. 2001). Indeed, yields of non-irrigated maize fall in range of 25% to 95% as compared to a normally irrigated crop effecting plant development and yield potential, the greatest harm varying according to maturity at the time of stress (Tacker et al. 2001).

Water requirements and water management

Maize has an excellent water use-yield response curve (Rogers 2007). Figure 2 is a smoothed out representation to reflect average weather and maize growth. Maize achieves its most rapid growing stage and reaches its peak following the uptake of 0.6 cm–0.7 cm of water per week (O'Gara 2007). This crop has a critical growth stage at the beginning of its reproductive stage. Failure to supply adequate water at tasselling and silking will drastically reduce yield. Therefore, water management and system capacity are important issues to assure good yields. The use of irrigation systems allows maintenance of adequate soil water content (Rogers 2007).

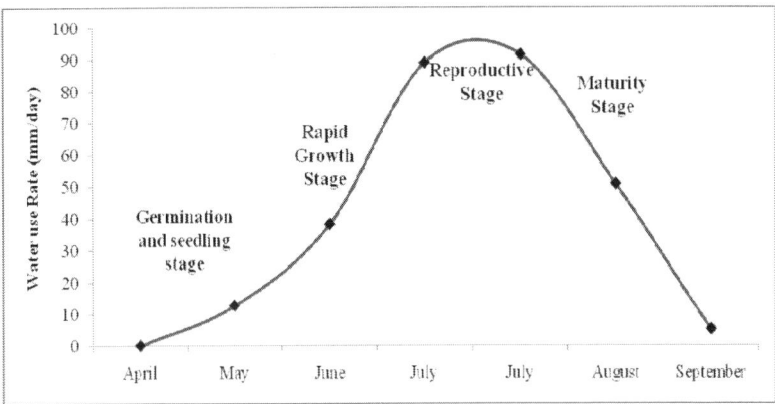

Figure 2 Characteristic water use pattern of maize. Adapted from: (Rogers 2007).

The amount of irrigation that a maize crop requires will vary depending on the soil moisture already present and the rainfall received during the growing season (Tacker et al. 2001). One ton of maize requires approximately 745 million liters of water (Wright et al. 2002) equivalent to 0.5–0.7 cm of water applied through the growing season (O'Gara 2007). An irrigated maize crop in a dry season will require 5–7 million liters per hectare of water. Typically, however, maize requires 0.25 cm of water every 7–10 days until it stands 38 cm, then 2.5 cm every 5–7 days until tassel emergence, and finally another 2.5 cm every three days until physiological maturity (Wright et al. 2002). In general, these requirements vary according to planting time, plant density, fertility, days to maturity, soil type, seasonal conditions, and irrigation efficiency (Tacker et al. 2001).

It is critical to establish pumping schedules and determine whether the crop can be irrigated sufficiently (O'Gara 2007). Maize is a relatively deep-rooted crop, but only 0.9 m–1.2 m of the root zone is usually monitored for irrigation management. In a uniformly wetted profile, 70% of water and nutrients are removed from the upper half of the root zone (Rogers 2007).

Irrigation methods

Selection of an irrigation method is a decision based on diverse factors such as geographic location, type of technology and cost-benefit ratio. Sandy soils have lower water storage capacity and high infiltration rate needing small irrigation applications. For this kind of soil, sprinkler irrigation is recommended. Otherwise, on loam or clay soils many irrigation methods can be used, but surface irrigation is the most common. Clay soils with low infiltration rates are ideally suited to surface irrigation. Strong winds can disturb the direction of the spray. Under very windy conditions, drip or surface irrigation methods are preferred. In areas of supplementary irrigation, sprinkler or drip irrigation may be more suitable than surface irrigation because of their flexibility and adaptability to varying irrigation demands on the farm.

The choice of irrigation method is most often determined by the technology. In general, drip and sprinkler irrigation are technically more complicated and require high capital investment per hectare. To maintain the equipment, local high-level technical expertise has to be available. In addition, a regular supply of fuel and spare parts must be maintained, which together with the purchase of equipment may require foreign currency. Surface irrigation systems, in particular small-scale schemes, usually require less sophisticated equipment for both construction and maintenance (Brouwer et al. 1987). A summary of systems used for irrigation can be found in Table 4.

Table 4 Common types of irrigation used in maize planting.

Type	Feature
Flood or Furrow Irrigation	This one of the most popular irrigation methods; here, water is pumped to fields and is allowed to flow among crops through furrows. It is a simple and inexpensive method. With this kind of irrigation, it is important to get the water across the field as quickly as possible. Also, it is recommended that water not be pooled in any area for longer than two days.
Border Irrigation	In border irrigation, a large volume of water is flushed over a relatively flat field surface in a short period of time. Borders are graded strips of land, separated by earth bunds. The objective is to release water into the area between the borders at the high end of the field. A clay soil that cracks is sometimes difficult to irrigate, but with border irrigation, the cracking actually helps as a distribution system between the borders. This factor also makes it possible to use borders on clay fields that have a slight side or cross slope.
Center Pivot Irrigation (Sprinkler Irrigation)	In this method, an irrigation device rotates around a pivot, and the circular area centered on the pivot is irrigated. Center pivots offer the ability to irrigate fields with surface slopes that make it impossible or impractical to irrigate with surface methods. The need for good surface drainage still exists with pivot irrigation and should not be overlooked. Pivots provide the ability to control the irrigation amount applied by adjusting the system's speed.

Source: (Omary 1997; Tacker 2010; USGS 2011)

Maize Harvesting

Maize is physiologically mature at about 60 days after pollination (Coulter et al. 2005), when a black layer appears at the base of the grain where it attaches to the cob. Grain moisture at this stage is around 15%–32%, which is too high for storage. Maize can be harvested at higher moisture, but will require artificial drying. According to variety, maize will usually take between 135 and 160 days from planting to harvest at the correct moisture content (O'Gara 2007). Currently, maize can be harvested either manually or mechanically. Manual harvesting is a labor intensive job, with a high impact over global production costs. This task can be performed when red maize cobs have been dried down, so they are handpicked, hand shelled and then dried in the sun (Hanna and Quick 2007; Belfield and Brown 2008).

Before maize can be machine harvested, several requirements must be met. The plants and the cobs should be at similar heights across the field and not be too green or too dry. Furthermore, the plants should be erect, expressing a strong rooting with fruits of sufficient size to expedite gathering (Coulter et al. 2005; Belfield and Brown 2008). Once these criteria are met, seed moisture content becomes critical. Harvesting by machine is made possible when seed moisture levels drop below 18%–24%. Delivery and subsequent storage requires a further reduction down to below 14%.

Mechanical harvesting of maize can be accomplished by one of three different methods. These include harvesters, which pick and thresh the cobs, the kernels being directly emptied into the truck, machines that pick the cob from the stem and dehusk, immediately collecting the ears in a truck, and appliances that cut the stems, cobs and all. Should this last technology be employed, the cobs must be manually removed from the stalk and later threshed (Belfield and Brown 2008).

Threshing grain is the action of impact, rubbing that separates the detached free grain from other plant material (Hanna and Quick 2007). Maize threshing can be done either manually or mechanically. Manually, maize kernels are separated from the cob by pressing on the grains with the thumbs. In order to increase operational performance, small disk shellers are usually employed. Alternatively, in tropical countries, threshing is driven by putting cobs into bags and beating them with a stick. Mechanical threshing is performed with effective and robust maize shellers equipped with a rotating cylinder of the peg or bar type. Major mechanical threshing mechanism styles move the crop between the surface of a rotating cylinder and an open-mesh concave receptacle. Mechanical threshing equipment has output ranges 3–15 times higher than manual ones. Significantly, when the crop is machine harvested, the harvesting and threshing can be carried out in one single operation (Proctor 1994; Belfield and Brown 2008).

Harvesting Machinery

Harvesting causes some damage to maize kernels, so the machinery used should be chosen to maximize grain separation while preserving product quality. Ideally, the crop should be planted and rows spaced to match the planter and harvester. Gathering losses can be as great as 158 kg/ha, if the collector opening is 10 cm–12 cm off the row. Maize heads aligned with combine wheels and matched with planters and row bedders improve combine performance (Huitink 2001). Table 5 summarizes main features of machinery used during maize harvesting.

Maize Drying and Storage

The seed expresses its highest quality at harvest, but its lifetime when stored is limited. Harvesting, drying and storage of the grain should be completed in less than 24 hours. Therefore, the way maize is handled during drying and storage will determine how much of its quality is retained (Gardisser 2001). Moisture level must remain below 11% to store maize for long periods; otherwise, the kernels may become infected with *Aspergillus* and produce aflatoxin (Belfield and Brown 2008). If the seed is expected to be immediately marketed, drying to 15.5% is sufficient (Gardisser 2001). It may be specifically concluded that good storage management that takes into consideration climate, length of storage and seed quality can greatly influence the storability of maize (Harner and Sloderbeck 2007).

Maize kernels can be dried in the sun using mats, plastic tarpaulins or on a cement pad until their moisture content is below 11% (Belfield and Brown 2008), although, industrially, different systems for maize drying are employed. On an industrial scale, grain moisture is reduced by passing large quantities of dry air over the seed. The drying unit should be selected according to its drying efficiency, thus significantly reducing the possibility of deterioration (Harner and Sloderbeck 2007).

When temperatures are too high and steeping rates drop, further milling is more difficult accompanied by a loss of protease activity and starch release. Endogenous proteases, along with sulfur dioxide contribute to starch release (Morris 2004). A given volume of air can hold a given amount of moisture; consequently, the final moisture content of the maize kernel is highly influenced by the quality of the air related to equilibrium moisture content (EMC) of the used air. For instance, if air with an EMC of 12% is used, the grain moisture will probably reach approximately 12% (Gardisser 2001). In order to speed up drying processes, pass or continuous dries can be employed. Devices with this configuration are able to pass a large volume of airflow at high temperature, allowing the reduction of 3–6 moisture points in the grain in a single pass. This quick drying operation

Table 5 Main machinery used in maize harvesting.

Machinery	Feature
Combiner	Combine used in maize harvesting must be modified with a specialized front or maize head. The head must gather all grain into the hopper. Fronts can be either "cutter bar" or "snapper bar" types with large crop dividers matching the crop row width. Cutter bar fronts cut the maize stalks and thresh the whole plant and ear; they are mainly used in dry land or low-yielding crops. Otherwise, snapper bar fronts strip the ear from the plant so that stalks are not taken into the harvester. The adjustment to the front of the harvester is critical for harvest efficiency and for minimizing grain loss. Gathering head capacity determines combine throughput. Because threshing, separating, and cleaning mechanisms must operate at constant speeds, travel speed is adjusted by a variable speed drive to the crop intake capacity of the head, while the engine throttle is fixed by requirements of the separator and shoe.
Maize Head	A maize head has individual row units designed to strip the ear from the stalk and gather it into the machine. Six, eight, and twelve row maize heads are common. The head should be matched to row spacing to avoid machine losses during gathering.
Feederhouse	The feederhouse transports grain from the head to the threshing area. Large chains with cross slats fastened to them pull material along the bottom of the housing of the feederhouse into the threshing area. Chain position, tension, and speed should be adjusted to uniformly take material from the head. The height position of the front drum around which the feeder chains operate should be adjustable to accommodate different crop sizes and crop volumes due to yield or the amount of plant material moved through the combine. Many combiners have a rock trap in the feederhouse or close to the top end of the feederhouse.
Cylinder or Rotor and Concave	Grain quality is directly impacted by rotor or cylinder speed and concave clearance settings. Typical peripheral speed for maize is 15m/s per. A general recommendation is to increase rotor speed to just below the point where grain quality is adversely affected. Concave clearance should be narrowed to just the point where threshing is satisfactory without adversely influencing grain damage. Although threshing increases with increased rotor/cylinder speed and decreased concave clearance, grain damage greatly increases in order to thresh the most resistant grain heads. Grain damage increases with the square of rotor/cylinder speed.
Straw Walkers and Rotary Separation	There are two separation processes commonly used for the grain not separated in the threshing zone: gravity-dependent straw walkers or rotary separation. Straw walkers are sieve sections that oscillate up and down to shake remaining grain. The number of sections across the width or along the length of the straw walkers is limited by the chassis size of the combine, which in turn is limited by road transport width and length. Grain is separated in the straw walkers by gravity as the individual shakers oscillate on crankshafts at about 200 rpm with a throw of 15 cm to 45 cm. Larger straw, stems, and other material "walk" toward the rear exit of the combine, while heavier grain falls through the sieves. Rotary separation is commonly used in conjunction with rotary threshing and accomplished in the mid- and rear-sections of the rotor or rotor pair. Rotary action results in centrifugal separation as heavier grain flies through the concave. Rotary separators require more power than straw walkers as they grind away on the straw but grain damage may be acceptable over a wider range of speeds, and separation forces are greater.

Source: (Coulter 2005; O'Gara 2007; Hanna 2007)

helps to reduce the risk of toxin production. Similarly, in-bin drying can be used when a more gradual drying is required (Gardisser 2001). However, dryers at high temperatures require more energy and may cause quality problems in final grain. Stress cracks can be the result of improper cooling of a dryer and cause fines and broken kernels during grain handling operations (Harner and Sloderbeck 2007). Energy use is related to dryer type, grain type, airflow per kilogram, drying temperature, and excess amount of air recirculation. Due to inefficiencies in dry equipments, the energy required can be increased 26%–63% with respect to theoretical energy required to evaporate a kilogram of water. Therefore, low temperature drying prevents overdrying and improves the quality of grain, and although maize must be dry enough to be safely stored, overdrying must be avoided because it is both costly and unnecessary (Harner and Sloderbeck 2007).

Maize Milling

Every year more than 200 million tons of maize is produced in the US alone. Maize not employed for direct human food or animal consumption is milled, in order to obtain valuable derivative products from different parts of the maize kernel. Dry milling and wet milling are two different alternatives for maize processing, and each generates unique co-products (Davis 2001). Dry milling produces primarily flour and meal, while wet milling produces primarily high fructose maize syrup, oil, starch, and some animal feed products. In addition each technology may produce additional value added by products, some of which overlap. Indeed, both manufacturing alternatives are able to produce high quality ethanol, although with different production rates (O'Brien and Woolverton 2009). However, the main difference between both processes relies on their product quality. On typical example is the protein based products obtained from these two alternatives. On the one hand, in wet milling process most of the protein content in maize is initially recovered. As a result, high quality protein products can potentially obtained. However, this protein removal, cause that products used for human consumption (i.e., sweeteners) contain extremely low levels of intact proteins (Safriet 1994; EPA 2001). On the other hand, in dry milling there is not a direct protein removal stage. Then, the proteins contained in maize are partially recovered in by products such as DDGS, intended exclusively for animal feed industry applications.

Dry Milling

Maize dry milling is a grinding process that uses one of three different methodologies. Each varies according to its complexity and the products generated are of different particle sizes. The most basic of these procedures generates flaking grits, large pieces of vitreous endosperm, as its main and most valuable product. The more complex options generate flour, meal, grits, maize bran, and feed mixtures (Morris 2004). Dry milling products are used in brewing, in foods, in building products, and in fermentation that results in pharmaceutical and ethanol production as well as in animal feeds. Indeed, dry milling of food products from both white and yellow maize in the US alone represents approximately 5 million tons annually (EPA 2001).

The primary quality criterion for dry milling is a high proportion of vitreous endosperm in a yellow dent variety, free of mold, and mycotoxins. Stress cracks from poor drying are undesirable because they reduce the recovery of grits. Even though flint maize is entirely vitreous, it is actually less desirable for dry milling because its more spherical kernel morphology is not conducive to producing grits (Morris 2004).

Given the number of dry milling strategies available, product recovery determines the technological option. Stone grinding, also known as whole kernel dry milling or full flat dry milling, is the option of choice if fractionating the maize kernel is not required. Here, the seed is ground into particles of uniform size as is typical for flour and meal (Eckhoff 2004). Among devices used in white kernel dry milling, are hammer mills, pin mills and disk mills.

The process of dry milling begins by cooking whole kernels in an alkaline solution (sodium hydroxide 1.0%) for 20 minutes (Fig. 3). The seed is then allowed to soak 8–12 hours. This step is known as steeping. The mixture is drained and washed with clean water to remove excess lime and the pericarp, which has been loosened. The washed maize has reached about 45%–50% moisture and is then grounded forming dough. If the dough is formed into strips and then fried, maize chips can be produced. Otherwise, if the dough is converted into thin pancake-like sheets and baked, maize

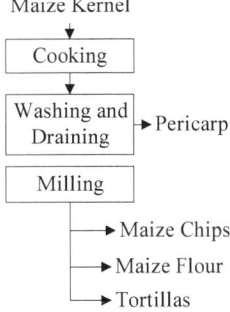

Figure 3 First dry milling option.

tortillas are produced. If the baked tortillas are subsequently fried, tortilla chips are produced (EPA 2001).

The second dry milling process is primarily oriented to ethanol production (Fig. 4). First, maize is ground using a hammer mill until the desired particle size is obtained prior to jet cooking, liquefaction, saccharification, and fermentation. This process is known as dry grind ethanol (Eckhoff 2004). In dry milling for ethanol production, the kernels, whose moisture had been previously increased is ground using a hammer mill that converts the mash into a medium-coarse to fine ground meal. This fine maize meal is mixed with fresh and recycled waters in known ratios to form slurry. Then, pH and temperature of this mixture are adjusted to 5–6 and 82°C–90°C, respectively as most convenient values. In this step, the slurry is hydrolyzed using an alpha amylase, in order to facilitate the conversion of starch to dextrin. This step is also known as liquefaction.

After complete liquefaction of the starch, the mash is cooked at 30°C to eliminate lactic acid, which can potentially produce a medium for the growth of contaminating bacteria. Subsequently, during saccharification the dextrin is converted to dextrose sugars using a glucoamylase enzyme. Then, these dextrose sugars are fermented using *Saccharomyces cerevisiae* to produce ethanol and carbon dioxide. This fermentative mixture is also known as beer. In this stillage, maize protein and recycled waters

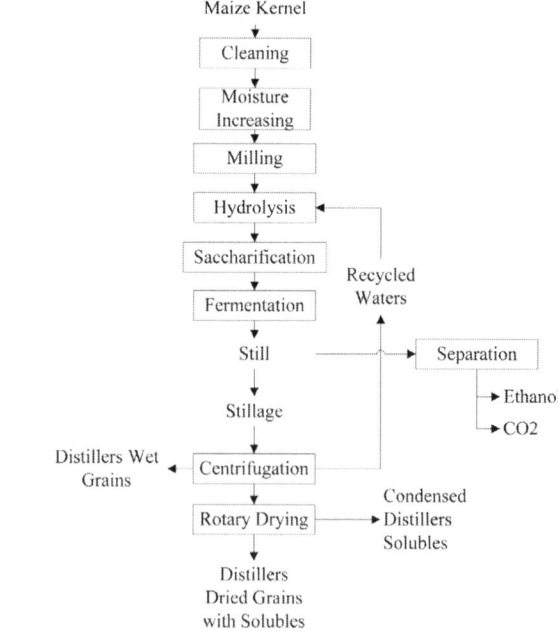

Figure 4 Second dry milling option.

provide nitrogen compounds that are metabolized by the yeast during the fermentation. The fats and fiber in the fermenter remain untouched and concentrate as the starch is converted to ethanol. Fermentation is completed in 40–60 hours. The beer is then processed to remove the ethanol. The water and all solids (protein, fat and fiber) are collected from the distillation bottom. This mixture is also known as whole stillage and then is centrifuged to separate the coarse solids from the liquid. The liquid referred to as thin stillage, is either recycled to initiate the process again or concentrated in the evaporator to become 'maize condensed distillers solubles'. The coarse solids collected from the centrifuge are called wetcake. Wetcake and condensed solubles are then combined and dried in a rotary dryer to form the feed coproduct 'maize distillers dried grains with solubles' (Davis 2001; O'Brien and Woolverton 2009).

The third choice is the degerminated dry milling process (Fig. 5). The objective here is to remove the germ and the pericarp, separating them from the grits, meal and flour that remain. Degermination improves the shelf life of the endosperm products by removing the bulk of the oil in the maize kernel (Eckhoff 2004). The degerminated dry milling begins with mechanical cleaning of the seed, where broken pieces of maize, weed seed, other grain, and additional contaminants are removed. Then, the moisture content is increased to about 20% in a tempering tank and subsequently moved to a degerminator unit where two streams are obtained. In this step the objective is to separate a mixture of detached germ, endosperm and pericarp fiber from the tails containing larger endosperm pieces. While the

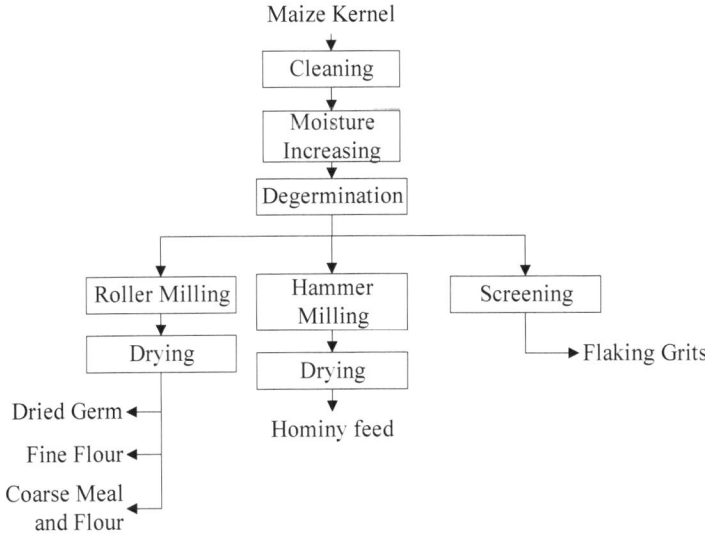

Figure 5 Third dry milling option.

germ is isolated for the production of maize oil, the large endosperm fraction is processed through the screening stage where flaking grits are recovered. The remaining products of the germinator are sieved into different fractions. These in turn are then either roller or hammer milled and dried according to their particle size. Both milling systems finally produce dried germ, fine flour, coarse meal, and flour, as well as hominy feed (EPA 2001; Eckhoff 2004). Otherwise, maize oil is obtained from the extracted germ as a co-product, the remaining tissue being rich in protein and used as animal feed. An example of a lower cost degerminator is a Beall-Type.

Wet Milling

Wet milling is a complex process, which includes a number of unitary operations and processes (Eckhoff 2004). In the US, most of the wet milling plants process about 2,800 tons/day, operating continuously 365 days per year, although 25% are smaller and have lesser capacities. The magnitude of the processing, in addition to the number of steps required, involve a large amount of expensive equipment requiring significant capital outlay to meet the sizeable economy of scale (Galitsky et al. 2003). The objective of wet milling is to efficiently separate maize into various products and parts aiming to find an optimum use and maximum value from each constituent (Davis 2001). These constituents are starch, germ, gluten, fiber, and steep liquor, which are then processed and recombined to yield products for paper, food, beverage, and biofuels, among other industries (Galitsky et al. 2003; O'Brien and Woolverton 2009).

The primary aim of wet milling is the recovery of starch products. These serve as refining products with specific applications in the food, paper, corrugating, and fuel industries. The main derivatives of maize starch are maize sweeteners, ethanol and unmodified industrial food starches (Morris 2004). Maize sweeteners can be found in either of three categories and include glucose syrup, dextrose and high fructose maize syrup. Ethanol is an important fuel additive in the United States. In 2008, about 20% of this alcohol came from wet milling of maize (Noureddini and Dang 2010). Other important products are starches and syrup. In addition to these goods, wet milling produces germ derived oil that when added to other byproducts has significant uses for the animal feed industry (Galitsky et al. 2003). Furthermore, from maize germ and gluten, hydrolyzed vegetable protein (HVP) can be obtained following acidification (EPA 2001). The basic steps of wet milling includes steeping, germ separation, grinding, starch separation, syrup conversion, and fermentation (see Fig. 6; EPA 2001). These steps are summarized in the next sections.

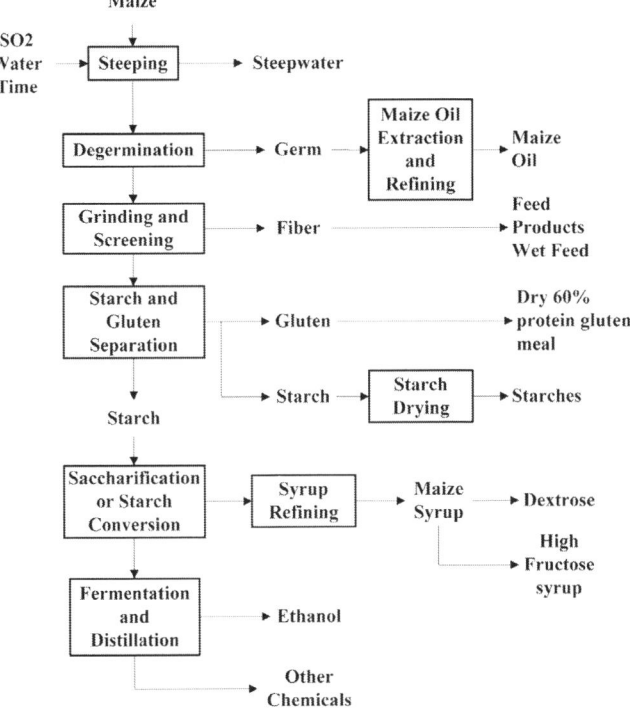

Figure 6 Overview of the process and products of wet maize milling. Sources: (EPA 2001; Galitsky, Worrell et al. 2003)

Cleaning

Maize is delivered by truck, by rail, or by river barge and is held in large silos before being cleaned. It should meet minimum physical quality standards such as high starch content and express low levels of mycotoxin. Moreover, the kernel should be well filled (high test weight indicates that kernels have a high vitreous to starchy endosperm ratio). Vitreous endosperm slows steeping and the processing time required Mechanically damaged and broken kernels are likewise undesirable (Morris 2004). In order to clean these impurities, maize is transferred through a cleaning system to grain handling/storage facilities attached to a milling plant. Then the maize feedstocks are soaked in heated chemical solutions, in order to soften the maize kernels and release soluble nutrients into solution (O'Brien and Woolverton 2009). During cleaning and inspection different elements such as debris, cob, dust, chaff and any foreign materials are removed (EPA 2001; AESSEAL 2009).

Steeping

This step is performed in order to soften and condition the maize kernel for further milling and to prevent premature germination and fermentation (AESSEAL 2009). Specifically, the grain is softened in stainless steel tanks, with side-entry agitator units, for 30–40 hours at 50°C in soaking water containing 0.1% sulfur dioxide. This prevents not only excessive bacterial growth but also reduces bonds between gluten and starch, resulting in the further release of soluble nutrients (EPA 2001). During steeping, moisture levels are increased 15%–45%, and the kernel size doubles. The resulting mixture contains ground maize and then goes to a cyclone separator where the water is separated from the milled maize (EPA 2001). This water is known as steep water and contains much of the soluble material from the maize, including a significant percentage of proteins and sugars (Galitsky et al. 2003). The protein content of the steep water ranges from 35%–45%. This water is further condensed to capture additional nutrients. The product is known as condensed maize fermented extractives and is further used for animal feed and in the fermentation process (EPA 2001).

Germ Separation

Following steeping, the germ is removed from the soaked kernel. The germ contains about 85% of the seed's total oil. Since this fraction contains most of the oil, it can be separated using hydrocyclones taking advantage of density differences between components. These cyclone separators are similar to centrifuges. They spin low density maize germ out of the slurry, while pumping it onto a series of screens where it is washed repeatedly until any residual starch and gluten is removed (EPA 2001). Afterwards, the germ is dewatered using a conical screw press and a perforated mesh system. The germ then has a water content of 50%–60% and is further dried using a countercurrent rotary steam tube dryer until it reaches a moisture content of 2%–4% (Galitsky et al. 2003). The dried germ goes to an extraction stage, where it is processed both mechanically and with solvents, in a series of steps to remove all impurities of refining and filtering. Degumming and alkaline treatment follow to remove fatty acids. Finally, the oil is deodorized to remove residual proteins/amino acids and color bodies. Refined maize oil is the main product of this stage. Extraction residues, a mixture of maize starch and husks, known as maize germ meal can be used as a component of animal feed after a treatment (Davis 2001).

Grinding and Screening

The remaining maize and water slurry from the degerminator are sent to an impact or attrition-impact mill. In this stage, starch and gluten are liberated from the fiber in the kernel by means of grinding. The resulting suspension flows over a concave screen where the fiber is caught, while remaining starch and gluten pass through (EPA 2001). These threads are collected and dewatered in two steps. First, moisture is reduced using a screen centrifuge until it reaches 65%–75% and reduced further to 10% using a screw press. The collected water is added to the steep liquor to recover any residual starch or protein by a further drying (Galitsky et al. 2003). This residual fiber, known as maize gluten feed, can be used as a major component of animal feed. Meanwhile, the starch-gluten suspension is sent to the starch separator (EPA 2001).

Starch and Gluten Separation

Using a set of hydrocyclons or filtering systems, starch and the remaining glutens are separated until the final starch slurry contains 2% protein in addition to a solids concentration of 33%–37%. The incoming starch slurry is diluted 8–14 times, rediluted and washed again in the hydrocyclon to remove the last traces of gluten until high quality starch is obtained (> 99.5%) (EPA 2001). This starch's moisture content varies between 33%–42%. Therefore, this product must be immediately dried and processed, in order to avoid microbial infections that will affect the color and odor of the final product. This starch is dried using a flash dryer, spray drier or a film drier, and can be either marketed as unmodified maize starch once it has been powdered in a compact granulator or further refined and converted to glucose syrup (EPA 2001; Galitsky et al. 2003; AESSEAL 2009).

The gluten, which had been previously separated, is dewatered using either filter or centrifuge technologies. Belt vacuum filters or rotary drum filters give rise to filter cakes with a solids content of 40%–43%. Alternatively, a decanter centrifuge may be employed giving rise to a gluten mud with a solids content ranging between 30%–40% (Galitsky et al. 2003). Regardless of methodology required for drying before marketing, maize gluten meal is a major byproduct containing 60% protein (Davis 2001).

Syrup Conversion

During saccharification, long chain starch molecules are broken down into smaller polysaccharide units when mixed with dilute hydrochloric acid temperature between 130°C–140°C and pressure between 10–50 bar. The resulting product passes through a vacuum flash cooler where temperature is reduced and pH is checked in the hydrolysate tank. Centrifugation follows, to separate soluble and insoluble proteins along with subsequent filtration in rotary vacuum filters, to remove particles combined of starch slurry and syrup. The liquor is further refined passing through columns containing activated charcoal; then, a clear colorless liquid is collected. The liquid is neutralized with sodium carbonate and pumped into an evaporator train, where syrup with different dextrose equivalent grades are obtained (AESSEAL 2009).

Fermentation

Maize starch not used to produce sweeteners is funneled into ethanol, amino acids, and other fermentation products (Galitsky et al. 2003). As with dry milling, wet milling uses either yeast or bacterial fermentation according to the fermentation product desired. Again, final products are purified using distillation and/or dehydration. The carbon dioxide byproduct of fermentation is sold to carbonated beverages manufacturers (EPA 2001).

Milling Products

Maize is an efficient factory for converting large amounts of radiant energy from the sun into a stable form of chemical energy stored as cellulose, oil, and starch. It has proven to be a very versatile grain and its end products are used daily (Davis 2001). During more than 150 years, maize milling factories have been perfecting the process of separating maize into its four basic components: starch, germ, fiber, and gluten (AESSEAL 2009). From different components of a maize kernel and its derivatives, assorted products are obtained that impact a broad range of industries.

Animal Feed

Maize and its derivatives are essentials for animal feed, providing vitamins, minerals and energy to different animals, including cattle, poultry and swine. The leading consumers of maize for animal feed are the US, China, the European Union and Brazil, sharing 70% of world production (Abbassian 2006). Four products are obtained from different combinations of steep water, maize germ residues, fiber and gluten and are dedicated to animal feed exclusively. These products include gluten meal, gluten feed, germ meal and condensed fermented maize extracts. All of them have significantly high protein content (EPA 2001).

Maize gluten meal is a dried high protein feed that typically contains 60% protein, 2.5% fat and 1% fiber. It is used as a medium protein ingredient in complete feeds for dairy and beef cattle, poultry, swine, and in pet food. It can be marketed either wet or dry as pellets to facilitate handling and contains 21% protein, 2.5% fat and 8% fiber. Maize zein is an insoluble protein that is also derived from maize gluten. It is used as a glazing and coating agent for the food and pharmaceutical industries. Additionally, gluten can be treated with acid resulting in hydrolyzed vegetable protein that is commonly produced from soybean and wheat (EPA 2001).

After the oil is removed from the germ, a meal is produced that contains 20% protein, 2% fat and 9.5% fiber. Its amino acid balance makes it valuable for poultry and swine rations and as a carrier of liquid feed nutrients (O'Brien and Woolverton 2009). Another important animal feed derived from maize is liquid feed syrup, which is a protein and energy source added to enhance animal feed quality (Davis 2001). Maize seed cake contains 17% protein and is used as the dietary source of choice when breeding chickens, duck, geese, pigs, and other livestock (Abbassian 2006). Finally, steep water may be used as a protein supplement for cattle while gluten feed provides fiber for beef cattle.

Other animal feed sources obtained during maize milling are those derived from ethanol production. The demand for maize based ethanol has both redirected and brought increases in the supply of dry grains, wet grains and condensed extractives to the biofuel industry. These products are mainly obtained by dry milling (Abbassian 2006). Condensed maize fermented extracts are partially removed by evaporation and result in a high-energy liquid feed ingredient having a protein content of 25%–50%. These can be combined with maize gluten feed or marketed separately as a liquid protein source. Maize condensed distillers solubles are a highly palatable feed, which can be used as a feed supplement containing 29% protein, 9% fat and 4% dry fiber. In 2008, about 29,000 tons per month were produced in the US (O'Brien and Woolverton 2009). Dry grain fermentation for alcohol production generates about one-third of the dry matter (DM) recovered from the wet cake and the condensed solubles in a rotary dryer as 'maize condensed distillers solubles' and 'maize distillers dried grains with solubles' (DDGS) (Liu 2009). In 2009, ethanol plants in the US generated over 25 million metric tons of DDGS, mainly used in animal feed (Winkler-Moser and Breyer 2011). These coproducts are rich in essential nutrients (Kingsly and Ileleji 2009), containing 27% protein, 11% fat, and 9% fiber. They can be used as a source of protein in ruminants and as feed ingredient for other livestock species.

Starch and its Derivative Products

Starch and its derivatives account for approximately 74% of products from maize wet milling (EPA 2001). This kind of milling starch is separated with a yield of 65%–68% (Morris 2004). Starch can be marketed directly as powder where it is mainly converted into other value added products such sweeteners and ethanol. Sweeteners from maize are marketed as syrup with different grades of sweetness expressed as dextrose equivalents. The main syrup products include dextrose, high fructose syrups and crystalline fructose (EPA 2001; AESSEAL 2009). High fructose syrup is a popular substitute for sucrose and is used in soft drinks and other processed foods (Abbassian 2006). Another indirect product from starch production is germ-derived oil that is used as cooking and salad products and accounts for nearly 25% of margarine production. Notably as well, starch developed fuel ethanol is a safe replacement for toxic octane enhancers of gasoline such as benzene, toluene and xylene (Gardisser 2001). Today in the US, most fuel ethanol is produced from maize by either dry mill (67%) or wet mill (33%) process. Additional information on ethanol production is further presented in the study case below.

Case Study: Simultaneous Production of Biodiesel and Ethanol from Maize Dry Milling

Today, most of the ethanol produced in the US comes from the dry milling process generating DDGS as a main byproduct (Winsness 2006). However, the EPA'S renewable fuel standard program 2 (RFS2) includes maize oil extraction as an advanced technology. This is designed to reduce energy consumption of maize ethanol production as all post-fermentation flows are sent for extraction, thus decreasing the mass of DDGS to be dried (EPA 2009; Krablin 2010). This technology boasts additional benefits too, by increasing the DDGS feed properties as a consequence of fat reduction. About one third of the oil content can be recovered using conventional solvent extraction (Krablin 2010). A newly patented process using green technology is able to recover up to 75% of maize oil contained in distiller grains (GreenShift 2009).

In dry milling for ethanol and biodiesel production, the finely ground meal is mixed with fresh and recycled water in known ratios to form slurry. The pH of this mixture is adjusted to 5–6 and temperature to 82°C–90°C. In this step, the slurry is hydrolyzed using an alpha amylase to facilitate the conversion of starch to dextrin. The last step is known as starch liquefaction. After complete liquefaction, the mash is cooked at 30°C to eliminate lactic acid, which can potentially stimulate the growth of contaminating bacteria.

During the saccharification stage the dextrin is converted to dextrose sugars using a glucoamylase. Subsequently these dextrose sugars are fermented using *Saccharomyces cerevisiae* to produce ethanol and carbon dioxide. This fermentative mixture is also known as beer. The output mixture of fermentation contains ethanol, water, protein, fiber, and maize oil. The water and all solids are collected from the distillation base and is known as whole stillage. In this stillage, maize protein and recycled waters provide nitrogen compounds absorbed by the yeast microorganism in the fermentation process.

The fats and fiber in the fermenter remain untouched. Fermentation is completed in 40–60 hours. The whole stillage output from the distillation stage of the ethanol process is centrifuged to remove water and soluble, leaving the thin stillage that contains the remaining water, fiber, and maize oil. This water content is reduced using evaporators to produce a concentrated thin stillage (Eckhoff 2004). During this evaporation process, the thin stillage is pretreated to break down possible emulsions and subsequently centrifuged to remove maize oils. These remaining solids are mixed with those from the centrifuged solids of whole tillage and returned to the evaporators where they are concentrated to produce distillers dried grains (GreenShift 2009). The residual mixture, with oil removed, is recycled to the beginning of the process or concentrated in the evaporator to become 'maize condensed distillers solubles'. Currently, 1 L ethanol also generates between 6–7 L thin stillage and the dry grind industry recycles 30°C–50% of thin stillage as backset (Arora et al. 2010). The coarse solids collected from the centrifuge are called wetcake. Wetcake and condensed solubles are then combined and dried in a rotary dryer to form the feed coproduct DDGS (Davis 2001; O'Brien and Woolverton 2009).

The biodiesel production from maize oil is defined by a series of steps, including pretreatment, reaction, purification and refining. Maize oil is first purified to remove different pigments, particle matter and other impurities, prior to entering the reaction stage. If the crude oil has water content higher than 0.06%, the oil should be dried first. Following this, maize-oil free fatty acids are esterified, in order to avoid undesirable saponification, which may result in unnecessary biodiesel yield reduction and an increase in soap formation. The remaining triglycerides in the maize oil are converted to methyl esters (biodiesel) using transesterification reaction. This is done by reacting the triglycerides with methanol using sodium hydroxide as a catalyst. Then, the biodiesel is purified by vacuum distilling at 0.5 bar, to remove methanol, followed by a final washing with warm water at 50°C, in a 7 stage liquid-liquid extractor in order to separate methyl esters from glycerin (Gerpen 2005; Haas 2006). The biodiesel enriched steam is purified using distillation, in order to remove excess catalyst, neutralization salts and soaps. These steps result in a final product having a desirable smell and color. Similarly, glycerin a value added byproduct may be purified, using an analogous procedure, according to its purity requirements. Glycerin purified to 70% is a raw product, whereas in concentrations 99% or higher may be funneled for use in the pharmaceutical industry as USP glycerin. This protocol is summarized in Fig. 7.

Simulation Procedure

The simulation of the dry milling to produce ethanol and biodiesel from maize was carried out using Aspen plus v 7.1 (Aspen Technology Inc, USA). Ethanol production was designed using a plant capacity of about 185 million L/year of anhydrous ethanol and the required raw material necessary to meet the desired output. This strategy obtains data specifying mass and energy balances as well as basic engineering estimates of equipment size along with equipment energy consumption. Specific compounds separated in the different processing stages include free fatty acids, triglycerides, alkyl esters, proteins, salts, cellular, and enzymes. Other complex molecules used and produced during dry milling process are not available on the Aspen Plus Database. Physicochemical data of the components required for simulations for those components not included in the database were obtained from Wooley and Putsche (1996) (Wooley and Putsche 1996a, 1996b) and calculated using the Marrero and Gani Method (Marrero and Gani 2001). The UNIFAC Dortmund model for liquid phase was utilized to calculate the activity coefficients in the liquid phase whereas the Soave RedlichKwong in conjunction with the Bosto Mathias was used to model the vapor phase. The Kinetic model used to model the biodiesel production reaction was a second order reaction reported by Granjo et al. (Granjo et al. 2009).

The economic analysis was performed using the Aspen Icarus Process Evaluator package (Aspen Technology, Inc, USA), to calculate a mean cost in US dollar per liter for biodiesel produced with the selected feedstocks. This analysis was performed using the design information provided by Aspen Plus and economic information of cost and taxes in Colombia, while feedstock prices were those reported by ICIS pricing (ICIS pricing 2011). Based on aforementioned data, a 12-year straight line depreciation method was considered and operating charges such as labor and management costs were defined at USD 2.14/h and USD 4.29/h, respectively. Electricity, potable water, low and high steam pressure costs were respectively USD 0.0304/kWh, USD 1.25/m^3, and USD 8.18/ton.

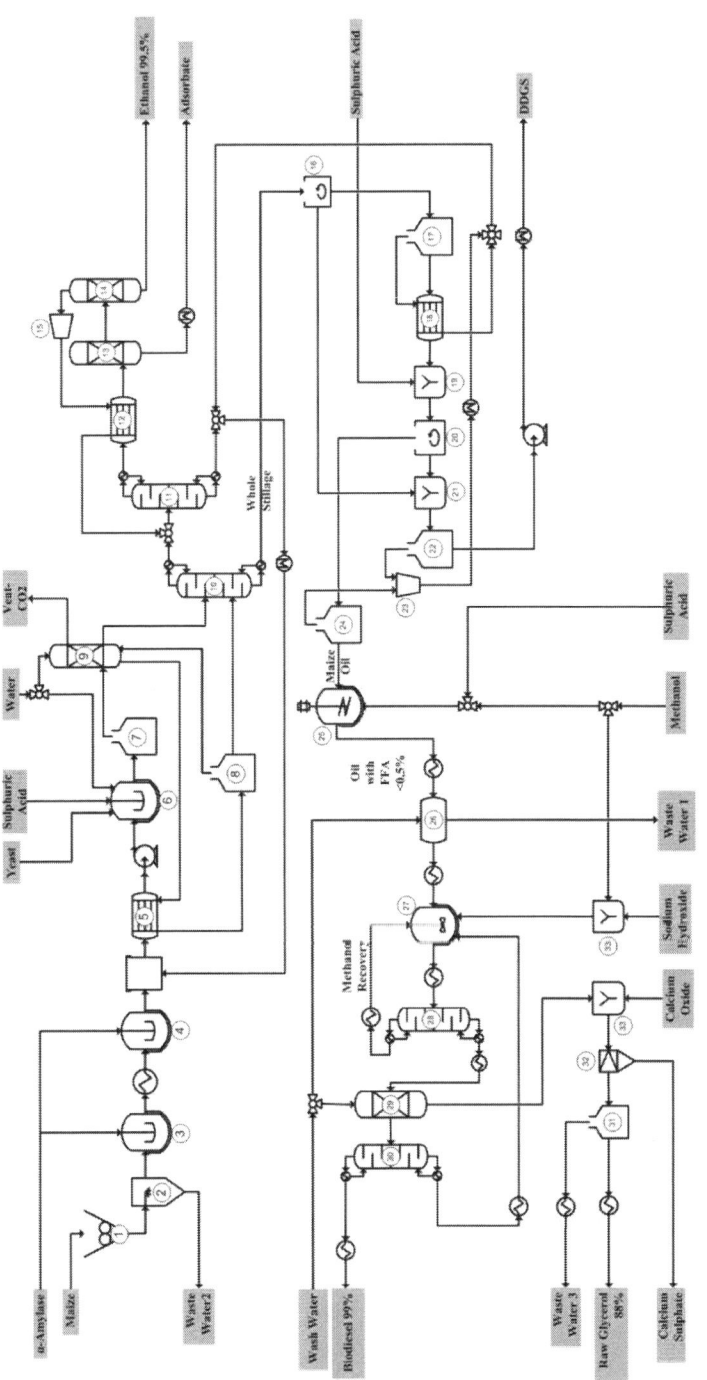

Figure 7 Integrated scheme for biodiesel and bioethanol production from maize.

Case Study Results

The simulation scenario clearly demonstrates how it is possible to obtain anhydrous ethanol (99.5%) and high quality biodiesel (> 99 wt) from single source maize, while taking no account of DDGS, raw glycerol (88%) and CO_2 that are also recovered. In addition, raw glycerol (88% wt) was produced. As result of the recreation of biodiesel and ethanol production from dry milling, production rates were observed with values that agree with those reported at industrial levels. Table 6 summarizes the results with regard to mass and energy balances.

The dry milling process was able to produce biofuels with yields of 0.45 L Ethanol/Kg maize and 0.02 L biodiesel/Kg maize, along with 0.554 kg DDGS/Kg maize with a protein composition of 36%. These results were higher than those reported, of 0.37 L Ethanol/Kg maize and 0.33 kg DDGS/Kg maize by Pimentel et al. (Pimentel et al. 2007). The protein composition of DDGS, however, was in accord with those values reported by Drapcho et al. (Kim et al. 2008). Dry milling yields and production rates were likewise in agreement with those reported for commercial technologies (Kwiatkowski et al. 2006; Szulczyk et al. 2010).

The value for biodiesel yield was low with respect to maize. If the amount of extracted oil is added to biodiesel recovery, the amount of biodiesel/kg of extracted oil increases to 1.061 L, a value that is slightly lower for those reported for rapeseed, soybean and palm (1.1 L) (Drapcho et al. 2008; Cushion et al. 2010). The above results were compared to previous published studies. Specifically, ethanol and DDGS yields were higher than the values reported by Pimentel and his co-authors as 0.37 L Ethanol/Kg Maize and 0.33 kg DDGS/Kg Maize (Pimentel et al. 2007).

With relation to production costs, these values were calculated as 0.687 USD/L fo subsidized maize (see Table 7). This production cost was higher than values between 0.29–0.40 USD/L reported for dry milling by Quintero, Montoya et al. and Crago, Khana et al. (Quintero et al. 2008; Crago et al. 2010). Apparently, the production costs associated with the simulation were high compared to conventional dry milling process. These production costs include two additional chemical processes: oil extraction and biodiesel production; therefore the number of units and the energy consumption are increased, and consequently the operational costs. However, in this case biodiesel is obtained as co-product, increasing the available products to be marketed. Then, traditional dry milling lessens production costs but only produces two products (ethanol and DDGS), while the simulated process has a higher production cost but produces three products (ethanol, biodiesel and DDGS). This fact allows improving the potential profit margin of the process as explained below.

Table 6 Simulation Results of simultaneous production of biodiesel and bioethanol from maize.

Bioethanol		Biodiesel	
Materials	Kg/h	Materials	Kg/h
Maize	50034.59	Thin Stillage	20891.51
Oil content (% wt)	3.68%	Oil Content (% wt)	0.09
Protein content (% wt)	8.80%	Protein Content (% wt)	0.21
Water content (% wt)	15.68%	Water Content (% wt)	0.48
Sugars content (% wt)	2.22%	Others Content (% wt)	0.01
Fiber content (% wt)	8.31%	Fiber Content (% wt)	0.20
Starch content (% wt)	61.31%	Crude oil extracted from whole stillage	1284.67
		FFA content (% wt)	10%
		Methanol	2277.01
Yeast	52.43	NaOH	12.33
Sulphuric acid@ 20%wt	1.00	Water	1798.54
Process water	16133.34	H2SO4 @98% wt	14.39
Energy	(Electricity. kW)	Energy	(Electricity. kW)
Milling	21.23	Pumping	4.36
Pumping	5.00	Agitation Reactors	1.55
Compression	2787.45		
Energy	(Heat. MW)	Energy	(Heat. MW)
Heating	109.08	Heating	67.83
Cooling	85.48	Cooling	87.15
Products	Kg/h	Products	Kg/h
Ethanol @99.5%wt	17836.83	Biodiesel @>99% wt	1088.93
CO2	17247.81	Glycerol @>88% wt	105.43
		DDGS	27744.13
Residues	Kg/h	Residues	Kg/h
Waste water	53544.35	Waste Water	23437.00
Carbon dioxide	16926.41	CaSO4	71.00

Thus, an additional analysis was included in order to analyze the feasibility of the dry milling process for ethanol and biodiesel production, using a selling price/total production ratio calculated that is summarized in Fig. 8. The costs were compared under different scenarios of products marketed for ethanol production. Scenario 1 considers ethanol as single

Table 7 Production cost of simultaneous production of biodiesel and bioethanol.

	USD/L	SHARE
Raw material cost	$ 0.36	67.97%
Total utilities cost	$ 0.05	9.34%
Operating labor	$ 0.00	0.69%
Maintenance	$ 0.03	5.29%
Operating charges	$ 0.00	0.17%
Plant overhead	$ 0.02	2.82%
General and administrative cost	$ 0.07	13.73%
Total operating cost	$ 0.54	100.00%
Depreciation of capital	$ 0.14	
Total project cost	$ 0.68	

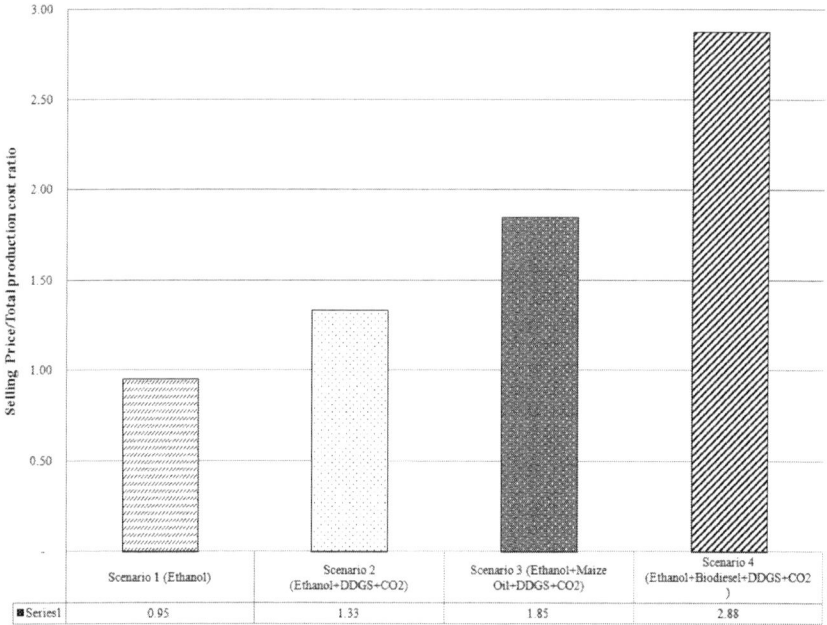

	Scenario 1 (Ethanol)	Scenario 2 (Ethanol+DDGS+CO2)	Scenario 3 (Ethanol+Maize Oil+DDGS+CO2)	Scenario 4 (Ethanol+Biodiesel+DDGS+CO2)
■ Series1	0.95	1.33	1.85	2.88

Figure 8 Comparison of potential selling Price/Total production cost under different product scenarios.

product from dry milling; scenario 2 considers the actual trend including DDGS and CO_2 as marketable products; scenario 3 includes the maize oil extraction; and finally scenario 4 includes the biodiesel production. In summary, the total operating cost is but a component of the total cost of operation. The revenue derived from the sale of the dry milling process must be considered in the value of the production process. Profit is the difference between the total cost of operation and the total revenue obtain from the process.

Figure 8 compares the ratio of the potential selling price divided by the production costs under different production scenarios. If the selling price is less than the selling cost, the ration is less than one, meaning that the scenario is not economically viable. A ratio of more than one signal profitability, with ethanol as its only product, scenario 1 is not feasible given its absolute reliance on government subsidy. This reliance diminishes in scenario 2, given the value associated with such by-products as DDGS and CO_2. Production schemes, which include maize oil extraction and biodiesel production as simulated in this work, were able to increase the sale price/ total production ratio to values higher than those reported in scenarios 1, 2 and 4. Thus a simple scenario considering just the maize oil extraction obtains a ratio of 1.85 (scenario 3), while a full scheme including biodiesel production achieves the maximum level of 2.88 (scenario 4).

Conclusion

Maize is one of the world's main cereals and plays a key role in the food system, for animal feed, for ethanol production, and for manufacture of a range of important food, feed, and industrial ingredients. In the US, Canada, France, Argentina and other major maize growing countries, dramatic rates of yields have been achieved by breeding, through addition of value-added genes by genetic engineering, and by superior agronomic management practices.

Two major outputs of maize wet milling are sweeteners and ethanol. Both maize sweeteners and ethanol are made from starch. Sweeteners fall into three major categories that include maize syrup or glucose syrup, dextrose and high fructose syrup. Ethanol is an increasingly important component of the US fuel supply. About 60% of the ethanol produced in the US currently comes from maize wet milling. Maize dry millers seek a maximum amount of endosperm recovered as large grits. Maize wet milling is a relatively sophisticated process producing a variety of products for the paper, food, beverage, and other industries.

The viability of maize-based ethanol production is a source of highly disputed opinion that touches everything from hard-core science to political and economic considerations. Indeed, some studies suggest that the production of maize ethanol would increase pollution contributing to environmental problems that include transforming the Gulf of Mexico into a dead zone. Moreover, there exists a serious concern, as to whether the total energy obtained from maize ethanol exceeds that required for its manufacture. The last part of this chapter has demonstrated that an integrated approach towards the concept of bio-refineries can be a good base to increase efficiency in this industry.

Acknowledgements

To the Colombian Institute for Development of Science and Technology (COLCIENCIAS) and Universidad Nacional de Colombia sede Manizales, for the financial support of the research that originated this work.

References

Abbassian A (2006) Maize: International Market Profile. Rome, Italy, FAO-Food and Agriculture Organization of the United Nations.

AESSEAL (2009) A Guide to Sealing WET CORN MILLING & REFINING. United Kingdom.

Arora A, Dien BS, Belea RL et al. (2010). Heat transfer fouling characteristics of microfiltered thin stillage from the dry grind process. Bioresource Tech 101: 6521–6527.

Belfield S and Brown C (2008) Field Crop Manual: Maize A Guide to Upland Production in Cambodia, NSW Department of Primary Industries.

Brouwer C, Prins K, Kay M et al. (1988) Irrigation Water Management: Irrigation Methods. Rome, Italy, FAO-Food and Agriculture Organization of the United Nations.

Coulter J, Sheaffer C, Moncada K et al. (2005) Corn Production. Risk Management Guide for Organic Producers. 9.1–9.21. Moncada KM and Sheaffer CC (eds) University of Minnesota.

Crago CL, Khanna M, Barton J et al. (2010) Competitiveness of Brazilian sugarcane ethanol compared to US corn ethanol. Ener Pol 38: 7404–7415.

Cushion E, Whiteman A, Dieterle G et al. (2010) Bioenergy Development. Washington D.C., USA, The World Bank.

Chaudhary AL, Miler M, Toley PS et al. (2008) Amylose content and chemical modification effects on the extrusion of thermoplastic starch from maize. Carbo Poly 74(4): 907–913.

Davis KS (2001) Corn Milling, Processing and Generation of Co-products. Minnesota Nutr. Conf.—Techn.l Symp., Minnesota, USA, Minnesota Corn Growers Association.

Drapcho C, Nghiem J and Walker T (2008) Biofuels Engineering Process Technology, McGraw Hill, 371 pp.

du Plessis J (2003) Maize Production. Pretoria, South Africa, Depart. Ag. in cooperation with ARC-Grain Crops Institute.

Eckhoff SR (2004) MAIZE | Dry Milling. In: Encyclopedia of Grain Science. Colin W, Cooke H and Walkers CE (eds) Oxford, Elsevier, pp 216–225.

Eckhoff SR (2004) MAIZE | Wet Milling. Encyclopedia of Grain Science. Colin W, Cooke H and Walker CE (eds) Oxford, Elsevier, pp 225–241.

EPA (2001) White Paper on the Possible Presence of Cry9C Protein in Processed Human Foods made from Food Fractions produced through the Wet Milling of Corn. Assessment of Additional Scientific Information Concerning StarLink Corn, Washington, USA. http://www.epa.gov/pesticides/biopesticides/cry9c/index.htm, http://www.epa.gov/scipoly/sap/index.html.

EPA (2009) EPA Proposes New Regulations for the National Renewable Fuel Standard Program for 2010 and Beyond. August 28th, from http://www.epa.gov/oms/renewablefuels/420f09023.htm.

Espinoza L (2001) Growth and Development. In: Corn Production Handbook. Espinoza L and Ross J (eds) Univ of Ark, USDA, and County Govermments Cooperating, pp 3–6.

FAOSTAT (2011) Crops Yields Statistics for 2007. Retrieved July 8th, 2011, from http://faostat.fao.org/.

FAOSTAT (2012) Crops Yields and Harvested Area Statistics for 2012. Retrieved Febraury 8th, 2012, from http://faostat.fao.org/.

Galitsky C, Worrell E and Ruth M (2003) LBNL-52307-Energy Efficiency Improvement and Cost Saving Opportunities for the Corn Wet Milling Industry—An ENERGY STAR Guide for Energy and Plant Managers. Berkeley, USA, U.S. EPA-Environ Energy Techno Div.

Gardisser DR (2001) On Farm Storage and Drying. In: Corn Production Handbook. Espinoza L and Ross J (eds) Univ Ark, USDA and County Govermments Cooperating, pp 73–77.

Gardisser DR (2001) Renewable Energy. In: Corn Production Handbook. Espinoza L and Ross J (eds) Univ Ark, USDA and County Govermments Cooperating, pp 95–97.

Gerpen JV (2005) Biodiesel processing and production. Fuel Process Tech 86: 1097–1107.

Granjo JFO, Duarte BPD and Olivera NMCl (2009) Kinetic Models for the Homogeneous Alkaline and Acid Catalysis in Biodiesel Production. In: Computer Aided Chemical Engineering. de Brito Alves R, do Nascimento CO and Biscaia E (eds) Elsevier 27: 483–488.

GreenShift (2009) Method of Freeing the Bound Oil Present in Whole Stillage and Thin Stillage. G. C. Corporation. United States. US Patent No. 7,608,729.

Haas MJ (2006) A process model to estimate biodiesel production costs. Bioresource Tech 97: 671–678.

Halm M, Hornbaek T, Arneborg N et al. (2004) Lactic acid tolerance determined by measurement of intracellular pH of single cells of Candida krusei and Saccharomyces cerevisiae isolated from fermented maize dough. Intl J Food Microbio 94: 97–103.

Hanna HM and Quick GR (2007) Grain Harvesting Machinery Design. In: Handbook of Farm, Dairy, and Food Machinery. Kutz M (ed) William Andrew, Inc, Norwich NY, pp 93–111.

Harner J and Sloderbeck P (2007) Drying and Storing. In: Corn Production Handbook, Kansas State University, Agricultural Experiment Station and Cooperative Extension Service, pp 38–41.

Huitink G (2001) Corn Harvesting. In: Corn Production Handbook. Espinoza L and Ross J (eds) Univ Ark, USDA and County Govermments Cooperating, pp 65 72.

ICIS pricing (2011) Ethanol Prices and Pricing Information. Retrieved June 6th, 2011, from http://www.icis.com/v2/chemicals/9075312/ethanol/pricing.html.

Kim Y, Mosier NS, Hendriksen R et al. (2008) Composition of corn dry-grind ethanol by-products: DDGS, wet cake, and thin stillage. Bioresource Tech 99: 5165–5176.

Kingsly ARP and Lleleji KE (2009) Sorption isotherm of corn distillers dried grains with solubles (DDGS) and its prediction using chemical composition. Food Chem 116: 939–946.

Krablin R (2010) RFS2 and the Impact of Corn Oil Extractionon the Ethanol Industry. Retrieved August 17th, 2011, from http://www.greenshift.com/pdf/RFS2_and_the_Impact_of_Corn_Oil_Extraction_Final.pdf.

Kwiatkowski JR, McAloon AJ, Taylor F et al. (2006) Modeling the process and costs of fuel ethanol production by the corn dry-grind process. Indus Crops Prod 23: 288–296.

Liu K (2009) Effects of particle size distribution, compositional and color properties of ground corn on quality of distillers dried grains with solubles (DDGS). Bioresource Tech 100: 4433–4440.

Marrero J and Gani R (2001) Group-contribution based estimation of pure component properties. Fluid Phase Equilib 183–184: 183–208.

Morris CF (2004) CEREALS | Grain—Quality Attributes. In: Encyclopedia of Grain Science. Colin W, Corke H and Walker CE (eds) Oxford, Elsevier, pp 238–254.

Motes JE, Roberts W and Cartwright BO (2001) Sweet Corn Production, Okla. St. U. Cooperative Extension Service.

Noureddini H and Dang J (2010) An integrated approach to the degradation of phytates in the corn wet milling process. Bioresource Tech 101: 9106–9113.

O'Brien D and Woolverton M (2009) Recent Trends in U.S. Wet and Dry Corn Milling Production. AgMRC Renewable Energy Newsletter

O'Gara F (2007) Irrigated maize production in the top end of the northern territory production guidelines and research results, Crops, Forestry and Horticulture Division. Australia.

O'Keeffe K and Byrne R (2010) Maize. Summer crop production guide 2010. L. Serafin, L. Jenkins and R. Byrne, I&I NSW management guide.www.dpi.nsw.gov.au./pubs/summer-crop-production-guide.

Pimentel D, Patzek DT and Cecil G (2007) Ethanol Production: Energy, Economic, and Environmental Losses. Rev. Environ. Contamin. Toxicol 189: 25–41.

Pirgozliev V, Murphy TC, Owens B et al. (2008) Fumaric and sorbic acid as additives in broiler feed. Res Vet Sci 84: 387–394.

Proctor DL (ed) (1994) Grain storage techniques. FAO AGRICULTURAL SERVICES BULLETIN No. 109. Rome, Italy, Food and Agriculture Organization of the United Nations (FAO).

Quintero JA, Montoya MI, Sanchez OJ et al. (2008) Fuel ethanol production from sugarcane and corn: Comparative analysis for a Colombian case. Energy 33: 385–399.

Regehr D, Olson B and Thompson C (2007) Weed Management. In: Corn Production Handbook, KS St U, Ag Expt Stat Coop Extens Serv, pp 21–22.

Robinson C (2005) Tillage. Retrieved July 26th 2011, from http://www.wtamu.edu/~crobinson/TILLAGE/Tillage.htm.

Rogers D (2007) Irrigation. Corn Production Handbook, KS St U Ag Expt Stat Coop Extens Serv pp 23–27.

Roozeboom K, Devlin D, Duncan K et al. (2007). Optimum Planting Practices, pp 10–13. Corn Production Handbook, Kans. S U, Ag Expt Stat Coop Exten Serv.

Ross J, Huitink G, Tacker P et al. (2001) Cultural Practices. In: Corn Production Handbook. L. Espinoza and J Ross, Univ Ark, USDA and County Govermments Cooperating, pp 7–11.

Safriet D (1994) Emission Factor Documentation for AP-42-Section 9.9.7-Corn Wet Milling-Final Report. North Carolina, USA, U.S. EPA Office of Air Quality Planning and Standards.

Singh N, Inouchi N, Nishinari K et al. (2006) Structural, thermal and viscoelastic characteristics of starches separated from normal, sugary and waxy maize. Food Hydrocolloids 20(6): 923–935.

Singh N, Singh S, Shevkani K et al. (2011) Maize: Composition, Bioactive Constituents, and Unleavened Bread. In: Flour and Breads and their Fortification in Health and Disease Prevention. Victor RP, Ronald Ross W and Vinood BP (eds) San Diego, Academic Press, pp 89–99.

Soil Science Glossary Terms Committee (2008) Tillage. Glossary of Soil Science Terms. Madison, WI, Soil Science Glossary Terms Committee 92.

Szulczyk KR, McCarl BA, Cornforth G et al. (2010) Market penetration of ethanol. Renewable and Sustainable Energy Reviews 14: 394–403.

Tacker P, Vories E and Huitink G (2001) Drainage and Irrigation. Corn Production Handbook. L. Espinoza and J Ross, Univ Ark, USDA and County Govermments Cooperating, pp 13–22.

van Soest JJG, Hulleman SHD, de Wit D et al. (1996) Crystallinity in starch bioplastics. Indus Crops Prod 5: 11–22.

Winkler-Moser JK and Breyer L (2011) Composition and oxidative stability of crude oil extracts of corn germ and distillers grains. Indus Crops Prod 33: 572–578.

Winsness D (2006) Increase Ethanol Industry Profits with Corn Oil Extraction Technology. Retrieved August 31th, 2011, from http://www.ethanolproducer.com/articles/1835/increase-ethanol-industry-profits-with-corn-oil-extraction-technology.

Wooley R, and Putsche V (1996a) Development of an ASPEN PLUS Physical Property Database for Biofuels Components. Report No. NREL/MP-425-20685. Golden, CO, USA, National Renewable Energy Laboratory.

Wooley R and Putsche V (1996b) Report NREL/MP-425-20685. Golden, CO, USA, National Renewable Energy Laboratory 38.

Wright D, Marois J, Rich J et al. (2002) Field Corn Production Guide, University of Florida.

CHAPTER 4

Bioenergy-Related Traits and Model Systems

Hugh Young,[1] George Chuck[2] and Ludmila Tyler[3,]*

ABSTRACT

Humans' use of maize for food has yielded extensive agricultural knowledge that is now accelerating the development of maize as a bioenergy feedstock. The first half of this chapter summarizes the traits that should be targeted to optimize maize's utility for bioenergy. The goal is to maximize the production of high-quality biomass while minimizing the required expenditure of resources. The kernels and the vegetative tissues of the stalk represent two distinct feedstocks whose production must be balanced for maximal efficiency. For kernels, controlling starch synthesis and degradation is key. For vegetative biomass, traits including plant architecture, cell wall composition, and flowering time are important. In general, plant feedstocks containing large amounts of an easily accessible glucose-based polymer are well-suited for current processing technologies. Improving photosynthetic capacity, disease and insect resistance, drought tolerance, and nutrient use efficiency can boost yield for the plant as a whole. The second half of the chapter describes tools and model systems for maize. New tools

[1]Department of Plant and Microbial Biology, University of California–Berkeley, Berkeley, CA 94720, USA.
Email: hugh.young@ars.usda.gov; georgechuck@berkeley.edu
[2]Plant Gene Expression Center/US Department of Agriculture (USDA) and University of California-Berkeley, Albany, CA 94710, USA.
Email: georgechuck@berkeley.edu
[3]Department of Biochemistry and Molecular Biology, University of Massachusetts–Amherst, Amherst, MA 01003, USA.
*Corresponding author: ltyler@biochem.umass.edu

such as the Intermated B73 MO17 and nested association mapping (NAM) populations are supplementing classical genetics to enable the identification of genes underlying complex biofuel traits. *Arabidopsis thaliana*, a small, diploid, dicotylendous weed, is the preeminent model plant for molecular research. Joining *Arabidopsis* are the emerging models *Brachypodium distachyon* (purple false brome) and *S. viridis* (green millet). As grasses, both *Brachypodium* and *Setaria* are more closely related to maize than *Arabidopsis* is. With its small physical stature, simple growth requirements, sequenced genome, amenability to genetic transformation, and diversity of germplasm, *Brachypodium* is useful for investigating questions such as cell wall composition and the identity of cell-wall-associated proteins in the grasses. *Setaria*, although having fewer research resources at present, still promises to be an excellent model for C_4 photosynthesis. This rich research toolbox will support the improvement of maize as a dual-use crop.

Keywords: feedstock traits, starch, vegetative biomass, model species

Introduction

Since its domestication from the wild grass teosinte thousands of years ago in Mexico (Hastorf 2009; van Heerwaarden et al. 2011), maize has been a staple food crop. Only recently has maize begun to be exploited for the production of biofuels, predominantly ethanol. Maize is currently the major feedstock converted to ethanol in the United States, accounting for more than 96% of the 2011 production capacity (Boundy et al. 2011). In 2000, the US produced only 1.6 billion gallons of ethanol for fuel purposes; a decade later, in 2010, production had jumped eight-fold to 13.3 billion gallons, with 4.9 billion bushels (or approximately one-third of the available harvest) of corn being used to make fuel ethanol (Boundy et al. 2011). Compared to the millennia of agricultural knowledge supporting the breeding, growth, and utilization of maize as a food source, research into the development of maize as a bioenergy feedstock is in its infancy. Nevertheless, given the finite nature of fossil fuels and the continuously expanding needs of the human population, accelerating the development of sustainable energy sources is imperative.

The first half of this chapter summarizes the traits that should be targeted to optimize maize's utility for bioenergy. The second half describes tools and model systems useful for elucidating these traits and their genetic basis.

Target Bioenergy Traits: High Biomass Quantity and Quality Achieved with Low Input

In developing a bioenergy feedstock, the goal is to maximize the quantity and quality of the harvested biomass, while minimizing the resources required for production and processing. In short, outputs must exceed inputs. The quantity of feedstock is the total amount of useable plant material harvested per unit of land. "Quantity" is intuitively comprehensible; it is directly related to traditional measures of yield. Feedstock "quality," however, is a more complex concept; it is being defined by the emerging technologies of the bioenergy field. In general, quality depends on the type, relative abundance, and accessibility of plant polymers, specifically those which can be converted to energy. Inputs are the resources—pesticides, water, fertilizer, electricity, etc.—required to grow the feedstock, to transport and process it, and then to release its stored energy. Inputs are subtracted from outputs: the higher the inputs, the lower the net energy gain. This energy equation determines not only the environmental sustainability of a bioenergy crop, but also its economic viability (West et al. 2009). Over the long term, then, optimizing the production of suitable plant biomasss, together with resource use efficiency, is essential for maize or any other species to be a successful bioenergy crop.

Maize represents two potential sources of bioenergy feedstock: the kernels and the vegetative tissues of the stalk. Making these structures during the reproductive and vegetative phases, respectively, involves interconnected processes. Depending on the circumstances, increased vegetative growth may boost or reduce grain yield. Thus, improving maize will require striking an appropriate balance—not to mention acquiring the biological understanding necessary to achieve this balance.

Also, the differing characteristics of corn kernels and stover necessitate different biorefining approaches. The readily accessible starch in kernels can be converted into ethanol using a few, defined enzymes and the well-studied baker's yeast, *Saccharomyces cerevisiae* (Lin and Tanaka 2006; Sánchez and Cardona 2008). This relatively mature technology now forms the foundation of a domestic ethanol industry. In contrast, the knowledge, techniques, and infrastructure for processing vegetative biomass are still developing. The bulk of stover consists of rigid plant cell walls. Releasing sugars from the cellulose and other carbohydrate components of the cell wall requires extensive pretreatment and enzymatic digestion (Lin and Tanaka 2006; Himmel et al. 2007; Sánchez and Cardona 2008). Furthermore, to supplement *S. cerevisiae*, researchers are investigating additional microbes for their abilities to express a myriad of wall-degrading enzymes, to ferment monosaccharides other than glucose, or to produce next-generation biofuels such as butanol (Demain et al. 2005; Fischer et al. 2008; Kumar et al. 2008).

Because of these distinctions, the next two parts of the chapter will discuss quantity and quality traits separately for kernels and vegetative biomass. The third portion will discuss traits, such as biotic and abiotic stress tolerance, which affect resource use efficiency for the plant as a whole.

Kernels

Currently, the US produces most of its biofuels from carbon sources stored in the plant, such as starches. The biosynthesis, as well as the degradation, of starch has been well-studied in plants, especially in the grasses that form the foundation for much of the starch industry. Starch quantity and quality traits have been under intense breeding selection for thousands of years, resulting in the maize ear, a near-perfect vessel for starch production. The focus of this section is to explore the metabolism of starch, its use as a fermentable substrate for biofuels, and early attempts to engineer its production. The lessons from these early attempts provide a roadmap for the development of a cellulosic biofuel that can be viable on a commercial scale.

Starch biosynthesis

In plants, most of the fixed carbon cannot be used immediately and is thus stored for later use in the form of starch. In *Arabidopsis* leaves, this stored carbon supply is used at different times, and the circadian clock controls its degradation (Graf and Smith 2011). Starch is a long polymer of glucose units that is synthesized and stored in specialized organelles called amyloplasts. Starch is stored as granules composed of amylose and amylopectins. Amylose consists of glucose linked by alpha 1–4 bonds; amylopectin has the same organization, but also contains branched alpha 1–6 bonds between glucose units. Both the synthesis and degradation of this specialized storage polymer are tightly regulated (Zeeman et al. 2010).

Starch synthesis begins with the formation of ADP-glucose, catalyzed by the multi-subunit enzyme ADP-glucose pyrophosphorylase (AGPase) (Orzechowski 2008). This enzyme consists of two large subunits and two small subunits. In maize, these subunits are encoded by the *shrunken2* (*sh2*) and *brittle2* (*bt2*) genes, respectively (Hannah and Nelson 1976); mutations in each of these genes lead to greatly reduced starch levels, and higher sugar levels. Mutant alleles of both *sh2* and *bt2* are found in many of the varieties of sweet corn consumed today. AGPase has at least two forms, a cytosolic version important for endosperm starch synthesis and a chloroplast version important for the formation of leaf starch. AGPase activity is positively regulated by certain sugars such as hexose phosphate and negatively regulated by inorganic phosphate (Smith 2008). Once ADP-glucose has been synthesized, several classes of starch synthase enzymes—together with branching and debranching enzymes—catalyze the transfer of new glucose units to ADP-glucose (Orzechowski 2008), ultimately forming amylose and amylopectin. In the plastid, this process also requires enzymes for importing and interconverting ATP, such as adenylate transporters and adenylate kinase (Smith 2008).

Starch degradation

Recent data from *Arabidopsis* indicate that regulation of starch degradation can be as important as biosynthesis in determining overall starch levels in the plant. For example, screens for plants that accumulate starch, such as the *starch excess* (*sex*) mutants, uncovered a new enzyme involved in initiating degradation of the starch granule (Yu et al. 2001). The *sex1* gene encodes glucan water dikinase, which phosphorylates glucan polymers—a prerequisite for degradation—and opens up the granule for degradation by amylases and starch debranching enzymes (Ritte et al. 2002). In *sex1* mutants, starch degradation is reduced: mutant plants have more than seven-fold higher starch levels than wild-type in the dark.

Attempts to enhance starch levels

Since starch synthesis is dependent on AGPase activity that catalyzes the first committed step in the formation of the starch granule, early attempts to control starch biosynthesis focused on altering AGPase expression. For example, in maize a variant of the *sh2* gene that was less sensitive to negative inhibition was isolated (Giroux et al. 1996). However, this mutant did not produce more starch as a percentage of seed weight in maize, or in rice or wheat under limiting field conditions (Meyer et al. 2007). A bacterial enzyme called GlgC, important for the synthesis of glycogen, a polymer similar to starch, was engineered to be resistant to negative regulation (Stark et al. 1992). Overexpression of this gene in maize (Sakulsingharoj et al. 2004), rice (Wang et al. 2007), and cassava (Ihemere et al. 2006) resulted in modest increases in weight, but increases in starch were not reported. Although there was no starch increase in cassava, root yield was two to three times greater in some transgenic lines. Clearly, starch yield is not as dependent on AGPase activity as previously thought. In cereal seed, starch already makes up 80% of the dry weight, while in cassava it is 45%, thus raising the question of whether the upper limit of starch levels has already been reached in these organs (Smith 2008).

Overexpression of *GlgC* in potato yielded slightly more promising results. Stark and colleagues reported increases in starch content of up to 60% in some potato lines, though these increases were not correlated with *GlgC* expression levels (Stark et al. 1992). A greater increase in potato starch, ranging from 65–85%, was reported when the adenylate cyclase gene was downregulated, ostensibly boosting starch accumulation by allowing higher levels of plastid ATP to exist (Regierer et al. 2002). In field trials, some plants contained up to two-fold more starch. Similar approaches to increase plastid levels of ATP by overexpressing adenylate translocators in potato also yielded positive—though less dramatic—results, with starch increases ranging from 16–36% (Geigenberger et al. 2001). Taken together, these experiments represent the most promising starch increases obtained by transgenic technology reported to date. In addition, the fact that these measurements were derived from field-grown plants, and not from plants grown in the greenhouse under near-ideal conditions, highlights the need for well-controlled field trials in assaying starch levels.

An alternative strategy for increasing starch levels may be to repress the starch degradation pathway. In *Arabidopsis*, this may include reducing the activities of glucan water dikinase (*SEX1*), beta amylase (*BAM4*), or starch debranching enzymes (*ISA3*) (Zeeman et al. 2010). Mutations in most of these genes appear to only affect leaf starch levels, and knockdowns of several *SEX* genes were not reported to compromise plant growth (Caspar et al. 1991). However, recent data indicate that blocking starch degradation may have negative consequences for plant development, resulting from premature chloroplast death (Stettler et al. 2009). This may explain why, to date, no field trial data have been reported using this approach in any plants. Elevating starch levels without significant deleterious effects may require either temporal or tissue-specific knockdown of the starch degradation pathway genes.

Plants that self process starch

An alternative to saccharification of starch using exogenous reagents is to express hydrolytic enzymes directly in plant tissues. Degradation of maize starch to fermentable sugars normally requires a hydrolysis step, either using acids or digestive enzymes such as alpha amylase at high temperatures. Recently, a new approach to reduce the cost of starch hydrolysis was commercialized and approved for release by the USDA. The Enogen line of genetically modified maize (Syngenta) contains a heat-stable, microbial α-amylase engineered directly into the plant, thus allowing the plant to self-process its own starch and reducing the cost of hydrolysis to fermentable sugars. A similar approach was also successfully implemented in sweet potato (Santa-Maria et al. 2011). The advantage of using a thermostable amylase is that it does not function at room temperature and only becomes active at high temperatures; therefore, it should not have adverse effects on the growth of plants in the field (Santa-Maria et al. 2011).

Vegetative Biomass: Quantity Traits

Making next-generation biofuels from lignocellulosic feedstocks will require huge supplies of vegetative biomass. The US Departments of Energy and Agriculture have, for example, set a goal of producing 1 billion dry tons of biomass annually to supplant petroleum usage (Perlack et al. 2005). However, traits specific to the production of vegetative biomass have traditionally not been the targets of human selection. Small-scale exceptions to this larger trend include the development of trees for forestry and plants for fodder; knowledge from these areas may help to improve bioenergy crops. Certainly, though, the unique demands of the emerging bioenergy sector present novel challenges and define new priorities for plant phenotypes. The following paragraphs discuss the traits and phenotypes relevant to two major aspects of feedstock production: the quantity and quality of vegetative biomass.

Traits advantageous for grain production—precisely those traits favored during the domestication and long, subsequent cultivation of cereals—can be disadvantageous for vegetative biomass production. Channeling resources to reproductive development often leaves fewer resources available for vegetative growth, and vice versa. The Green Revolution of the 1960s and 1970s is an excellent illustration of this concept. During the Green Revolution, tremendous gains in the grain yields of rice and wheat were achieved, primarily by introducing cultivars carrying dwarfing alleles (Evans 1996; Conway 1997; Hedden 2003). In rice, the causative mutation disrupted a gene encoding GA 20-oxidase, an enzyme involved in synthesizing the plant hormone gibberellin (Sasaki et al. 2002; Spielmeyer et al. 2002). In wheat, a gain-of-function mutation caused a repressor of gibberellin signaling, *Rht-B1/Rht-D1*, to be constitutively active (Peng et al. 1999). Gibberellin promotes vegetative growth, including stem elongation (Fleet and Sun 2005; Sun 2011). Therefore, in Green Revolution rice and wheat, the reduction in gibberellin levels or responses, respectively, resulted in plants with shorter stems. Genetically restricting the crops' stature, in turn, maximized the impact of applying nitrogen fertilizers: The new cultivars channeled the increased nutrients to seed production, whereas older varieties, lacking the dwarfing alleles, simply grew taller (Evans 1996; Sasaki et al. 2002; Hedden 2003).

Conversely, to meet bioenergy needs, researchers have suggested increasing vegetative growth by upregulating gibberellin biosynthetic and signaling pathways, or those of another growth-promoting phytohormone, brassinosteroid (Salas Fernandez et al. 2009). In fact, to minimize the plant's investment in seed set and thus maximize the resources available for incorporation into vegetative biomass, T. Sang has proposed that dedicated bioenergy crops should be sterile, or at least have reduced fertility (Sang 2011). Clearly, however, in the case of maize, decreasing grain yield and negating centuries of breeding efforts are not desirable. A new paradigm is necessary: For maize, the aim is to maintain grain production while simultaneously increasing vegetative growth. In fact, recent work suggests that it is possible to design breeding programs to select for greater stover yield and composition without adversely affecting grain yield (Lewis et al. 2010).

Plant architecture, morphology, and secondary cell walls

Numerous aspects of plant architecture and morphology influence the total amount of harvestable biomass. One approach to understanding this trait is to measure the quantity of biomass directly. In phenotyping a population of sorghum recombinant inbred lines (RILs), Murray and colleagues found that total dry biomass was negatively correlated with grain harvest index and positively correlated with stand density, tillering, flowering time, and especially plant height (Murray et al. 2008). These results hint at the bigger picture: Total biomass is the product of a myriad of internal and external factors—developmental programs, phytohormones, light conditions, abiotic stresses, diseases, etc.—and their complex interactions throughout the growing season.

The gibberellin (GA) and brassinosteroid (BR) phytohormone pathways are particularly attractive targets for breeding and engineering to increase biomass. These hormones promote vegetative growth in a variety of organs, through mechanisms which are well-conserved among flowering plants (Fleet and Sun 2005; Müssig 2005; Salas Fernandez et al. 2009; Sun 2011). For example, GA forms a complex with its receptor (GID1) and a repressor of GA responses (a DELLA protein), targeting the repressor for ubiquitin-mediated degradation (reviewed in Sun 2011). This GA-GID1-DELLA signaling module is conserved across the angiosperms: GID1 was first identified in rice, and DELLA proteins, named for a characteristic N-terminal motif, were first characterized in the dicotyledenous model species *Arabidopsis thaliana* (Peng et al. 1997; Silverstone et al. 1998; Ueguchi-Tanaka et al. 2005). The maize *d8* gene —like the wheat *Rht-1* Green Revolution gene—was found to encode an ortholog of the *Arabidopsis* DELLA repressor proteins (Peng et al. 1999). This conservation increases the likelihood that elevating the levels of, or responses to, GAs and BRs could increase plant growth in maize, as has been found in other species (Salas Fernandez et al. 2009). At the same time, both the GA and BR pathways integrate a number of signals, including those important for responses to other hormones, light, and pathogen attack (Salas Fernandez et al. 2009; Sun 2011). This consideration, together with taller plants' greater susceptibility to lodging, or falling over, means that breeders must exercise caution in order to generate high-biomass lines without deleterious side effects.

Because the aboveground vegetative tissues arise from the shoot apical meristem and the vascular cambium, increasing the activity of these meristems has been proposed as another way to boost vegetative biomass accumulation (Demura and Ye 2010). Possible mechanisms include altering the activities of *KNOX* homeobox genes, raising cytokinin levels or signaling in response to this hormone, and delaying flowering to prolong the activities of vegetative meristems (Demura and Ye 2010).

Other strategies entail increasing the formation of secondary cell walls, for example by upregulating the glycosyl transferases or accessory proteins involved in synthesizing cell wall carbohydrates (Demura and Ye 2010). One accessory protein is sucrose synthase (SuSy), which cleaves sucrose into UDP-glucose and fructose and thus provides the building blocks for cellulose synthesis (Amor et al. 1995). Overexpressing a cotton sucrose synthase gene in hybrid poplar increased cellulose content by 2–6% and also resulted in thicker secondary cell walls in the xylem and in higher wood density (Coleman et al. 2009). Additionally, it may be possible to upregulate the entire secondary cell wall synthesis program by elevating the activity of specific NAC domain transcription factors (Demura and Ye 2010). A suite of secondary wall NACs serves as top-level, master regulators of a network of transcription factors, including second-level MYB transcription factors, to control the production and deposition of cellulose, hemicellulose, and lignin (reviewed in Zhong et al. 2011). Although the existence of putative secondary wall transcription factors in diverse vascular plant taxa suggests conservation of this regulatory mechanism (Zhong et al. 2011), the identity and functionality of maize orthologs must still be confirmed. A further advantage of these approaches is that increasing the amount of cellulose or total cell wall material could yield not just more, but denser biomass; higher density would enhance sustainability by reducing the cost of transporting plant material from the field to the refinery (Richard 2010).

Finally, optimal leaf angle and an extended vegetative growth phase can increase biomass by improving the plant's ability to capture light. For example, a loss-of-function mutation in the rice brassinosteroid biosynthetic gene, *OsDWARF4*, resulted in erect leaves and greater biomass at high planting densities, presumably by allowing more light to reach lower leaves (Sakamoto et al. 2006). The presence of another rice cytochrome P450 gene that is functionally redundant with *OsDWARF4* probably prevented the pleiotropic effects of severe BR deficiency (Sakamoto et al. 2006). Weak loss-of-function alleles of the rice BR receptor *OsBRI1* similarly result in more acute leaf angles (Morinaka et al. 2006). However, disrupting BR signaling in this way causes additional phenotypes such as semi-dwarfism and small grain size; it was only by cosuppression of the *OsBRI1* gene and careful selection of transgenic lines that Morinaka and colleagues were able to identify plants with erect leaves but without other, undesirable phenotypes (Morinaka et al. 2006).

Longer vegetative growth can likewise improve light capture, as illustrated by a comparison of maize and the tall bioenergy grass *Miscanthus x giganteus*: In a side-by-side study in the Midwest, *Miscanthus* yielded 59% more biomass than maize (Dohleman and Long 2009). Maize had a biochemical advantage in that it exhibited significantly higher photosynthesis at the leaf level during midsummer. However, the *Miscanthus* leaf canopy developed several weeks earlier and persisted several weeks longer than the maize canopy, extending the *Miscanthus* growing season by an average of 59%. On average, *Miscanthus* had a green leaf area index 2.6 times that of maize and, correspondingly, intercepted light 61% more efficiently (Dohleman and Long 2009).

Photosynthetic capacity

Overall yield is a function of the amount of light intercepted by a plant, the efficiency with which the light is used to fix carbon, and the portion of this carbon which is partitioned into harvested biomass (Zhu et al. 2010). In the case of maize, the third component of the yield equation—the partitioning of biochemical energy into various parts of the plant—requires balancing the dual uses of maize: grain for food and stover for fuel. Funneling photosynthate to one part of the plant means diverting photosynthate from the rest of the plant. In contrast, improving the first two components of the equation—the capture of light and its conversion into biomass—can simultaneously benefit the production of reproductive and vegetative tissues. As discussed above, optimizing plant architecture and extending the length of the leafy, vegetative growth phase are two ways to maximize light interception. This section examines the potential for optimizing carbon fixation, i.e., photosynthetic efficiency *per se*.

Photosynthetic efficiency depends in large part on the type of photosynthetic pathway present in an organism. Maize utilizes the C_4 photosynthetic pathway, which evolved from the more ancient C_3 pathway (Vicentini et al. 2008). In both pathways, the enzyme Rubisco (ribulose 1,5-bisphosphate carboxylase oxygenase) plays a key role. Rubisco catalyzes the reaction between CO_2 and ribulose 1,5-bisphosphate (RuBP) to form 3-phosphoglycerate, which is ultimately reduced to produce carbohydrates. Some of the triose phosphate resulting from the reduction of 3-phosphoglycerate is used to regenerate the starting substrate, RuBP (reviewed in Zhu et al. 2010; Byrt et al. 2011; Sage and Zhu 2011).

In C_3 plants, Rubisco catalyzes the carboxylation of RuBP in mesophyll cells. These cells absorb both CO_2 and O_2 from the intercellular airspaces, which are connected to ambient air via stomata. Although Rubisco exhibits higher affinity for CO_2 than O_2, the amount of CO_2 within the mesophyll cells is relatively low, and Rubisco can catalyze not only carboxylation but also oxygenation of RuBP. This competing oxygenation reaction, which begins a cycle known as photorespiration, reduces the efficiency of carbon fixation in C_3 species (Zhu et al. 2010; Sage and Zhu 2011).

In C_4 plants, Rubisco acts in bundle sheath cells, specialized cells arranged around the vasculature. This Kranz, or "wreath," anatomy allows for separation between the activity of Rubisco in the bundle sheath cells and a CO_2-concentrating mechanism in the mesophyll cells. In the initial step of carbon fixation, the mesophyll-localized enzyme phosphoenol pyruvate carboxylase (PEPC) catalyzes the carboxylation of phosphoenol pyruvate, using CO_2 in the form of HCO_3^-. This reaction ultimately leads to the formation of a four-carbon acid (hence the name C_4). This acid—malate in the case of maize—is transported to the bundle sheath cells, where the C_4 acid is decarboxylated to a C_3 acid by one of three enzymes: NADP-dependent malic enzyme (NADP-ME, utilized in maize), NAD-dependent malic enzyme (NAD-ME), or phosphoenolpyruvate carboxykinase (PEP-CK). The resulting C_3 acid is then transported back to the mesophyll to regenerate phosphoenol pyruvate, in a reaction catalyzed by pyruvate Pi dikinase (PPDK) (reviewed in Matsuoka et al. 2001).

As a C_4 plant, maize has an inherent photosynthetic advantage. Particularly at high temperatures, which favor photorespiration, C_4 photosynthesis is more efficient than C_3 at converting light energy into chemical energy stored in carbon compounds (Long et al. 2006). The calculated theoretical maximum for conversion efficiency in a C_4 plant is 6.0%, compared to 4.6% in a C_3 plant, with observed efficiencies of around 4.2% and 2.9% for C_4 and C_3 plants, respectively (Zhu et al. 2010). In general, this increased photosynthetic efficiency translates into higher biomass yields for C_4 versus C_3 crops (Byrt et al. 2011).

Although most efforts to boost photosynthetic efficiency have focused either on minimizing inefficiencies in C_3 photosynthesis or engineering C_4 pathways into C_3 species (Matsuoka et al. 2001; Sage and Zhu 2011), there may also be opportunities for improving C_4 photosynthesis itself. Studies in the C_4 plants *Flaveria bidentis* and *Amaranthus edulis* indicate that the reactions catalyzed by PEPC, Rubisco, and PPDK are potentially rate-limiting and may thus represent targets for genetic improvement (Dever et al. 1997; Furbank et al. 1997; Bailey 2000; Matsuoka et al. 2001). In comparison to the biochemical processes of C_4 photosynthesis, the specification of Kranz anatomy is poorly understood. However, anatomical features critical to the C_4 pathway represent an active area of research, which may provide additional opportunities for gains in efficiency (Nelson 2011; Sage and Zhu 2011). In summary, given the complexity of photosynthetic pathways and their interconnectedness with other metabolic and developmental pathways, modeling and systems biology approaches will likely be needed to engineer comprehensive improvements in photosynthetic efficiency (Zhu et al. 2010).

Flowering time: using corngrass1 *to put the starch into cellulosic biomass*

In general, the earlier that plants flower, the less biomass will be made. Flowering also causes a remobilization of fermentable substrates from vegetative portions of the plant, reducing the amount of energy that can be extracted from the vegetative biomass. Thus, altering the timing of the floral transition is an important consideration in the design and breeding of new biofuel crop plants. Genes involved in regulating flowering time are discussed in more detail below. Here, manipulation of the *Corngrass1* (*Cg1*) gene will be discussed as an example of the possibilities for simultaneously improving biomass quantity and quality.

Recently, it has been shown that both flowering time and starch levels can be altered through overexpression of a novel grass microRNA called *Corngrass1* (*Cg1*) (Chuck et al. 2007). *Corngrass1* is a dominant maize mutation that fixes plant development in the juvenile phase, leading to late flowering and increased early-vegetative biomass through the constant initiation of axillary branches, or tillers. In addition, juvenile maize biomass has reduced amounts of lignin and, in theory, should be more easily digestible compared to adult biomass (Abedon et al. 2006). The *Cg1* gene is a tandem microRNA of the miR156 class that functions to negatively regulate *SPL* transcription factors which have diverse functions, controlling processes such as flowering time (Wang et al. 2009), leaf initiation (Wang et al. 2008), and bract suppression (Chuck et al. 2010).

Cg1 was overexpressed in a variety of plants, including the bioenergy crop plant switchgrass (Chuck et al. 2011). Such plants were fixed in the juvenile phase of development and displayed excessive vegetative branching. Interestingly, none of the switchgrass transformants flowered after several years in the greenhouse or field, likely as a result of downregulation of *AP1* MADS box genes that have been shown to be important for floral meristem initiation in *Arabidopsis* (Kobayashi and Weigel 2007). An interesting consequence of *Cg1* overexpression in switchgrass is the fact that transgenic stems have an excess of starch, up to 250% more compared to wild-type (Chuck et al. 2011). This is probably a result of the lack of flowering, since the stored carbon reserves, including starch, which are normally used to produce reproductive structures remain unused in these plants. Treatment with a standard mix of saccharification enzymes plus alpha amylase and amyloglucosidase could release the starch from the transgenic tissue. Using these enzymes, almost the same amount of glucose could be extracted as from wild-type tissue pretreated with acid, but without the requirement for pretreating the *Cg1* tissue (Chuck et al. 2011). As discussed below, eliminating the need for pretreatment represents a significant cost savings, because pretreating biomass requires high levels of energy in the form of heat and expensive reagents. This study presents a new solution to the difficulty of trying to control starch levels in organs where the amount of starch is already high. By simply preventing the formation of inflorescences, carbon reserves such as starch remain unused, thereby increasing the amount of fermentable substrates in plant tissue normally valued only for its cellulose.

Vegetative Biomass: Quality Traits

For plant feedstocks, the process used to convert biomass to bioenergy directly determines the definition of "quality." In the case of combustion, i.e., burning biomass to generate heat, high-quality feedstocks are those containing large amounts of compounds with high energy density. One such compound is lignin, a phenylpronanoid polymer which releases substantial amounts of heat upon burning. Thus, in developing feedstocks for combustion, a breeder might select for plants with high levels of lignin. The opposite is true for biofuel production: High-quality biofuel feedstocks are more likely to have low levels of lignin. This discrepancy can be understood by examining biofuel production in detail.

In general, plant material is converted into liquid fuels (or other specialty chemicals proposed as replacements for petroleum-derived compounds) via a four step process: 1. mechanical and thermochemical pretreatment of feedstocks to increase the accessibility of constituent carbohydrates, 2. enzymatic digestion of these carbohydrate polymers to release simpler sugars, 3. microbial fermentation of the released sugars, and 4. recovery of the desired fermentation products from the culture medium (Fig. 1A). The specific nature of each particular conversion process dictates which characteristics improve or detract from the quality of the feedstock. For example, maize kernels are commonly converted to ethanol by employing amylases to digest the starch stored in the kernels, followed by *S. cerevisiae*-mediated fermentation of the glucose subunits of starch into ethanol (Fig. 1B). In this case, as discussed in Section 4.2.1, higher levels of starch

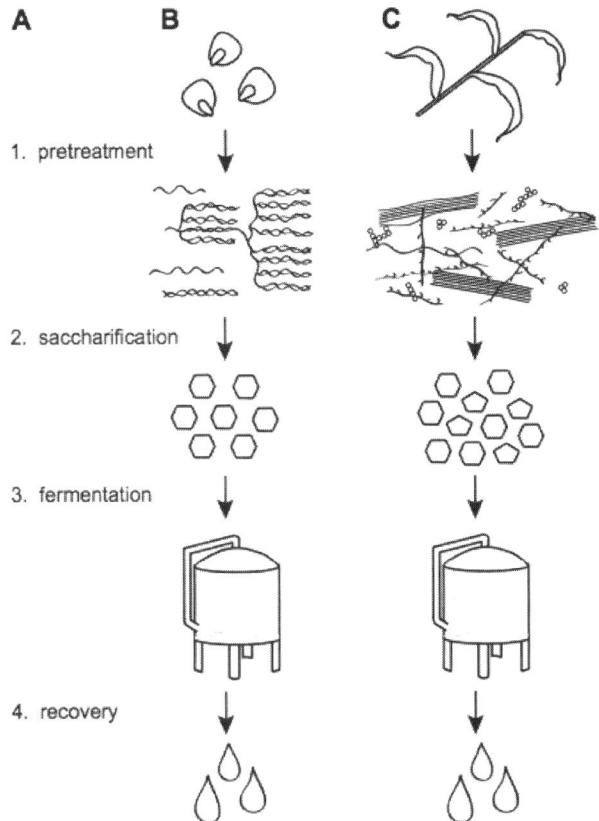

Figure 1 Overview of the biofuels production process for maize feedstocks. Maize represents two major feedstocks: starch-rich kernels and lignocellulosic vegetative tissues, such as leaves and stems, in which diverse carbohydrate polymers are locked within the cell walls. (A) In general, biofuel production proceeds via four steps, 1. pretreatment of plant material, to make the constituent carbohydrates more accessible; 2. saccharification, the process of enzymatically digesting the carbohydrate polymers to release simpler sugars; 3. microbial fermentation of the sugars to produce biofuels; and 4. recovery of the liquid biofuels, e.g., ethanol. (B) Maize kernels can be milled as a pretreatment to access starch, which consists of the glucose polymers amylose (unbranched) and amylopectin (branched). Saccharification with amylases releases glucose, represented as hexagons. *S. cerevisiae*-mediated fermentation of the glucose yields ethanol, which can be recovered via distillation. (C) Resource-intensive pretreatment, for example exposure to hot acid, is required to loosen the carbohydrate polymers in vegetative biomass from the cell wall matrix, which is extensively cross-linked and also contains lignin. Mixtures of enzymes, often including combinations of cellulases and xylanases, break the wall carbohydrates (cellulose, hemicellulose, and mixed-linkage glucan) into hexose and pentose subunits, depicted as hexagons and pentagons, respectively. Various microbes may be used for fermentation. The choice of microbe(s) determines whether hexoses, pentoses, or both will be fermented and also dictates the type of fermentation product. As in the case of processing kernels, if *S. cerevisiae* is utilized, the yeast will convert glucose to ethanol, followed by distillation to recover the biofuel. Diagrams are simplified for illustrative purposes and are not drawn to scale.

would contribute to increased feedstock quality. However, for vegetative biomass (Fig. 1C), the composition and the conversion processes—and thus the determinants of quality—are more complex.

Overview of grass cell wall composition

Vegetative biomass consists of cell walls. Thus, understanding the molecular basis of cell wall polymers will be key to generating stable and sustainable bioenergy feedstocks (US Dept. of Energy 2006; Pauly and Keegstra 2010). As described in Chapter 2, the cell walls of grasses—such as maize—are primarily comprised of cellulose, hemicellulose, lignin, and mixed-linkage glucan (see also Pauly and Keegstra 2008 and Vogel 2008 for reviews). In contrast to eudicots, in which the primary hemicellulose is xyloglucan, grasses contain glucuronoarabinoxylan (GAX) as the predominant hemicellulose (Vogel 2008). Grasses also contain relatively little pectin and low levels of structural protein (Vogel 2008). Additionally, grass cell walls are characterized by the presence of the hydroxycinnamate compounds ferulic acid and ρ-coumaric acid (Vogel 2008). Ferulate residues can cross-link GAX polymers to each other and to lignin (Grabber 2005; Vogel 2008). However, the exact composition of the wall varies with cell and tissue type, as well as with developmental stage (Knox 2008; Pauly and Keegstra 2010). Consequently, improved quality could be achieved via alterations within a particular type of cell wall or by changing the abundance of different cell types.

Contributors to cell wall recalcitrance

Plant cell walls are inherently strong: they provide mechanical support and protection against disease and herbivory. Reducing this strength can diminish the plant's fitness. At the same time, the cell walls' resistance to degradation, also known as "recalcitrance," represents the largest obstacle to sustainably producing biofuels from lignocellulosic feedstocks (US Dept. of Energy 2006; Himmel et al. 2007; Wyman 2007). Overcoming recalcitrance is the goal of the first two steps in the production process, pretreatment and enzymatic digestion (Fig. 1A).

Pretreatments typically encompass mechanical disruption (e.g., cutting, chipping, or grinding) together with the application of heat and chemicals. The treatments can take many forms including exposure to hot sulfuric acid or water, ammonia, extreme pressures, lime, ionic liquids, or a combination of these (Wyman et al. 2005; Brodeur et al. 2011). A common feature of pretreatments is their requirement for large amounts of energy, time, chemicals, water, or other resources. Nevertheless, pretreatments are necessary to make the carbohydrate polymers in the cell walls accessible to the hydrolytic enzymes used in the second step of the conversion process. For example, Zhang and colleagues have shown that extrusion-pretreated corn stover releases 2.2 times more glucose and 6.6 times more xylose than untreated biomass (Zhang et al. 2011b).

Saccharification, the enzymatic release of sugars following pretreatment (Fig. 1A), is also resource-intensive. Whereas starch degradation can be achieved with α-amylase and glucoamylase, the heterogeneous composition of cell walls dictates the use of multicomponent enzyme cocktails that can cleave a myriad of polymers. Despite advances in the identification of efficient wall-degrading enzymes (see Chapter 2), their production remains expensive (Kumar et al. 2008). Together, the costs of pretreatment and saccharification drastically reduce the economic competiveness of biofuel production (Wyman 2007; West et al. 2009). Consequently, a reduction in biomass recalcitrance would reduce production costs and improve both the economic and environmental sustainability of plant-derived fuels.

A number of factors contribute to recalcitrance. Among these are the presence of lignin, cellulose crystallinity, and cross-linkages between cell wall polymers. Although desirable for combustion, lignin is detrimental to biofuel production, because lignin cross-linkages make cell wall carbohydrates less accessible to digestive enzymes (Grabber 2005). Additionally, pretreatments can trigger the release of lignin-derived compounds that inhibit the growth of fermentative microbes (Klinke et al. 2004). Modifying lignin content and composition to increase feedstock digestibility is an important target of bioenergy breeding, but one which will need to be monitored in the context of maintaining crop fitness (Zhang et al. 2011a; Zhang et al. 2011c). One potential strategy is to reduce lignin levels in specific cell types—such as parenchyma—where maximal cellulose and hemicellulose levels are found, while maintaining lignin synthesis in plant support structures (Byrt et al. 2011).

Amounts of hexoses versus pentoses

Most of the microbes currently being considered for biofuel applications ferment six-carbon sugars, or hexoses. The most abundant hexose in plant cell walls is glucose, which forms the building blocks of cellulose. The yeast *S. cerevisiae* readily converts glucose to ethanol under anaerobic conditions and can also ferment the hexoses mannose, fructose, and galactose (van Maris et al. 2006). However, *S. cerevisiae* cannot ferment pentoses such as xylose (van Maris et al. 2006). Similarly, *Clostridium thermocellum*—an ethanogenic bacterial species proposed for biofuel production—ferments glucose but not xylose (McBee 1954; Lynd et al. 1989; Demain et al. 2005). Thus, xylans in plant cell walls are typically not utilized in the production of bioethanol. For these production processes, it would be desirable to shift feedstock composition away from xylose-based polymers toward higher amounts of glucose-based polymers.

At the same time, hemicellulose, primarily in the form of the xylan polymer GAX, can comprise as much as a quarter to half of the dry weight of grass cell walls (Vogel 2008). Therefore, recent research efforts have focused on engineering or identifying microbes which can ferment five-carbon sugars. Through extensive genetic engineering, multiple groups have created *Saccharomyces* strains that can convert xylose into ethanol, albeit less efficiently than converting glucose into ethanol (Ho et al. 1999; Matsushika et al. 2009; Bera et al. 2011). Co-culturing *C. thermocellum* with the pentose-fermenting bacterium *C. thermosaccharolyticum* has been investigated (Demain et al. 2005). Additionally a new species of *Clostridium*, *C. phytofermentans*, which can ferment both glucose and xylose, has been discovered (Warnick et al. 2002). In consolidated bioprocessing of pretreated corn stover under optimal conditions, *C. phytofermentans* produced approximately 72% as much ethanol as an engineered, xylose-fermentating *S. cerevisiae* strain used together with commercial cell wall degrading enzymes (Jin et al. 2011). As these and other conversion technologies develop, ideas about ideal feedstock characteristics are likely to evolve. Currently, high quantities of an easily accessible glucose-based polymer (e.g., cellulose) are considered advantageous.

Genetic engineering of cell walls

Various transgenic approaches for changing the properties of cell walls are being investigated. For instance, heterologous carbohydrate-binding modules can be introduced to reduce the crystallinity of cellulose (reviewed in Abramson et al. 2010). Expression of wall degrading glycosyl hydrolases (GHs) from bacterial or plant sources is also being tested (Lopez-Casado et al. 2008; Taylor et al. 2008; Pauly and Keegstra 2010; Abramson et al. 2010). However, in contrast to the successful transgenic expression of α-amylase to degrade starch, other attempts to produce self-processing plants have failed to produce healthy, viable plants. For example, maize engineered to overproduce xylanase enzymes from *Bacillus* and *Clostridium* species, to help degrade the cell walls in corn stover, exhibited severe phenotypic defects (Gray et al. 2011). These findings point to the need for clever engineering of self-processing enzymes before they can be successfully expressed in plants.

Indeed, any strategy for lowering biomass recalcitrance must do so without compromising the plant's structural integrity, disease resistance, or reproductive fitness. One approach is to achieve precise spatial and/ or temporal regulation of GH expression, lignin modifications, or other recalcitrance-reducing mechanisms. For instance, one could employ tissue- and developmental-stage-specific promoters to drive the expression of cell wall modifying enzymes: Using a senescence-specific promoter to express a GH in vegetative tissues just before harvest could improve conversion efficiency without impacting the health of the plant during the growing season. Another approach is to search for naturally occurring variation in recalcitrance, as variants with high conversion efficiencies represent evolutionarily successful solutions to the recalcitrance problem. As an example, a survey of 1,100 *Populus trichocarpa* trees identified extensive natural variation in the amount of glucose and xylose released from the pretreated wood; compared to the most recalcitrant wood cores, the least recalcitrant samples released between 2.7 and 3.4 times as much sugar (Studer et al. 2011).

Reducing Required Inputs

For both grain and vegetative biomass, the resources needed to grow, harvest, and process the feedstock detract from the economic and environmental attractiveness of the final biofuel product. Therefore, any discussion of desirable traits for a bioenergy crop plant must include an emphasis on minimizing the use of resources. For example, breeding for disease and insect resistance, drought tolerance, and nutrient use efficiency can reduce the requirement for expensive "inputs" such as pesticides, water, and fertilizer. This reduction will, in turn, lead to a net gain in sustainability.

Expansions in maize crop production and changes in crop utilization are influencing biotic and abiotic stress factors that affect plant growth. The increase in corn acreage necessary to balance future bioenergy and food demands will result in greater pressure from pests and disease-causing pathogens that ultimately reduce crop yields. Climate change may also impact biotic stresses like pests and pathogens and increase the influence of abiotic stresses such as drought and heat. In addition, land use concerns are likely to expand maize cultivation onto marginal soils with low nutrient levels. Tolerance to one or, more optimally, complex combinations of these stress factors is critical for maize productivity and yield. Therefore, in a world with increasing food and fuel demands, improved resistance to biotic and abiotic stresses is a priority in maize breeding strategies.

Biotic stress tolerance

Since the advent of maize cultivation, growers have had to compete with a multitude of biotic pests, including animals (insects, rodents, birds), plant pathogens (fungi, bacteria, viruses), and other competing weed plants.

Protection from these biotic invaders has and continues to be critical for the prevention of crop losses in maize. In the context of bioenergy needs, emphasis is placed on foliar parts (leaves, stocks, ears) of the maize plant for both cellulosic and starch-based biofuel production. As a result, tolerance to foliar pathogens is critical to the bioenergy future of maize production.

Breeding for disease resistance

Global losses in maize due to disease-causing pathogens have been estimated at 9% for the 2001–2003 growing seasons (Oerke 2006). This number varies from 8 to 13% based on regional and climatic differences and does not include animal or viral pests. In industrialized nations, investment in resistant germplasm and chemical crop protection has attenuated losses in maize production. However, in less industrialized parts of Central Africa and Southeast Asia, disease pressure and economic constraints can lead to a loss of nearly one third of potential production (Oerke 2006). Historical disease data serve to highlight the importance of breeding efforts for pathogen resistance. Even with the advances of molecular genetics and genomics (discussed in Chapters 5 and 8), great challenges exist in ensuring production of stable, disease-resistant maize for bioenergy needs.

Plant pathogenic resistance is genetically divided into two major categories of "qualitative" and "quantitative" resistance. Major gene, or qualitative resistance, generally provides race-specific resistance to a particular pathogen at a high level of protection and is governed by host resistance genes (*R* genes). R gene-mediated resistance leads to pathogen recognition and rapid cell death called the hypersensitive response (HR) around the site of pathogen invasion. This qualitative resistance is classically described as the gene-for-gene hypothesis in which a host *R* gene recognizes a single pathogen avirulence (*avr*) gene product, resulting in the HR and disease resistance (Flor 1971; Jones and Dangl 2006). Such a host response is typically effective against biotrophic pathogens that require living host cells to survive. Although it is pathogen-specific, gene-for-gene resistance is often short-lived in an agronomic setting because changes in *avr* genes can lead to newly virulent pathogens. As a result, only a few resistance genes have been regularly used in maize, including the *Rp* genes for resistance to common rust (Smith and Hulbert 2005) and the *Ht* genes for resistance to northern leaf blight (Welz and Geiger 2000).

In contrast to qualitative resistance, quantitative resistance provides broad-spectrum protection at intermediate levels of resistance and is governed by multigenic regions called quantitative trait loci (QTLs). The multigene basis of quantitative resistance typically results in more durable protection against a wide range of pathogenic invaders (Parlevliet 2002; Kou and Wang 2010). However, quantitative traits are typically more difficult to analyze in genetic studies: QTLs controlling quantitative resistance often confer varying extents of disease protection, and the genetic basis is still poorly understood. Nevertheless, the vast majority of disease resistance used by maize breeders is quantitative in nature (Wisser et al. 2006).

The expanded use of maize grain and stover for bioenergy needs will likely result in new growth environments and exposure to novel pathogens. Thus, the potential for complex loci to govern sustainable and broad-spectrum disease resistance holds great promise for future maize production. Multiple disease resistance (MDR) conditioned by genes at the same locus to protect against multiple pathogens is critically important for maize fitness. Although still poorly understood, evidence that MDR genes exist in plants has been described for *Arabidopsis* (Cao et al. 1998; Nurmberg et al. 2007), wheat (Krattinger et al. 2009), rice (Manosalva et al. 2009), and maize (Wisser et al. 2006). In addition, multivariate analysis of resistance to three globally important fungal diseases (discussed below) suggests that a glutathione S-transferase gene is responsible for levels of resistance to all three pathogens (Wisser et al. 2011). Quantitative resistance is more often associated with resistance to necrotrophic pathogens that derive nutrition from dead cells, and this is a shared characteristic among the three pathogens described by Wisser et al. (2011). MDR in the host plant may therefore respond to shared aspects of pathogenesis, leading to broad-spectrum protection against a wide array of potential invaders.

Among the most damaging diseases of maize worldwide are southern leaf blight (SLB) caused by *Cochliobolus heterostrophus*, gray leaf spot (GLS) caused by *Cercospora zeae-maydis*, and northern leaf blight (NLB) caused by *Exserohilum turcicum*. Although these and several other foliar diseases will be mentioned only briefly in this chapter, there are excellent reviews on this topic (Pratt and Gordon 2006; Wisser et al. 2006; Balint-Kurti and Johal 2009). Ultimately, these three fungi derive nutrition as necrotrophic pathogens from dead host cells. Necrotrophs cause disease by directly suppressing the induction of plant defenses or by nullifying the consequences of defense (van Kan 2006). Another strategy employed by these fungi is the production of phytotoxins that can enable or enhance infection and disease development. Therefore, breeding efforts to ensure resistance to these pathogens have and continue to be directed toward both pathogen invasion strategies and maize defense mechanisms.

Disease risk associated with agricultural practices must also be considered when developing germplasm for biofuel feedstocks. For example, maize carrying the Texas cytoplasm for male sterility (cms-T) was used extensively for hybrid seed production in the 1950s and 1960s. This monoculture system resulted in a severe epidemic of southern corn leaf blight (SLB) in the early 1970s. A new race (race T) of *Cochliobolus heterostrophus* was highly pathogenic on cms-T maize plants, causing the SLB infestation and massive crop loss. It was discovered that race T produces a polyketide called T-toxin that binds to URF13, a peptide in the inner membrane of maize mitochondria, resulting in membrane permeability and host cell death (Levings 1990; Levings and Siedow 1992). Gray leaf spot (GLS) is also a significant yield-limiting disease worldwide and a serious threat to maize in the US and sub-Saharan Africa. The incidence of GLS has greatly increased over the past 25 years due to increased use of conservation tillage or no-till practices, which allow inoculum to overwinter in plant debris (Ward et al. 1999). Resistance to GLS is mostly additive in nature and ranges from moderately to highly heritable (Gordon et al. 2006).

Another important pathogen, *Exserohilum turcicum*, causes moderate to severe epidemics of northern leaf blight (NLB) globally. NLB is of significant importance for the northeastern US and areas all over the world where high humidity and moderate temperatures persist. The genetics of NLB resistance has been previously reviewed (Welz and Geiger 2000) and demonstrates a complex genetic architecture with multiple QTLs (Zwonitzer et al. 2010). Recently, genome-wide nested association mapping has identified multiple gene candidates governing resistance to NLB (Poland et al. 2011).

Collectively, biotic pathogens pose a major threat to the bioenergy future of maize production. Of particular importance will be the need to closely monitor the tradeoff between disease resistance and desired bioenergy traits. Biomass development, sugar content, and plant digestibility are all key factors for efficient biofuel production, but advances in any one of these phenotypes may result in greater pathogen susceptibility. To balance potentially conflicting phenotypes, geneticists and breeders have the advantage of substantial genetic diversity in maize (Buckler et al. 2006). Adapted germplasms of maize carry much more genetic diversity than those of wheat or rice, allowing maize breeders to effectively achieve quantitative disease resistance while simultaneously improving germplasm for bioenergy needs. In light of disease resistance strategies, maize is also an effective model for biofuel research because of the incomparable body of knowledge; the large community of physiology, genetics, and molecular biology researchers; and the extensive history of value-added breeding.

Breeding for insect resistance

Insect damage continues to be a major concern for maize production worldwide. As a consequence, breeding strategies have focused on enhancing and/or developing novel host resistance mechanisms to guard against insect pests. These strategies have taken two major directions, which will be briefly summarized here, but are discussed in more detail in several reviews (Christou et al. 2006; Ferry et al. 2006; McMullen et al. 2009; Gatehouse et al. 2011). Native plant resistance involves chemical, biochemical, and genetic defenses found endogenously within the maize plant. The genetic basis of the host plant response to insects tends to be quantitative in nature; therefore QTL analysis has been the standard approach toward understanding native resistance. A secondary tactic that has gained extensive use in modern maize production is transgenic resistance mediated by the expression of foreign proteins. The use of transgenic biotechnology to combat insect pests has been enthusiastically adopted by breeders and farmers alike (James 2010).

Maize utilizes various natural defense mechanisms at different life stages to defend itself against insect pests. Several secondary metabolites are synthesized independently of insect herbivory as phytoanticipins, to function as a basal defense before pest damage. Upon insect infestation, many of these constitutive chemicals become modified, or new ones synthesized, to actively combat the insect menace. One constitutive defense chemical found at high levels in maize seedlings and young plants is the hydroxamic acid DIMBOA (Klun et al. 1970). DIMBOA acts as a feeding deterrent for the European corn borer (Robinson 1982). Hydroxamic acids and other defense-related metabolites derived from the shikimic biosynthetic pathway (chlorogenic acid, phenolics, maysin) contribute to broad-spectrum insect resistance in maize. Recent transcriptional analyses using microarrays identified several genes with potential insect defense functions in maize seedlings (Johnson et al. 2011). These putative defense proteins include defensin, hydroxyproline, cystatin, proteases, protease inhibitors, thaumatin-like protein, and lipase, demonstrating the wide variety of defense mechanisms employed by maize. While some of these defense molecules are insect-species-specific, maize possesses an endogenous level of nonhost resistance to a broad array of insect pests. These nonhost defenses may be structural in nature (i.e., waxy cuticle or trichomes) or biochemical products that are derived from multigenic QTLs. The synergistic activity of host and nonhost defenses can make maize tolerant or resistant to multiple insect invaders.

The use of biotechnology to introduce novel defense strategies into maize began in 1996 with the commercialization of lines carrying transgenes from the naturally occurring soil bacterium *Bacillus thuringiensis* (*Bt*) (Mendelsohn et al. 2003). These transgene products are highly effective at controlling lepidopteran pests like the European corn borer, the corn earworm, and the southwestern corn borer. *Bt* corn hybrids actively expressing insecticidal proteins have been adopted extensively worldwide to control insect pests (Head and Ward 2009; James 2010). However, to sustain durable pest resistance, "stacking" of multiple transgenes will likely be necessary to deter the development of resistant insect populations (Christou et al. 2006).

Another concern for maize breeding of insect resistance is the effect of changing environmental parameters. Climate change could cause shifts in the geographical distribution of insect species that come into contact with maize plants. New pressures or new insect pests could arise when both resident and exotic species occur in a given ecosystem. A changing environment could also alter selection pressure, favoring certain genes or genotypes within a pest population to make the insects stronger invaders. Quarantine and early intervention are important measures to deter fluctuating insect populations and to prevent potential invasions. Additionally, changes in seasonal weather patterns may require altered management practices involving rotation, intercropping, and crop diversification. In light of the effects of climate change, breeding for genetic host resistance will play an even more important role in future maize production.

Abiotic stress tolerance

Environmental change is also a key factor when considering the effects of drought stress on maize growth. Early analyses of crop yield data in the United States, including that of maize, found that only a fraction of genetic potential was being realized (Boyer 1982). The most significant constraint on yield potential was the abiotic stress of water deficiency. In general, maize is highly adapted to hot, dry environments due to its C_4 biology, which enables the plant to concentrate and fix CO_2 in bundle sheath cells. This process allows for reduced stomatal conductance, greater transpiration efficiency, and reduced water loss during periods of drought, as well as improved nitrogen use efficiency (Sage and Zhu 2011). However, a substantial challenge for maize production in the context of climate change will be to stabilize yield performance in years that are drier than average (Reynolds et al. 2010). One promising approach to improve yield in drought and other abiotic stress conditions is the identification and pyramiding of naturally occurring alleles that confer tolerance. For example, the 'stay-green' trait in sorghum improves grain yield and lodging resistance under drought conditions, but functions as a multigene QTL (Sanchez et al. 2002). Recent studies have focused on fine mapping and utilization of the sorghum stay-green trait in maize (Harris et al. 2007). Several other efforts are summarized in excellent reviews (Umezawa et al. 2006; Araus et al. 2008; Mullet 2009; Reynolds et al. 2010).

The increasing demand for maize production as a biofuel crop will require greater land acreage in order to minimize competition with food corn. This will lead to growth on marginal lands that are not suitable for food crops and will impose greater abiotic limitations on growth and development. Poor soil quality (saline or acidic), suboptimal temperatures (hot/cold), and greater nutrient deficiencies will ultimately affect rates of biomass and grain yield. Adaptation to these and other abiotic factors are notoriously quantitative, complicating breeding methodologies. Fortunately, several studies have successfully identified mechanisms of abiotic stress tolerance in maize, including heat tolerance regulated by acetylcholinesterase (Yamamoto et al. 2011), acclimation to frost through CBF/DREB transcription factors (Tondelli et al. 2011), and the involvement of nitric oxide (NO) in enhancing salt tolerance and controlling H_2O_2 levels (Bai et al. 2011). In addition, there are complex regulatory gene networks that will provide resources for future analysis and improvement of stress tolerance in maize (Yamaguchi-Shinozaki and Shinozaki 2005).

Nutrient use efficiency

Nutrient availability and use efficiency are essential, and often limiting, factors for plant growth. Nearly all cultivated maize worldwide receives nitrogen (N) and phosphorus (P) fertilizer applications. Fertilizer use for maize cropping, including for cellulosic ethanol, leads to greater input costs and negative impacts on soil, water, and air quality. For example, excess fertilizer use has resulted in hypoxic "dead zones" (Turner et al. 2008) and enhanced greenhouse gas emissions. Therefore, reducing the necessary amount of supplemental nutrient (especially N) and increasing nutrient use efficiency in maize will have positive effects on both the biofuel economy and the environment.

Genetic improvement of nitrogen use efficiency (NUE) in maize has been complicated by the large number of genes and interacting pathways associated with this trait. An integrated approach, using QTL mapping, expression profiling, transgene analysis, and functional genomics will be required to identify improvement strategies (Hirel et al. 2007). The ideal biofuel feedstock will require low N inputs, have a high NUE, and a high C:N ratio in the biomass. In addition, as discussed in Section 4.2.3, the feedstock should have high carbohydrate and low lignin levels. In maize cropping systems, N fertilization often leads to increased protein and decreased concentrations of carbohydrates (Zhang et al. 1993). High levels of N fertilization result in vigorous grain yields, but corn residues (used for cellulosic ethanol production) show only modest increases in carbohydrate yields (Gallagher et al. 2011). Nitrogen application also increases the lignin content of corn residue, diminishing its quality as a biofuel feedstock. Increasing utilization of maize for cellulosic ethanol will thus require a shift in the analysis of NUE from grain yield only to the physiology of the whole plant.

Plant growth habit and lifecycle also have a dramatic impact on nutrient use and resource allocation. Maize exhibits an annual growth pattern, resulting in heavy resource and nutrient allocation into seed production.

This scenario is certainly ideal for grain/food corn, but a perennial habit may be much more valuable in a biofuel feedstock system. Perennial plants spread both by seed and by vegetative organs like rhizomes, enabling them to overwinter and rapidly regrow at the start of the next season. Rhizomatous grasses translocate nutrients from aboveground organs to belowground organs, minimizing the need for fertilizer application. In addition, perennial grasses like switchgrass and *Miscanthus* remove less N, P, and K from the environment than high-yielding annual crops (Propheter and Staggenborg 2010). For these and other reasons, perennialization of maize is a target for efforts to improve maize as a bioenergy crop. Recent analyses of wild relatives of maize have identified multiple QTLs controlling perennial phenotypes (Westerbergh and Doebley 2004). Also, new teosinte species (ancestral to maize) that may provide a resource for genes governing perennial growth have been discovered (Sánchez et al. 2011).

Using New Populations and Classical Genetics to Map the Loci Underlying Bioenergy Traits

Maize has been studied as a genetic system for over 100 years and has been instrumental in the analysis of several classic biochemical pathways, including anthocyanin and starch biosynthesis. The wealth of genetic tools available for maize includes large mutant collections; excellent genetic and cytologic maps; transposon insertion alleles that can be used for reverse genetics; molecular markers for rapid mapping; and several fully sequenced genomes, which can be accessed at www.maizesequence.org. Maize carries the benefit of a diverse and well-established germplasm (Buckler et al. 2006), a long history of genetic improvement, and multiple genetic and cytogenetic stocks (Vasal et al. 2006). For plant breeding, many of the diverse germplasm resources can be obtained through the National Germplasm Repository; mutants can be ordered through the Maize Genetics Cooperation. All of these resources are consistently updated and curated through the Maize Genetics and Genomics Database (www.maizegdb.org). Each of these genetic tools will need to be utilized throughout the process of improving bioenergy traits in maize, and these are discussed in more detail in several recent reviews (Yuan et al. 2008; Byrt et al. 2011; Vermerris 2011).

Maize Quantitative Trait Loci and the Intermated B73 MO17 Population

Since most ethanol production in the US utilizes maize starch as a feedstock, there has not been a concerted effort to extract additional energy from residual corn stover until recently. With the advent of new methods to both generate and analyze large datasets, it is now feasible to apply complex statistical methods to identifying groups of maize genes responsible for enhancing biofuel production in corn stover. Resources to support these new methods include the maize IBM (Intermated B73 MO17) lines, a well-characterized mapping population derived from two widely used maize inbred lines (Lee et al. 2002). This population has undergone four generations of self-pollination, thus increasing recombination frequency four-fold over simple F_2 mapping and allowing greater map resolution. Furthermore, the existence of a large set of DNA markers for the entire maize genome makes rapid genetic mapping possible.

Many bioenergy traits are multigenic and can only be measured in populations as assigned values. When mapped, these traits localize to several locations, i.e. quantitative trait loci (QTLs). The maize IBM population has been screened for several beneficial biofuel properties measured as quantitative traits and then used to identify loci. For example, one goal in the creation of easily digestible biomass is to reduce lignin content. However, maize lignin mutants such as *brown midrib3* (Coors et al. 1994) are not useful because of unpredictable, negative agronomic effects of the mutation. Because the lignin pathway is still not completely understood in grasses, attempts to alter lignin levels through targeted mutagenesis of putative lignin genes may not give the intended results. Using the QTL approach, there is no preselection of genes responsible for the phenotype being assayed. Genes are only identified as associations with the phenotype after fine mapping of the bioenergy traits. For example, analysis of fiber quality and lignin in maize revealed several associations with lignin biosynthetic genes such a *brown midrib1* and 2, as well as a cellulose synthase gene (Cardinal et al. 2003). While these results were somewhat expected, these genes were identified only after the map location of the trait was found. In addition, the IBM population was used to correlate several maize biomass QTLs with the activity of ten genes involved in carbon and nitrogen metabolism, including glutamine synthetase and AGPase (Zhang et al. 2010). Since complex traits such as biomass quality are assumed to be controlled by multiple genes, this QTL mapping approach is an improvement over standard mutagenesis approaches, which can only identify gene function one-at-a-time. Fortunately, our increasing ability to sequence plant genomes, together with the development of bigger mapping populations with greater resolution, will make QTL mapping even easier and faster in the future.

The Maize Nested Association Mapping Population

The maize NAM population has emerged as a powerful tool for dissecting the genetic basis of quantitative traits (Yu et al. 2008). The NAM population was derived from 26 diverse maize germplasms, each of which has been genotyped with 1,106 molecular markers, enough to cover almost every centiMorgan of DNA in the maize genome. Having 26 diverse inbreds allows researchers to capture and assay the complete range of phenotypic variation present in maize. Each of the 26 founder lines was phenotypically characterized, crossed to the B73 reference genome that has been fully sequenced, and then selfed for six generations. From each self-cross, 200 recombinant inbred lines (RILs) were selected, thus forming a total population of 5,000 lines. Each of the 5,000 lines was then genotyped with the same molecular markers used to characterize the parents, allowing the identification of all the blocks of recombination that had occurred between the NAM parental genome and the B73 reference genome.

From a practical standpoint, if one wants to follow a complex trait and identify which genes are associated with it, one only needs to assay that trait in the NAM founder lines. If variation is found between two or more of the lines, the genes responsible for that variation can be quickly identified by growing the RILs from those founders, determining which ones display the variation and which do not, and then analyzing the suite of recombinations within those individuals. Not only will these recombinations identify a genetic locus, but having the large number of RILs will also—in most cases—simultaneously allow fine mapping and identification of the putatively associated genes. Previous methods of QTL mapping required generations of continuous selfing and recombination mapping in order to follow and fine map loci associated with single QTLs; in contrast, in the NAM population, this substantial, initial investment of effort has already been made for researchers on a large scale.

The NAM population has been shown to be a valuable tool for dissecting complex biofuel phenotypes such as maize flowering. While many genes have been discovered that control flowering time in different grasses (Distelfeld et al. 2009), in maize only one flowering time QTL corresponding to the *AP2* gene *vegetative to generative transition* (*Vgt*) has been cloned (Salvi et al. 2007). Clearly, other flowering genes must exist in maize. To find them, the 5,000 RILs from the NAM population and their progeny were screened for differences in flowering time, totaling over one million plants assayed in 8 different environments (Buckler et al. 2009). This study resulted in a broad picture of the genetic basis for maize flowering and showed that flowering time in maize is controlled not by a few genes with large effects, but by many genes with small additive effects (Buckler et al. 2009).

Additional studies of maize flowering QTLs under different photoperiodic environments identified different flowering loci, such as a QTL on chromosome 10 responsible for nearly 40% of the phenotypic variance (Wang et al. 2010a). To date, however, none of these flowering genes has been shown to cause a substantial difference in either the grain yield or the biofuel properties of maize. These results, along with the NAM results for flowering time, indicate that while QTL and association analysis can paint a useful picture of the genes involved in a biological process, there is still a place for traditional genetics to uncover gene function.

Classical Genetics

The power of traditional genetics has been demonstrated recently in sorghum. Closely related to maize, sorghum is a short-day tropical grass whose drought tolerance and high biomass and sugar yields make it an ideal biofuel crop in areas with unpredictable water supply. Sorghum varieties grown for forage and energy have been selected for delayed flowering, a trait that increases the duration of vegetative growth. Several of the loci responsible for flowering time phenotypes, i.e., maturity loci, have been identified (Pao and Morgan 1986). One of them, Ma_1, was cloned by chromosome walking and shown to correspond to a *PSEUDORESPONSE REGULATOR* (*PRR*) gene that is modulated by both photoperiod and the circadian clock (Murphy et al. 2011). In wheat and barley, both long-day plants, a different *PRR* gene (Turner et al. 2005) called *PHOTOPERIOD1* (*Ppd1*) regulates flowering time in response to photoperiod. Taken together, these studies indicate that the grass *PRR* gene family may be a good target for engineering flowering time for biofuel production from grass feedstocks.

Model Systems as Tools for Generating Knowledge to Improve Maize

In spite of the many advantages of maize, its large size and long generation time make high-throughput experiments impractical. To address questions which require growing large numbers of plants under controlled conditions, e.g., in a greenhouse or growth chamber, researchers often turn to model species. The choice of a model depends in large part on the question, but also on the characteristics of the model and the resources available for it. Experimental tractability—the defining feature of a successful plant model system—often encompasses small physical stature; simple growth requirements; self-compatibility; a compact, sequenced genome; amenability to mutagenesis and genetic transformation; and a diversity of germplasm. Good models are immensely useful for elucidating the relationships between genotype and phenotype, and this is especially true for complex, bioenergy-relevant phenotypes. Fortunately for plant biology, the model species toolkit is both deep and expanding.

The preeminent model plant is *Arabidopsis thaliana*, a small, diploid weed in the mustard family. *Arabidopsis* is short, easy to grow, and inbreeding (Table 1; Fig. 2). Detailed protocols for cultivating, characterizing, crossing, and mutagenizing *Arabidopsis* are readily available (Weigel and Glazebrook 2002). An efficient, straightforward, floral-dip method for transforming *Arabidopsis* using *Agrobacterium tumefaciens* (Clough and Bent 1998) has enabled the testing of transgenic constructs, the complementation of mutations, and the generation of extensive mutant collections, many of which are publically available. In addition, *Arabidopsis* was the first plant to have its genome fully sequenced (Arabidopsis Genome Initiative 2000). Over the course of the next decade, careful annotation and curation by countless researchers have yielded a gold-standard genome sequence (accessible at www.arabidopsis.org). Building upon this foundation, scientists are now sequencing the genomes of 1,001 diverse *Arabidopsis* accessions from across Eurasia and North Africa (Weigel and Mott 2009; Cao et al. 2011a)

Figure 2 Comparison of maize and the model species *Arabidopsis thaliana* and *Brachypodium distachyon*. Flowering, adult plants of maize (A), *Arabidopsis* (B), and *Brachypodium* (C) are shown. For all three panels, the scale bar represents 10 cm.

Table 1 Comparison of the characteristics of maize and three model plant species.

	Zea mays	*Arabidopsis thaliana*	*Brachypodium distachyon*	*Setaria viridis*
Common name	Maize	*Arabidopsis* (Mouse-ear cress)	*Brachypodium* (Purple false brome)	Green millet (or Green foxtail)
Height (cm)[a]	120–300	15–30	15–30	10–30
Growth requirements	demanding, space-intensive	simple	simple	simple
Generation time (weeks)[b]	8–15	8–12	8–12	6–9
Number of seeds per plant	200–1,000	> 1,000	100–200	> 2,000
Germplasm collections	extensive	extensive	extensive	limited
Reproductive Strategy	outcrossing, self compatible	selfing	selfing	selfing
Genome size (Mbp)	2,300	125	272	510
Crossing method	optimized	optimized	optimized	optimized but unpublished (T. Brutnell)
Transformation method	relatively inefficient, labor-intensive	optimized, *Agrobacterium*-mediated floral dip	optimized, *Agrobacterium*-mediated, requires tissue culture	optimized protocol unpublished (T. Brutnell), *Agrobacterium*-mediated
Photosynthetic pathway	C_4, NADP-malic enzyme type	C_3	C_3	C_4, NADP-malic enzyme type
Cell wall type	type II	type I	type II	type II

[a]Height is strongly dependent on growth conditions and genotype.
[b]For reference lines under standard laboratory or greenhouse growth conditions.

(www.1001genomes.org). The wealth of resources available for *Arabidopsis*, together with a vibrant research community, continues to place this model species at the forefront of plant biology (Koornneef and Meinke 2010).

Although knowledge gained from *Arabidopsis* informs nearly every molecular biological study of plants, *Arabidopsis* is not a universally appropriate model. As a eudicot, *Arabidopsis* is in an entirely different phylogentic clade from monocotyledonous plants, which include grasses such as maize. The monocot and dicot lineages diverged approximately 140 to 150 million years ago (Chaw et al. 2004), and it is thus not surprising that many aspects of grass biology—e.g., development, morphology, cell wall composition, and biotic interactions—are distinct from those of *Arabidopsis*. These distinctions are also reflected at the genome level. For instance, analysis of the maize genome revealed that maize shares 92% and 93% of its 11,892 predicted gene families with rice and sorghum, respectively; in contrast, maize shares only 73% of its gene families with *Arabidopsis* (Schnable et al. 2009). It is therefore advisable to complement the power of *Arabidopsis* with grass model species. Two such models are *Brachypodium distachyon* (purple false brome) and *S. viridis* (green millet), described below.

Brachypodium distachyon

Introduction to Brachypodium distachyon

The grass family, the Poaceae, is divided into several subfamilies, including the agriculturally important Panicoideae (of which maize is a representative), the Ehrhartoideae (represented by rice), and the Pooideae (represented by wheat, oat, and barley) (Kellogg 2001). *Brachypodium distachyon* (also known as purple false brome and hereafter referred to as *Brachypodium*) is a small annual in the Pooideae subfamily. As a grass, *Brachypodium* is much more closely related to maize than is *Arabidopsis*: The lineages leading to *Brachypodium* and maize diverged approximately 52 million years ago (Vicentini et al. 2008). In agreement with the existence of a relatively recent, common monocot ancestor, 92% of the 25,179 genes predicted in *Brachypodium* have significant similarity to sequences from other monocots (International Brachypodium Initiative 2010).

First proposed as a model in 2001 (Draper et al. 2001), *Brachypodium* has many of the characteristics that have made *Arabidopsis* an incredibly powerful research tool (Garvin et al. 2008; Opanowicz et al. 2008; Brkljacic et al. 2011) (Table 1; Fig. 2). *Brachypodium* is short, has a rapid lifecycle, and is undemanding its growth requirements (Vogel and Bragg 2009). As is the case for *Arabidopsis*, these features enable high-throughput experiments by allowing researchers to grow multiple generations and thousands of individual plants per year in the controlled environments of growth chambers or greenhouses. *Brachypodium* is also a self-fertilizing diploid, a fact which simplifies genetic studies and the maintenance of inbred lines. Even in the wild or when flowering plants of different genotypes are intentionally tied together in the laboratory, outcrossing appears to be extremely rare: When 25 wild-type *Brachypodium* plants were each surrounded by up to 20 transgenic plants and over 2,200 offspring from the wild-type plants were examined, none of the progeny showed signs of inheriting the transgene (Vogel et al. 2009). The numerous and expanding resources for *Brachypodium*—from bacterial artificial chromosome libraries and gene expression data to protocols and bioinformatic tools—have been described in excellent reviews (Garvin et al. 2008; Opanowicz et al. 2008; Vogel and Bragg 2009; Brkljacic et al. 2011; Mur et al. 2011) and will be summarized briefly here, with an emphasis on *Brachypodium*'s relevance to maize research.

Germplasm

Brachypodium is native to the Mediterranean region and Middle East, an area which remains a center of *Brachypodium* diversity (Garvin et al. 2008; Opanowicz et al. 2008). Originally, *Brachypodium distachyon* was thought to encompass accessions with varying genome sizes and chromosome numbers of $2n = 10$, $2n = 20$, and $2n = 30$ (Draper et al. 2001; Hasterok et al. 2004). However, cytogenetic analyses revealed that, instead of representing a polyploidy series, the accessions actually represent two diploid taxa of $2n = 10$ and $2n = 20$, which hybridized to give rise to $2n = 30$ allotetraploids (Idziak and Hasterok 2008; Idziak et al. 2011). Recently, the $2n = 20$ and $2n = 30$ accessions were assigned new species names: *B. stacei* and *B. hybridum*, respectively (Catalán et al. 2012). The species name *B. distachyon* was retained for the $2n = 10$ cytotypes. Although *B. stacei* and *B. hybridum* are valuable for studying genome evolution and speciation, the *Brachypodium distachyon* accessions ($2n = 10$) are the focus of both this section and resource development for *Brachypodium* as a model system.

Historical collections of *Brachypodium* include population samples dating to the 1940s and housed at the United States Department of Agriculture's National Plant Germplasm System (NPGS, www.ars-grin.gov/npgs/). Several of these samples, originating in Turkey and Iraq, were used to generate the diploid inbred lines Bd1-1, Bd2-3, Bd3-1, Bd18-1, Bd21, and Bd21-3 (Vogel et al. 2006; Vogel and Hill 2008). Bd21 was the source of the reference nuclear and chloroplast genome sequences (Bortiri et al. 2008; International Brachypodium Initiative 2010), while Bd21-3 was selected for its amenability to efficient *Agrobacterium*-mediated transformation (Vogel and Hill 2008). Both Bd21 and Bd21-3 are commonly used as wild-type, reference lines, and both can be readily crossed to the genetically divergent Bd3-1 line for mapping experiments. For example, RILs have been made from a cross of Bd21 and Bd3-1, and further RIL populations are also being produced (Garvin et al. 2008; Brkljacic et al. 2011). Additional *Brachypodium* lines were developed by John Draper and colleagues at the University of Wales, Aberystwyth. The initial nine "ABR" lines include accessions from Turkey, France, Spain, Italy and Croatia and were first used to explore disease responses (Draper et al. 2001).

The last several years have seen an explosion of *Brachypodium* germplasm resources. Seeds harvested at 51 locations across Turkey were used to generate over 170 diploid inbred lines, divided into two collections: The "BdTR" collection was generated by Hikmet Budak's group (Filiz et al. 2009). The second collection was generated and characterized by Metin Tuna, John Vogel, and colleagues (Vogel et al. 2009). Names of lines in this second collection begin with a three-letter prefix designating the town closest to the original harvest site (e.g., "Adi" for "Adiyaman"). Together, the Bd and Turkish lines exhibit extensive natural variation in numerous traits, including several relevant to bioenergy, such as flowering time, growth habit, height, average biomass, and drought tolerance (Filiz et al. 2009; Vogel et al. 2009; Schwartz et al. 2010; Luo et al. 2011). Furthermore, analysis of the lines using simple sequence repeat (SSR) markers revealed extensive genetic diversity underlying the phenotypic diversity (Vogel et al. 2009). Inbred lines are also being generated from seeds collected from a variety of habitats at 46 sites in northern Spain; preliminary SSR analysis suggested that the genetic diversity in these Spanish lines may be even greater than in the Turkish accessions (Mur et al. 2011). Researchers continue to collect *Brachypodium* in the wild, and new accessions originating in France, Portugal, and Georgia could soon contribute to the growing wealth of *Brachypodium* germplasm (Brkljacic et al. 2011). With over 150 diploid inbred lines, primarily from the Turkish collections, already freely available through the NPGS (www.ars-grin.gov/npgs/), *Brachypodium* is a highly accessible system for mining natural variation.

Genome sequence and resequencing

Due to a low abundance of repetitive elements, *Brachypodium* has a remarkably compact—and thus experimentally manageable—genome (International Brachypodium Initiative 2010). The 272 Mb *Brachypodium* genome is only about twice the size of the *Arabidopsis* genome and more than eight times smaller than the maize genome (Arabidopsis Genome Initiative 2000; Schnable et al. 2009; International Brachypodium Initiative 2010). The genome sequence of the *Brachypodium* reference line Bd21 was published in 2010 and is of very high quality: The final assembly yielded five pseudo-molecules corresponding to the five *Brachypodium* chromosomes and incorporated 99.6% of the sequence information, for 9.4x coverage of the genome (International Brachypodium Initiative 2010). Gene models were computationally predicted and then refined using transcriptome data and manual annotation (International Brachypodium Initiative 2010). The *Brachypodium* genome sequence is accessible via several Web sites, including www.brachypodium.org and through the Munich Information Center for Protein Sequences (mips.helmholtz-muenchen.de/plant/brachypodium/).

Building upon the success of the initial *Brachypodium* sequencing project, the Joint Genome Institute, in collaboration with John Vogel's group at the USDA and others, has undertaken the resequencing of 56 additional *Brachypodium* lines. These lines are being chosen to maximize genotypic and phenotypic diversity and will be freely available to the research community (brachypodium.pw.usda.gov). Inbreeding and the inherently high levels of homozygosity in *Brachypodium* (Vogel et al. 2009; Mur et al. 2011) should reduce the complexities associated with distinguishing haplotypes. Also, the availability of the Bd21 genome sequence allows for reference-guided, as well as *de novo*, sequence assembly. Although 57 genomes, in total, constitute an insufficient sample size for genome-wide association studies, the re-sequencing data will support traditional mapping, the production of RIL populations, and the investigation of candidate genes, while also laying the foundation for large-scale studies in the future.

Genetic transformation

Efficient transformation is indispensable for a modern, model genetic system. Although biolistic transformation of *Brachypodium* has been reported (Christiansen et al. 2005), *Agrobacterium*-mediated transformation is generally preferred, because it results in lower-copy-number and less complex DNA insertions (Dai et al. 2001; Travella et al. 2005). *Agrobacterium*-mediated transformation inserts transferred DNAs (T-DNAs) essentially randomly into the plant genome (Krysan et al. 2002) and is thus useful for insertional mutagenesis, promoter trapping, activation tagging, and the introduction of transgenes.

High-throughput, *Agrobacterium*-mediated transformation protocols have been developed for the *Brachypodium* reference lines Bd21 and Bd21-3 (Vain et al. 2008; Vogel and Hill 2008). Both protocols begin with the dissection of tiny, immature embryos from seeds; followed by the induction of embryogenic callus from these embryos; subculturing to increase the amount of callus; transformation of callus pieces via cocultivation with the hyper-virulent *Agrobacterium* strain AGL1; selection of transformed callus tissue; and regeneration of transgenic plants from the transformed callus (Vain et al. 2008; Vogel and Hill 2008). Linking a green-fluorescent protein (GFP) gene to the selectable marker within the T-DNA can aid in the identification of Bd21 transformed callus, whereas Bd21-3 forms embryogenic callus that is readily identifiable by its yellow color (Vain et al. 2008; Vogel and Hill 2008). Since their first publication, *Brachypodium* transformation protocols have been continually updated. Improvements include incorporating copper sulfate into the callus induction medium and, particularly for Bd21-3, performing cocultivation under dessicating conditions (Alves et al. 2009; Bragg et al. 2012). Under optimized conditions, the entire transformation process—from the dissection of embryos to the harvesting of T1 transgenic seeds—can take as little as 21 to 28 weeks (Thole and Vain 2012) (brachypodium.pw.usda.gov). Transformation efficiencies, calculated as the percentage of callus pieces producing transgenic plants, are impressive: Average efficiencies of 20% for Bd21 or upwards of 50% for Bd21-3 are routine, and efficiencies for individual transformation experiments can be much higher (Alves et al. 2009; Bragg et al. 2012).

Mutant collections

Brachypodium can be mutagenized with ethyl methanesulfonate (EMS) or fast neutron irradiation to generate point mutations or large deletions, respectively. An EMS mutagenesis protocol is available at brachypodium. pw.usda.gov. By far the largest publically available collections of *Brachypodium* mutants, however, consist of T-DNA lines generated at the United Kingdom's John Innes Centre (www.brachytag.org) and the USDA's Western Regional Research Center (brachypodium.pw.usda.gov). The T-DNAs inserted via *Agrobacterium*-mediated transformation serve as tags which allow sequencing, and thus identification, of the plant genomic regions flanking the insertions. As demonstrated in *Arabidopsis*, such tagged, insertional mutant collections are tremendously powerful tools for forward and reverse genetics.

The John Innes Centre Collection of T-DNA Mutants

This collection consists of 5,000 T-DNA lines in the Bd21 genetic background (www.brachytag.org). The lines were generated with vectors for either insertional mutagenesis (BrachyTAG lines) or promoter trapping (BrachyTRAP lines). The BrachyTAG T-DNA contains a GFP coding sequence under the control of a constitutive promoter. In contrast, the BrachyTRAP T-DNA contains a promoter-less GFP construct, such that insertion of this construct downstream of a plant promoter, in the correct orientation, will place GFP expression under the control of the plant promoter. The GFP reporter can thus reveal the endogenous pattern of activity for the adjacent plant promoter. Both the BrachyTAG and BrachyTRAP T-DNAs contain a hygromycin phosphotransferase gene for antibiotic-based selection of transformed callus tissue and transgenic plants (www.brachytag.org). When the T-DNA carries an intact GFP gene, as in the BrachyTAG lines, green fluorescence provides additional confirmation of transformation (Vain et al. 2008).

To date, 1,005 flanking sequence tags have been recovered for T-DNA lines in the John Innes collection (Thole et al. 2009) (www.brachytag.org). Of these tags, a subset correspond to insertions in or near (i.e., within 500 bp of) a coding sequence. At least 365 genes with predicted or putative functions have been tagged in this way. The Modelcrop.org website (www.modelcrop.org) displays alignments between all the flanking sequence tags and the *Brachypodium* genome, allowing scientists to search for insertions in genes of interest. Seeds from the T-DNA mutants can be requested online (www.brachytag.org).

The USDA collection of T-DNA Mutants

Researchers at the USDA have generated over 8,700 T-DNA lines in the Bd21-3 genetic background, and there are plans to expand the collection to 37,500 lines (brachypodium.pw.usda.gov). Vectors for insertional mutagenesis alone, promoter trapping, and activation tagging have been used to generate the transformants. The USDA promoter trapping vectors incorporate a promoterless β-glucuronidase (GUS) construct and a promoterless GFP construct just inside the left and right T-DNA borders, respectively (brachypodium.pw.usda.gov). Therefore, depending on the orientation of the T-DNA, either GUS or GFP expression might reflect the activity of the promoter adjacent to the insertion site. The activation tagging T-DNAs contain four enhancer sequences from the constitutive cauliflower mosaic virus *35S* promoter; these enhancers cause the overexpression of plant genes near the T-DNA insertion sites. Although a phosphinothricin acetyltransferase gene conferring resistance to the herbicide BASTA was utilized for some of the early transformations, most of the transgenic lines carry T-DNAs with a hygromycin resistance cassette, because selection on hygromycin was found to be more effective (brachypodium.pw.usda.gov).

Over 7,200 flanking sequence tags for the USDA lines have been aligned to the *Brachypodium* genome, and more than 2,200 genes contain an insertion within an exon, an intron, or the adjacent region—defined as the 1,000 base pairs upstream and the 500 base pairs downstream of the gene. The brachypodium.pw.usda.gov website offers multiple options to search for insertions in genes of interest and also includes a link for requesting seeds.

Additionally, the Gbrowse viewer accessible through www.brachpodium. org has a track for displaying flanking sequence tag information. Whether the goal is to disrupt a particular gene, cause its overexpression, or study the activity of its promoter, the sizable *Brachypodium* T-DNA collections are an excellent starting point for these investigations. As the collections grow, more and more *Brachypodium* genes will be associated with T-DNA tags.

Crossing

Like mutagenesis, crossing is fundamental to genetic studies. A major challenge in crossing *Brachypodium* arises from the fact that the flowers usually remain tightly closed from early floral development through pollination and seed maturation. The need to gently separate the palea and lemma without damaging the reproductive organs represents one of the largest obstacles to successful crosses. Nevertheless, two optimized crossing protocols (available at the websites brachypodium.pw.usda. gov and www.ars.usda.gov/pandp/docs.htm?docid=18531) have been developed. Although *Brachypodium* flowers are considerable larger than *Arabidopsis* flowers, the use of a dissecting microscope or a jeweler's loupe can facilitate the necessary manipulations. Also, waiting a day between emasculating flowers and pollinating them can help to ensure that the pistil is at the correct developmental stage for crossing. The success rate of pollinations is strongly dependent on the time of day, often being the highest a few hours after dawn; the optimal timing may need to be empirically determined for each particular laboratory. Finally, because the viability of *Brachypodium* pollen declines rapidly, pollen should be applied to pistils within 15 minutes of its release from anthers (brachypodium.pw.usda.gov). With these considerations and some practice, it is quite possible to cross a number of *Brachypodium* genotypes.

Utility of Brachypodium as a grass model species

By revealing commonalities between *Brachypodium* and other grasses, several studies support the utility of *Brachypodium* as a functional model for the grasses in bioenergy research. For example, cell wall composition is broadly similar across the grass family (see also Chapter 2). Gomez et al. found that when monosaccharides were extracted from cell wall matrix carbohydrates, the profiles for *Brachypodium* and three other grasses—including *Miscanthus*, in the Panicoideae subfamily with maize—were all alike, showing relatively large amounts of arabinose, presumably because of the prominence of arabinosylated xylan (GAX) in grass cell walls (Gomez et al. 2008). In contrast, the monosaccharide profile of *Arabidopsis* exhibited substantially lower levels of arabinose but relatively high levels of rhamnose, probably derived from rhamnose-containing pectins (Gomez et al. 2008).

Analyses of genes for cell-wall-associated proteins have identified additional parallels between *Brachypodium* and other grasses. For example, *Brachypodium*, rice, and sorghum all have substantially more expansin genes—involved in loosening cell walls (Cosgrove 2000)—than either *Arabidopsis* or poplar (International Brachypodium Initiative 2010). Glycosyl transferases (GTs) and glycosyl hydrolases (GHs) are also relevant to cell wall dynamics, because certain GT and GH families have been implicated in cell wall synthesis and remodeling, respectively. GTs form and GHs break bonds linking carbohydrates to other moieties, including other carbohydrates (www.cazy.org). Angiosperms appear to share common sets of GT and GH families; however, within individual enzyme families, there are clear distinctions between eudicots and grasses. Compared to *Arabidopsis*, the GT37, GT43, and GT61 families in *Brachypodium* and rice are expanded, with the *Brachypodium* and rice sequences often clustering into orthologous pairs (International Brachypodium Initiative 2010). In the case of the GH28 family, which is associated with pectin degradation, *Arabidopsis* has approximately 60–70% more family members than *Brachypodium*, rice, or sorghum (Tyler et al. 2010). In the GH28 family—as well as in the GH5 and GH51 families of cell-wall-modifying enzymes, the GH18 and GH19 chitinase families, and the GH13 family of starch-modifying enzymes—*Brachypodium* sequences consistently cluster with those from other grasses (Tyler et al. 2010), further supporting the use of *Brachypodium* as a general model for the Poaceae. As an illustration of *Brachypodium*'s potential to contribute specifically to biofuel research, *Brachypodium* was utilized as a feedstock in developing a simultaneous saccharification and fermentation assay based on the cell-wall-degrading and ethanol-producing capabilities of the microbe *Clostridium phytofermentans* (Lee et al. 2012).

Increasingly, *Brachypodium* is also informing maize research. For instance, when maize *St* genes were found to be important for resistance to northern leaf blight, the sequence conservation of the encoded nucleotide-binding, leucine-rich-repeat proteins in maize, sorghum, foxtail millet, and *Brachypodium* allowed the genes' evolutionary history to be traced (Martin et al. 2011). In another example, *Brachypodium* is being developed as a grass model for iron homeostasis (Yordem et al. 2011). Cataloging *Brachypodium* genes encoding proteins similar to the maize Yellow Stripe1 iron transporter helped to identify a grass-specific clade of Yellow-Stripe-Like proteins in *Brachypodium*, rice, and maize (Yordem et al. 2011). Furthermore, the high-efficiency transformation and short lifecycle of *Brachypodium* mean that maize sequences can be rapidly tested in this model system. Conversely, *Brachypodium* has been mined for promoters that are functional in maize (Coussens et al. 2012).

Brachypodium's ability to bridge the gap between *Arabidopsis* and the grasses was elegantly demonstrated by Cao et al. Building upon work in *Arabidopsis*, which had revealed a role for Nuclear Factor Y (NF-Y) transcription factors in drought resistance, photosynthesis, and flowering, Cao and colleagues identified 36 NF-Y genes in *Brachypodium* and analyzed their expression (Cao et al. 2011b). Then, the researchers showed that overexpression of *Brachypodium* NF-YB6—predicted to be an ortholog of flowering-time-regulating NF-Y proteins from *Arabidopsis*—could largely rescue the late-flowering phenotype of *Arabidopsis nf-yb2 nf-yb3* double mutants (Cao et al. 2011b). In summary, *Brachypodium* provides both a test bed for investigating grass-specific phenomena and a conduit for translating a treasure-trove of biological knowledge from *Arabidopsis* into the agronomically important Poaceae.

Setaria viridis *(green millet)*

Introduction to Setaria viridis

Setaria viridis (green millet) is an emerging grass model system. Like maize, *S. viridis* belongs to the Panicoideae subfamily of the grasses. Within the Panicoideae, *Setaria* is a member of the Paniceae tribe, while maize is a member of the Andropogoneae (Kellogg et al. 2009); the last common ancestor of these two tribes is estimated to have lived between 24 and 29 MYA (Vicentini et al. 2008). *S. viridis* is a small, diploid, weedy species thought to be the ancestor of the domesticated foxtail millet (*S. italica*) (Le Thierry d'Ennequin et al. 2000; Benabdelmouna et al. 2001; Dekker 2003). To date, most research into *S. viridis* has focused on its weedy characteristics or its role in domestication. In fact, a study conducted to inform weed control strategies measured traits directly relevant to bioenergy: Li and Yang confirmed a positive correlation between the length of the vegetative growth phase (ending with the onset of flowering) and tiller number, height, and biomass accumulation for *S. viridis* (Li and Yang 2008). More recently, *S. viridis* has begun to emerge as a model for C$_4$ grasses (Brutnell et al. 2010; Li and Brutnell 2011).

Setaria viridis (also known as green millet) is short in both stature and generation time and has undemanding growth requirements, comparable to the model plants *Arabidopsis* and *Brachypodium* (Table 1). Grown in the laboratory using conditions similar to those for *Arabidopsis*, *S. viridis* can start flowering when it is approximately 10 cm tall and yield mature seeds at about 6 weeks of age (Brutnell et al. 2010). An immense advantage for large-scale genetic studies is that a single *S. viridis* plant can produce many thousands of seeds (Li and Yang 2008; Li and Brutnell 2011).

However, from the standpoint of making crosses for genetic studies, the huge number and especially the tiny size of *S. viridis* flowers are disadvantageous. *S. viridis* is naturally self-pollinating, with a low frequency of outcrossing. Under standard growth conditions, *S. viridis* exhibited 1–2% outcrossing; when spikes from different parents were brought into close contact with each other to favor outcrossing, 2–4% of the offspring were hybrids (Till-Bottraud et al. 1992). Intentional crossing of *S. viridis* to generate hybrids is reportedly possible, but difficult (Brutnell et al. 2010). Although improvements to the crossing protocol are being developed, they have not yet been widely publicized.

The lack of an optimized and highly efficient transformation method for *S. viridis* is likewise an obstacle to the full utilization of this species as a model genetic system. Initial success using *Agrobacterium* to transform *S. viridis* callus tissue, followed by the regeneration of transgenic plants, has been reported (Brutnell et al. 2010). These efforts promise to pave the way for routine, stable transformation of *S. viridis* in the near future.

Germplasm and genome characteristics

Originating probably in Africa and giving rise to foxtail millet through domestication in China more than 5,900 years ago, *S. viridis* now has a global range (Li and Wu 1996; Dekker 2003; Barton et al. 2009). In spite of the wide natural distribution of *S. viridis*, germplasm resources for research are limited. The few accessions that are currently available are, as yet, poorly characterized.

One germplasm resource that has been studied is a mapping population generated from a cross between *S. italica* accession B100 and *S. viridis* accession A10 (Wang et al. 1998). Using an F_2 mapping population from this interspecific cross, Doust and colleagues identified a number of QTLs underlying differences in vegetative branching and inflorescence architecture between *S. italica* and its weedy ancestor (Doust et al. 2005; Doust and Kellogg 2006). Interestingly, some of the QTLs were linked to candidate genes identified in maize. For example, *teosinte branched1* (*tb1*) and *barren stalk1* (*ba1*) both mapped to *Setaria* QTLs for tillering and axillary branching (Doust and Kellogg 2006). Also, *zea floricaula leafy1* (*zfl1*) and *tasselseed4* (*ts4*) appear to be good candidate genes for *Setaria* QTLs affecting inflorescence traits (Doust et al. 2005). Other QTLs were not associated with obvious candidate genes, suggesting that at least some of these loci could represent novel regulators of grass architecture (Doust et al. 2005; Doust and Kellogg 2006). Together, these results support the idea that *Setaria* species will be informative functional genomic models for other grasses, including maize.

The full utilization of *S. viridis* as a model will require a sequenced, assembled, and annotated genome. At 510 Mbp, the *S. viridis* genome is relatively small (Brutnell et al. 2010). It is twice the size of the *Brachypodium* genome, and four times as large as that of *Arabidopsis*, but still nearly five times smaller than the maize genome (Table 1). The power of next-generation sequencing will facilitate the rapid acquisition of sequence data. Numerous factors, including the extent of repetitive elements in the genome and the degree of homozygosity in the sequenced accession, will determine whether or not the sequences assemble with ease. In any case, the *S. viridis* genome sequencing effort currently underway (Brutnell et al. 2010) will greatly strengthen this emerging model system.

Utility of Setaria viridis *as a model for* C_4 *grasses*

Because *Brachypodium* is a C_3 grass, investigations of C_4 photosynthesis in *Brachypodium* are limited to attempts to engineer the C_4 pathway into a C_3 plant. While such engineering is useful, it is more applicable to improving rice than maize. In contrast, *S. viridis* is perfectly poised to become a model for C_4 photosynthesis in maize and other biofuel crops (Sage and Zhu 2011; Wang et al. 2011). *S. viridis*, like maize, has an NADP-malic enzyme type of C_4 pathway (Vicentini et al. 2008). Thus, research into the biochemistry, anatomy, and regulation of photosynthesis in *S. viridis* is likely to be directly applicable to maize.

Setaria species have also been proposed as models for abiotic stress tolerance (Li and Brutnell 2011). The local adaptation of *S. viridis* to a wide range of habitats around the world (Dekker 2003) suggests that *S. viridis* accessions could be good resources for investigating natural variation in responses to the environment. Additionally, a survey of nine loci in 50 and 34 accessions of *S. italica* and *S. viridis*, respectively, revealed higher nucleotide diversity and lower linkage disequilibrium in *S. viridis* than in its domesticated relative (Wang et al. 2010b). Although these findings are not unexpected for a wild species and its domesticate, the results—particularly the decay of linkage disequilibrium within 150 bp in *S. viridis*—bode well for the feasibility of future association mapping studies (Wang et al. 2010b).

Finally, in some circumstances, utilizing two or more models may be advantageous. Because the biological knowledge derived from different models can be complementary, multispecies comparisons are sometimes the most informative. Thus, the strengths of several systems—*Arabidopsis*, *Brachypodium*, and *S. viridis*—can be combined to gain broadly applicable biological knowledge for improving maize as a bioenergy feedstock.

Conclusion

Undoubtedly, maize will continue to be a key food crop. However, the additional use of maize as a bioenergy crop holds great promise. In particular, the improvement of corn stover for conversion to biofuels could make food production (from kernels) and fuel production (from vegetative biomass) compatible, rather than competing, objectives. As evident in the discussions of kernel characteristics and genetic tools, the long history of maize cultivation provides researchers with a wealth of knowledge and resources. In this sense, *Z. mays* has a tremendous advantage over the largely undomesticated species being developed as dedicated bioenergy crops, e.g., switchgrass (*Panicum virgatum*) and *Miscanthus* (*Miscanthus x giganteus*). The current challenge is to leverage this knowledge, together with new tools including model species, to improve the characteristics of vegetative biomass. Especially in the context of resource utilization, efforts to improve maize for food and fuel can be synergistic. Breeding for enhanced water use efficiency, for example, can result in higher yields of both grain and lignocellulosic biomass. Such advances will support the cultivation of maize as a sustainable, dual-use crop. In light of human population growth and pressing environmental and social concerns, meeting this goal quickly is critical.

References

Abedon BG, Hatfield RD, Tracy WF et al. (2006) Cell wall composition in juvenile and adult leaves of maize (*Zea mays* L.). J Agri Food Chem 54: 3896–3900.

Abramson M, Shoseyov O and Shani Z (2010) Plant cell wall reconstruction toward improved lignocellulosic production and processability. Plant Sci 178: 61–72.

Alves SC, Worland B, Thole V et al. (2009) A protocol for *Agrobacterium*-mediated transformation of *Brachypodium distachyon* community standard line Bd21. Nat Protoc 4: 638–649.

Amor Y, Haigler CH, Johnson S et al. (1995) A membrane-associated form of sucrose synthase and its potential role in synthesis of cellulose and callose in plants. Proc Natl Acad Sci USA 92: 9353–9357.

Arabidopsis Genome Initiative (2000) Analysis of the genome sequence of the flowering plant *Arabidopsis thaliana*. Nature 408: 796–815.

Araus JL, Slafer GA, Royo C et al. (2008) Breeding for yield potential and stress adaptation in cereals. Crit Rev Plant Sci 27: 377–412.

Bai X, Yang L, Yang Y et al. (2011) Deciphering the protective role of nitric oxide against salt stress at the physiological and proteomic levels in maize. J Proteome Res 10: 4349–4364.

Bailey KJ (2000) Control of C_4 photosynthesis: effects of reduced activities of phosphoenolpyruvate carboxylase on CO_2 assimilation in *Amaranthus edulis* L. J Exp Bot 51: 339–346.

Balint-Kurti PJ, Johal GS (2009) Maize disease resistance. In: Bennetzen JL and Hake S (eds) Handbook of Maize: Its Biology. Springer, New York, NY, USA, pp 229–250.

Barton L, Newsome SD, Chen F-H et al. (2009) Agricultural origins and the isotopic identity of domestication in northern China. Proc Natl Acad Sci USA 106: 5523–5528.

Benabdelmouna A, Abirached-Darmency M and Darmency H (2001) Phylogenetic and genomic relationships in *Setaria italica* and its close relatives based on the molecular diversity and chromosomal organization of 5S and 18S-5.8S-25S rDNA genes. Theor Appl Genet 103: 668–677.

Bera AK, Ho NWY, Khan A et al. (2011) A genetic overhaul of *Saccharomyces cerevisiae* 424A(LNH-ST) to improve xylose fermentation. J Ind Microbiol Biotechnol 38: 617–626.

Bortiri E, Coleman-Derr D, Lazo GR et al. (2008) The complete chloroplast genome sequence of *Brachypodium distachyon*: sequence comparison and phylogenetic analysis of eight grass plastomes. BMC Res Notes 1: 61.

Boundy B, Diegel S, Wright L et al. (2011) Biomass Energy Data Book, edn 4. Oak Ridge National Laboratory, US Department of Energy, Oak Ridge, TN, USA.

Boyer JS (1982) Plant productivity and environment. Science 218: 443–448.

Bragg JN, Tyler L and Vogel JP (2012) *Brachypodium distachyon*, a model for bioenergy grasses. In: Kole C, Joshi CP and Shonnard D (eds) Handbook of Bioenergy Crop Plants. Taylor and Francis, Boca Raton, FL, USA, pp 593–618.

Brkljacic J, Grotewold E, Scholl R et al. (2011) *Brachypodium* as a model for the grasses: today and the future. Plant Physiol 157: 3–13.

Brodeur G, Yau E, Badal K et al. (2011) Chemical and physicochemical pretreatment of lignocellulosic biomass: a review. Enzyme Res 2011: 787532.

Brutnell TP, Wang L, Swartwood K et al. (2010) *Setaria viridis*: a model for C$_4$ photosynthesis. Plant Cell 22: 2537–2544.

Buckler ES, Gaut BS and McMullen MD (2006) Molecular and functional diversity of maize. Curr Opin Plant Biol 9: 172–176.

Buckler ES, Holland JB, Bradbury PJ et al. (2009) The genetic architecture of maize flowering time. Science 325: 714–718.

Byrt CS, Grof CPL and Furbank RT (2011) C4 plants as biofuel feedstocks: optimising biomass production and feedstock quality from a lignocellulosic perspective. J Integr Plant Biol 53: 120–135.

Cao H, Li X and Dong X (1998) Generation of broad-spectrum disease resistance by overexpression of an essential regulatory gene in systemic acquired resistance. Proc Natl Acad Sci USA 95: 6531–6536.

Cao J, Schneeberger K, Ossowski S et al. (2011a) Whole-genome sequencing of multiple *Arabidopsis thaliana* populations. Nat Genet 43: 956–963.

Cao S, Kumimoto RW, Siriwardana CL et al. (2011b) Identification and characterization of NF-Y transcription factor families in the monocot model plant *Brachypodium distachyon*. PLoS ONE 6: e21805.

Cardinal AJ, Lee M and Moore KJ (2003) Genetic mapping and analysis of quantitative trait loci affecting fiber and lignin content in maize. Theor Appl Genet 106: 866–874.

Caspar T, Lin TP, Kakefuda G et al. (1991) Mutants of *Arabidopsis* with altered regulation of starch degradation. Plant Physiol 95: 1181–1188.

Catalán P, Müller J, Hasterok R et al. (2012) Evolution and taxonomic split of the model grass *Brachypodium distachyon*. Ann Bot 109: 385–405.

Chaw S-M, Chang W-H, Chen H-L et al. (2004) Dating the monocot-dicot divergence and the origin of core eudicots using whole chloroplast genomes. J Mol Evol 58: 424–441.

Christiansen P, Andersen CH, Didion T et al. (2005) A rapid and efficient transformation protocol for the grass Brachypodium distachyon. Plant Cell Reports 23: 751–758.

Christou P, Capell T, Kohli A et al. (2006) Recent developments and future prospects in insect pest control in transgenic crops. Trends Plant Sci 11: 302–308.

Chuck G, Whipple C, Jackson D et al. (2010) The maize SBP-box transcription factor encoded by *tasselsheath4* regulates bract development and the establishment of meristem boundaries. Dev 137: 1243–1250.

Chuck GS, Tobias C, Sun L et al. (2011) Overexpression of the maize *Corngrass1* microRNA prevents flowering, improves digestibility, and increases starch content of switchgrass. Proc Natl Acad Sci USA 108: 17550–17555.

Clough SJ and Bent AF (1998) Floral dip: a simplified method for *Agrobacterium*-mediated transformation of *Arabidopsis thaliana*. Plant J 16: 735–743.

Coleman, HD, Yan J and Mansfield SD (2009) Sucrose synthase affects carbon partitioning to increase cellulose production and altered cell wall ultrastructure. Proc Natl Acad Sci USA 106: 13118–13123.

Conway G (1997) The doubly green revolution: food for all in the twenty-first century. Cornell University Press, Ithaca, NY, USA, 360 pp.

Coors JG, Carter PR and Hunter RB (1994) Silage corn. In: Hallauer AR (ed) Specialty Corns. CRC Press, Boca Raton, FL, USA, pp 305–340.

Cosgrove DJ (2000) Loosening of plant cell walls by expansins. Nature 407: 321–326.

Coussens G, Aesaert S, Verelst W et al. (2012) *Brachypodium distachyon* promoters as efficient building blocks for transgenic research in maize. J Exp Bot 63: 4263–4273.

Dai S, Zheng P, Marmey P et al. (2001) Comparative analysis of transgenic rice plants obtained by *Agrobacterium*-mediated transformation and particle bombardment. Mol Breed 7: 25-33.

Dekker J (2003) The foxtail (*Setaria*) species-group. Weed Sci 51: 641–656.

Demain AL, Newcomb M and Wu JHD (2005) Cellulase, clostridia, and ethanol. Microbiol Mol Biol Rev 69: 124–154.

Demura T and Ye ZH (2010) Regulation of plant biomass production. Curr Opin Plant Biol 13: 299–304.

Dever LV, Bailey KJ, Leegood RC et al. (1997) Control of photosynthesis in *Amaranthus edulis* mutants with reduced amounts of PEP carboxylase. Functl Plant Biol 24: 469–476.

Distelfeld A, Li C and Dubcovsky J (2009) Regulation of flowering in temperate cereals. Curr Opin Plant Biol 12: 178–184.

Dohleman FG and Long SP (2009) More productive than maize in the Midwest: How does *Miscanthus* do it? Plant Physiol 150: 2104–2115.

Doust AN, Devos KM, Gadberry MD et al. (2005) The genetic basis for inflorescence variation between foxtail and green millet (Poaceae). Genet 169: 1659–1672.

Doust AN and Kellogg EA (2006) Effect of genotype and environment on branching in weedy green millet (*Setaria viridis*) and domesticated foxtail millet (*Setaria italica*) (Poaceae). Mol Ecol 15: 1335–1349.

Draper J, Mur LA, Jenkins G et al. (2001) *Brachypodium distachyon*. A new model system for functional genomics in grasses. Plant Physiol 127: 1539–1555.

Evans LT (1996) Crop Evolution, Adaptation and Yield. Cambridge University Press, Cambridge, UK.

Ferry N, Edwards M, Gatehouse J et al. (2006) Transgenic plants for insect pest control: A forward looking scientific perspective. Transgen Res 15: 13 19.

Filiz E, Ozdemir BS, Budak F et al. (2009) Molecular, morphological, and cytological analysis of diverse *Brachypodium distachyon* inbred lines. Genome 52: 876–890.

Fischer CR, Klein-Marcuschamer D and Stephanopoulos G (2008) Selection and optimization of microbial hosts for biofuels production. Metab Engg 10: 295–304.

Fleet CM and Sun T (2005) A DELLAcate balance: the role of gibberellin in plant morphogenesis. Curr Opin Plant Biol 8: 77–85.

Flor HH (1971) Current status of the gene-for-gene concept. Annu Rev Phytopathol 9: 275–296.

Furbank RT, Chitty JA, Jenkins CLD et al. (1997) Genetic manipulation of key photosynthetic enzymes in the C_4 plant *Flaveria bidentis*. Functl Plant Biol 24: 477–485.

Gallagher ME, Hockaday WC, Masiello CA et al. (2011) Biochemical suitability of crop residues for cellulosic ethanol: Disincentives to nitrogen fertilization in corn agriculture. Environ Sci Technol 45: 2013–2020.

Garvin DF, Gu Y-Q, Hasterok R et al. (2008) Development of genetic and genomic research resources for *Brachypodium distachyon*, a new model system for grass crop research Crop Sci 48: S69–S84.

Gatehouse AMR, Ferry N, Edwards MG et al. (2011) Insect-resistant biotech crops and their impacts on beneficial arthropods. Phil Trans Roy Soc Lond Sr B 366: 1438–1452.

Geigenberger P, Stamme C, Tjaden J et al. (2001) Tuber physiology and properties of starch from tubers of transgenic potato plants with altered plastidic adenylate transporter activity. Plant Physiol 125: 1667–1678.

Giroux MJ, Shaw J, Barry G et al. (1996) A single mutation that increases maize seed weight. Proc Nat Acad Sci USA 93: 5824–5829.

Gomez LD, Bristow JK, Statham ER et al. (2008) Analysis of saccharification in *Brachypodium distachyon* stems under mild conditions of hydrolysis. Biotechnol Biofuels 1: 15.

Gordon SG, Lipps PE and Pratt RC (2006) Heritability and components of resistance to *Cercospora zeae-maydis* derived from maize inbred VO613Y. Phytopathology 96: 593–598.

Grabber JH (2005) How do lignin composition, structure, and cross-linking affect degradability? A review of cell wall model studies. Crop Sci 45: 820.

Graf A, Smith AM (2011) Starch and the clock: the dark side of plant productivity. Trends Plant Sci 16: 169–175.

Gray BN, Bougri O, Carlson AR et al. (2011) Global and grain-specific accumulation of glycoside hydrolase family 10 xylanases in transgenic maize (*Zea mays*). Plant Biotechnol J 9: 1100–1108.

Hannah LC and Nelson OE Jr (1976) Characterization of ADP-glucose pyrophosphorylase from *shrunken-2* and *brittle-2* mutants of maize. Biochem Genet 14: 547–560.

Harris K, Subudhi PK, Borrell A et al. (2007) Sorghum stay-green QTL individually reduce post-flowering drought-induced leaf senescence. J Exp Bot 58: 327–338.

Hasterok R, Draper J and Jenkins G (2004) Laying the cytotaxonomic foundations of a new model grass, *Brachypodium distachyon* (L.) Beauv. Chrom Res 12: 397–403.

Hastorf CA (2009) Rio Balsas most likely region for maize domestication. Proc Natl Acad Sci USA 106: 4957–4958.

Head G and Ward D (2009) Insect resistance in corn through biotechnology. In: Kriz AL, Larkins BA (eds) Molecular Genetic Approaches to Maize Improvement. Springer, Berlin, Heidelberg, Germany, pp 31–40.

Hedden P (2003) The genes of the Green Revolution. Trends Genet 19: 5–9.

van Heerwaarden J, Doebley J, Briggs WH et al. (2011) Genetic signals of origin, spread, and introgression in a large sample of maize landraces. Proc Natl Acad Sci USA 108: 1088–1092.

Himmel ME, Ding S-Y, Johnson DK et al. (2007) Biomass recalcitrance: engineering plants and enzymes for biofuels production. Science 315: 804–807.

Hirel B, Le Gouis J, Ney B et al. (2007) The challenge of improving nitrogen use efficiency in crop plants: Towards a more central role for genetic variability and quantitative genetics within integrated approaches. J Exp Bot 58: 2369–2387.

Ho NW, Chen Z, Brainard AP et al. (1999) Successful design and development of genetically engineered *Saccharomyces* yeasts for effective cofermentation of glucose and xylose from cellulosic biomass to fuel ethanol. Adv Biochem Engg Biotechnol 65: 163–192.

Idziak D, Betekhtin A, Wolny E et al. (2011) Painting the chromosomes of *Brachypodium*-current status and future prospects. Chromosoma 120: 469–479.

Idziak D and Hasterok R (2008) Cytogenetic evidence of nucleolar dominance in allotetraploid species of *Brachypodium*. Genome 51: 387–391.

Ihemere U, Arias-Garzon D, Lawrence S et al. (2006) Genetic modification of cassava for enhanced starch production. Plant Biotechnol J 4: 453–465.

International Brachypodium Initiative (2010) Genome sequencing and analysis of the model grass *Brachypodium distachyon*. Nature 463: 763–768.

James C (2010) A global overview of biotech (GM) crops: adoption, impact and future prospects. GM Crops 1: 8–12.

Jin M, Balan V, Gunawan C et al. (2011) Consolidated bioprocessing (CBP) performance of *Clostridium phytofermentans* on AFEX-treated corn stover for ethanol production. Biotechnol Bioengg 108: 1290–1297.

Johnson ET, Dowd PF, Liu ZL et al. (2011) Comparative transcription profiling analyses of maize reveals candidate defensive genes for seedling resistance against corn earworm. Mol Genet Genom 285: 517–525.

Jones JDG and Dangl JL (2006) The plant immune system. Nature 444: 323–329.

van Kan JAL (2006) Licensed to kill: the lifestyle of a necrotrophic plant pathogen. Trends Plant Sci 11: 247–253.

Kellogg EA, Aliscioni SS, Morrone O et al. (2009) A phylogeny of *Setaria* (Poaceae, Panicoideae, Paniceae) and related genera based on the chloroplast gene *ndhF*. Int J Plant Sci 170: 117–131.

Kellogg EA (2001) Evolutionary history of the grasses. Plant Physiol 125: 1198–1205.

Klinke HB, Thomsen AB and Ahring BK (2004) Inhibition of ethanol-producing yeast and bacteria by degradation products produced during pre-treatment of biomass. Appl Microbiol Biotechnol 66: 10–26.

Klun JA, Guthrie WD, Hallauer AR et al. (1970) Genetic nature of the concentration of 2,4-dihydroxy-7-methoxy 2H-1,4-benzoxazin-34H)-one and resistance to the European corn borer in a diallel set of eleven maize inbreds. Crop Sci 10: 87–90.

Knox JP (2008) Revealing the structural and functional diversity of plant cell walls. Curr Opin Plant Biol 11: 308–313.

Kobayashi Y and Weigel D (2007) Move on up, it's time for change—mobile signals controlling photoperiod-dependent flowering. Genes Dev 21: 2371–2384.

Koornneef M and Meinke D (2010) The development of *Arabidopsis* as a model plant. Plant J 61: 909–921.

Kou Y and Wang S (2010) Broad-spectrum and durability: understanding of quantitative disease resistance. Curr Opin Plant Biol 13: 181–185.

Krattinger SG, Lagudah ES, Spielmeyer W et al. (2009) A putative ABC transporter confers durable resistance to multiple fungal pathogens in wheat. Science 323: 1360–1363.

Krysan PJ, Young JC, Jester PJ et al. (2002) Characterization of T-DNA insertion sites in *Arabidopsis thaliana* and the implications for saturation mutagenesis. OMICS 6: 163–174.

Kumar R, Singh S and Singh OV (2008) Bioconversion of lignocellulosic biomass: biochemical and molecular perspectives. J Ind Microbiol Biotechnol 35: 377–391.

Lee M, Sharopova N, Beavis WD et al. (2002) Expanding the genetic map of maize with the intermated B73 x Mo17 (IBM) population. Plant Mol Biol 48: 453–461.

Lee SJ, Warnick TA, Pattathil S et al. (2012) Biological conversion assay using *Clostridium phytofermentans* to estimate plant feedstock quality. Biotechnol Biofuels 5: 5.

Levings CS 3rd (1990) The Texas cytoplasm of maize: cytoplasmic male sterility and disease susceptibility. Science 250: 942–947.

Levings CS 3rd and Siedow JN (1992) Molecular basis of disease susceptibility in the Texas cytoplasm of maize. Plant Mol Biol 19: 135–147.

Lewis MF, Lorenzana RE and Jung H-JG (2010) Potential for simultaneous improvement of corn grain yield and stover quality for cellulosic ethanol. Crop Sci 50: 516–523.

Li H and Yang Y (2008) Phenotypic plasticity of life history characteristics: Quantitative analysis of delayed reproduction of green foxtail (*Setaria viridis*) in the Songnen Plain of China. J Integr Plant Biol 50: 641–647.

Li P and Brutnell TP (2011) *Setaria viridis* and *Setaria italica*, model genetic systems for the Panicoid grasses. J Exp Bot 62: 3031–3037.

Li Y and Wu S (1996) Traditional maintenance and multiplication of foxtail millet (*Setaria italica* (L.) P. Beauv.) landraces in China. Euphytica 87: 33–38.

Lin Y and Tanaka S (2006) Ethanol fermentation from biomass resources: current state and prospects. Appl Microbiol Biotechnol 69: 627–642.

Long SP, Zhu X-G, Naidu SL et al. (2006) Can improvement in photosynthesis increase crop yields? Plant Cell Environ 29: 315–330.

Lopez-Casado G, Urbanowicz BR, Damasceno CMB et al. (2008) Plant glycosyl hydrolases and biofuels: a natural marriage. Curr Opin Plant Biol 11: 329–337.

Luo N, Liu J, Yu X et al. (2011) Natural variation of drought response in *Brachypodium distachyon*. Physiol Plant 141: 19–29.

Lynd LR, Grethlein HE and Wolkin RH (1989) Fermentation of cellulosic substrates in batch and continuous culture by *Clostridium thermocellum*. Appl Environ Microbiol 55: 3131–3139.

Manosalva PM, Davidson RM, Liu B et al. (2009) A germin-like protein gene family functions as a complex quantitative trait locus conferring broad-spectrum disease resistance in rice. Plant Physiol 149: 286–296.

van Maris AJA, Abbott DA, Bellissimi E et al. (2006) Alcoholic fermentation of carbon sources in biomass hydrolysates by *Saccharomyces cerevisiae*: current status. Antonie Van Leeuwenhoek 90: 391–418.

Martin T, Biruma M, Fridborg I et al. (2011) A highly conserved NB-LRR encoding gene cluster effective against *Setosphaeria turcica* in sorghum. BMC Plant Biol 11: 151.

Matsuoka M, Furbank RT, Fukayama H et al. (2001) Molecular engineering of C4 photosynthesis. Annu Rev Plant Physiol Plant Mol Biol 52: 297–314.

Matsushika A, Inoue H, Kodaki T et al. (2009) Ethanol production from xylose in engineered *Saccharomyces cerevisiae* strains: current state and perspectives. Appl Microbiol Biotechnol 84: 37–53.

McBee RH (1954) The characteristics of *Clostridium thermocellum*. J Bacteriol 67: 505–506.

McMullen MD, Frey M and Degenhardt J (2009) Genetics and biochemistry of insect resistance in maize. In: Bennetzen JL and Hake S (eds) Handbook of Maize: Its Biology. Springer, New York, NY, USA, pp 271–289.

Mendelsohn M, Kough J, Vaituzis Z et al. (2003) Are *Bt* crops safe? Nat. Biotechnol 21: 1003–1009.

Meyer FD, Talbert LE, Martin JM et al. (2007) Field evaluation of transgenic wheat expressing a modified ADP-glucose pyrophosphorylase large subunit. Crop Sci 47: 336.

Morinaka Y, Sakamoto T, Inukai Y et al. (2006) Morphological alteration caused by brassinosteroid insensitivity increases the biomass and grain production of rice. Plant Physiol 141: 924–931.

Mullet J (2009) Traits and genes for plant drought tolerance. In: Kriz AL, Larkins BA (eds) Molecular Genetic Approaches to Maize Improvement. Springer Berlin Heidelberg, Berlin, Heidelberg, Germany, pp 55–64.

Mur LAJ, Allainguillaume J, Catalán P et al. (2011) Exploiting the *Brachypodium* tool box in cereal and grass research. New Phytol 191: 334–347.

Murphy RL, Klein RR, Morishige DT et al. (2011) Coincident light and clock regulation of pseudoresponse regulator protein 37 (PRR37) controls photoperiodic flowering in sorghum. Proc Natl Acad Sci USA 108: 16469–16474.

Murray SC, Rooney WL, Mitchell SE et al. (2008) Genetic improvement of sorghum as a biofuel feedstock: II. QTL for stem and leaf structural carbohydrates. Crop Sci 48: 2180–2193.

Müssig C (2005) Brassinosteroid-promoted growth. Plant Biol (Stuttg) 7: 110–117.

Nelson T (2011) The grass leaf developmental gradient as a platform for a systems understanding of the anatomical specialization of C4 leaves. J Exp Bot 62: 3039–3048.

Nurmberg PL, Knox KA, Yun B-W et al. (2007) The developmental selector AS1 is an evolutionarily conserved regulator of the plant immune response. Proc Natl Acad Sci USA 104: 18795–18800.

Oerke E-C (2006) Crop losses to pests. J Agri Sci 144: 31–43.

Opanowicz M, Vain P, Draper J et al. (2008) *Brachypodium distachyon*: making hay with a wild grass. Trends Plant Sci 13: 172–177.

Orzechowski S (2008) Starch metabolism in leaves. Acta Biochim Pol 55: 435–445.

Pao CI and Morgan PW (1986) Genetic regulation of development in *Sorghum bicolor*: I. Role of the maturity genes. Plant Physiol 82: 575–580.

Parlevliet J (2002) Durability of resistance against fungal, bacterial and viral pathogens; present situation. Euphytica 124: 147–156.

Pauly M and Keegstra K (2008) Cell-wall carbohydrates and their modification as a resource for biofuels. Plant J 54: 559–568.

Pauly M and Keegstra K (2010) Plant cell wall polymers as precursors for biofuels. Curr Opin Plant Biol 13: 305–312.

Peng J, Carol P, Richards DE et al. (1997) The Arabidopsis *GAI* gene defines a signaling pathway that negatively regulates gibberellin responses. Genes Dev 11: 3194–3205.

Peng J, Richards DE, Hartley NM et al. (1999) "Green revolution" genes encode mutant gibberellin response modulators. Nature 400: 256–261.

Perlack RD, Wright LL, Turhollow AF et al. (2005) Biomass as feedstock for a bioenergy and bioproducts industry: The technical feasibility of a billion-ton annual supply. Oak Ridge National Lab, Oak Ridge, TN, USA.

Poland JA, Bradbury PJ, Buckler ES et al. (2011) Genome-wide nested association mapping of quantitative resistance to northern leaf blight in maize. Proc Natl Acad Sci USA 108: 6893–6898.

Pratt RC and Gordon SG (2006) Breeding for resistance to maize foliar pathogens. In: Janick J (ed) Plant Breeding Reviews. Wiley, Hoboken, NJ, USA, pp 119–173.

Propheter JL and Staggenborg S (2010) Performance of annual and perennial biofuel crops: Nutrient removal during the first two years. Agron J 102: 798.

Regierer B, Fernie AR, Springer F et al. (2002) Starch content and yield increase as a result of altering adenylate pools in transgenic plants. Nat Biotechnol 20: 1256–1260.

Reynolds MP, Hays D and Chapman S (2010) Breeding for adaptation to heat and drought stress. In: Reynolds MP (ed) Climate Change and Crop Production. CABI, Wallingford, UK, pp 71–91.

Richard TL (2010) Challenges in scaling up biofuels infrastructure. Science 329: 793–796.

Ritte G, Lloyd JR, Eckermann N et al. (2002) The starch-related R1 protein is an alpha-glucan, water dikinase. Proc Natl Acad Sci USA 99: 7166–7171.

Robinson JF, Klun JA, Guthrie WD et al. (1982) European corn borer (Lepidoptera: Pyralidae) leaf feeding resistance: Dimboa bioassays. J Kansas Entomol Soc 55: 357–364.

Sage RF and Zhu X-G (2011) Exploiting the engine of C4 photosynthesis. J Exp Bot 62: 2989–3000.

Sakamoto T, Morinaka Y, Ohnishi T et al. (2006) Erect leaves caused by brassinosteroid deficiency increase biomass production and grain yield in rice. Nat Biotechnol 24: 105–109.

Sakulsingharoj C, Choi S-B, Hwang S-K et al. (2004) Engineering starch biosynthesis for increasing rice seed weight: the role of the cytoplasmic ADP-glucose pyrophosphorylase. Plant Sci 167: 1323–1333.

Salas Fernandez MG, Becraft PW, Yin Y et al. (2009) From dwarves to giants? Plant height manipulation for biomass yield. Trends Plant Sci 14: 454–461.

Salvi S, Sponza G, Morgante M et al. (2007) Conserved noncoding genomic sequences associated with a flowering-time quantitative trait locus in maize. Proc Natl Acad Sci USA 104: 11376–11381.

Sanchez AC, Subudhi PK, Rosenow DT et al. (2002) Mapping QTLs associated with drought resistance in sorghum (*Sorghum bicolor* L. Moench). Plant Mol Biol 48: 713–726.

Sánchez GJJ, De La Cruz LL, Vidal M. VA et al. (2011) Three new teosintes (*Zea* Spp., Poaceae) from México. Am J Bot 98: 1537–1548.

Sánchez ÓJ and Cardona CA (2008) Trends in biotechnological production of fuel ethanol from different feedstocks. Bioresour Technol 99: 5270–5295.

Sang T (2011) Toward the domestication of lignocellulosic energy crops: learning from food crop domestication. J Integr Plant Biol 53: 96–104.

Santa-Maria MC, Yencho CG, Haigler CH et al. (2011) Starch self-processing in transgenic sweet potato roots expressing a hyperthermophilic α-amylase. Biotechnol Prog 27: 351–359.

Sasaki A, Ashikari M, Ueguchi-Tanaka M et al. (2002) Green revolution: A mutant gibberellin-synthesis gene in rice. Nature 416: 701–702.

Schnable PS, Ware D, Fulton RS et al. (2009) The B73 maize genome: Complexity, diversity, and dynamics. Science 326: 1112–1115.

Schwartz CJ, Doyle MR, Manzaneda AJ et al. (2010) Natural variation of flowering time and vernalization responsiveness in *Brachypodium distachyon*. BioEnergy Res 3: 38–46.

Silverstone AL, Ciampaglio CN and Sun T (1998) The Arabidopsis *RGA* gene encodes a transcriptional regulator repressing the gibberellin signal transduction pathway. Plant Cell 10: 155–169.

Smith AM (2008) Prospects for increasing starch and sucrose yields for bioethanol production. Plant J 54: 546–558.

Smith SM and Hulbert SH (2005) Recombination events generating a novel Rp1 race specificity. Mol Plant-Microbe Interact 18: 220–228.

Spielmeyer W, Ellis MH and Chandler PM (2002) Semidwarf (sd-1) "Green Revolution" rice contains a defective gibberellin 20-Oxidase gene. Proc Natl Acad Sci USA 99: 9043–9048.

Stark DM, Timmerman KP, Barry GF et al. (1992) Regulation of the amount of starch in plant tissues by ADP glucose pyrophosphorylase. Science 258: 287–292.

Stettler M, Eicke S, Mettler T et al. (2009) Blocking the metabolism of starch breakdown products in *Arabidopsis* leaves triggers chloroplast degradation. Mol Plant 2: 1233–1246.

Studer MH, DeMartini JD, Davis MF et al. (2011) Lignin content in natural *Populus* variants affects sugar release. Proc Natl Acad Sci USA 108: 6300–6305.

Sun T-P (2011) The molecular mechanism and evolution of the GA-GID1-DELLA signaling module in plants. Curr Biol 21: R338–345.

Taylor LE 2nd, Dai Z, Decker SR et al. (2008) Heterologous expression of glycosyl hydrolases *in planta*: a new departure for biofuels. Trends Biotechnol 26: 413–424.

Le Thierry d'Ennequin M, Panaud O et al. (2000) Assessment of genetic relationships between *Setaria italica* and its wild relative *S. viridis* using AFLP markers. Theor Appl Genet 100: 1061–1066.

Thole V, Alves SC, Worland B et al. (2009) A protocol for efficiently retrieving and characterizing flanking sequence tags (FSTs) in *Brachypodium distachyon* T-DNA insertional mutants. Nat Protoc 4: 650–661.

Thole V, Vain P (2012) *Agrobacterium*-mediated transformation of *Brachypodium distachyon*. In: Dunwell JM, Wetten AC (eds) Transgenic Plants, Methods in Molecular Biology. Humana Press, New York, NY, USA, pp 137–149.

Till-Bottraud I, Reboud X, Brabant P et al. (1992) Outcrossing and hybridization in wild and cultivated foxtail millets: consequences for the release of transgenic crops. Theor Appl Genet 83: 940–946.

Tondelli A, Francia E, Barabaschi D et al. (2011) Inside the *CBF* locus in Poaceae. Plant Sci 180: 39–45.

Travella S, Ross SM, Harden J et al. (2005) A comparison of transgenic barley lines produced by particle bombardment and *Agrobacterium*-mediated techniques. Plant Cell Rep 23: 780–789.

Turner A, Beales J, Faure S et al. (2005) The pseudo-response regulator Ppd-H1 provides adaptation to photoperiod in barley. Science 310: 1031–1034.

Turner RE, Rabalais NN and Justic D (2008) Gulf of Mexico hypoxia: Alternate states and a legacy. Environ Sci Technol 42: 2323–2327.

Tyler L, Bragg JN, Wu J et al. (2010) Annotation and comparative analysis of the glycoside hydrolase genes in *Brachypodium distachyon*. BMC Genomics 11: 600.

Ueguchi-Tanaka M, Ashikari M et al. (2005) *Gibberellin Insensitive Dwarf1* encodes a soluble receptor for gibberellin. Nature 437: 693–698.

Umezawa T, Fujita M, Fujita Y et al. (2006) Engineering drought tolerance in plants: discovering and tailoring genes to unlock the future. Curr Opin Biotechnol 17: 113–122.

US Dept. of Energy (2006) Breaking the biological barriers to cellulosic ethanol: A joint research agenda, A research roadmap resulting from the Biomass to Biofuels Workshop sponsored by the US Department of Energy, December 7-9, 2005, Rockville, MD, USA.

Vain P, Worland B, Thole V et al. (2008) *Agrobacterium*-mediated transformation of the temperate grass *Brachypodium distachyon* (genotype Bd21) for T-DNA insertional mutagenesis. Plant Biotechnol J 6: 236–245.

Vasal SK, Riera-Lizarazu O, Jauhar PP et al. (2006) Genetic enhancement of maize by cytogenetic manipulation, and breeding for yield, stress tolerance, and high protein quality. In: Singh RJ, Jauhar PP (eds) Genetic Resources, Chromosome Engineering, and Crop Improvement: Cereals. CRC Press, Boca Raton, FL, USA, pp 159–197.

Vermerris W (2011) Survey of genomics approaches to improve bioenergy traits in maize, sorghum and sugarcane. J Integr Plant Biol 53: 105–119.

Vicentini A, Barber JC, Aliscioni SA et al. (2008) The age of the grasses and clusters of origins of C4 photosynthesis. Glob Change Biol 14: 2963–2977.

Vogel J (2008) Unique aspects of the grass cell wall. Curr Opin Plant Biol 11: 301–307.

Vogel J and Bragg J (2009) *Brachypodium distachyon*, a new model for the Triticeae. In: Muehlbauer GJ and Feuillet C (eds) Genetics and Genomics of the Triticeae. Springer, New York, NY, USA, pp 427–449.

Vogel J and Hill T (2008) High-efficiency *Agrobacterium*-mediated transformation of *Brachypodium distachyon* inbred line Bd21-3. Plant Cell Rep 27: 471–478.

Vogel JP, Garvin DF, Leong OM et al. (2006) *Agrobacterium*-mediated transformation and inbred line development in the model grass *Brachypodium distachyon*. Plant Cell Tiss Org Cult 84: 100179–100191.

Vogel JP, Tuna M, Budak H et al. (2009) Development of SSR markers and analysis of diversity in Turkish populations of *Brachypodium distachyon*. BMC Plant Biol 9: 88.

Wang C, Chen Y, Ku L et al. (2010a) Mapping QTL associated with photoperiod sensitivity and assessing the importance of QTL×environment interaction for flowering time in maize. PLoS ONE 5: e14068.

Wang C, Chen J, Zhi H et al. (2010b) Population genetics of foxtail millet and its wild ancestor. BMC Genet 11: 90.

Wang J-W, Schwab R, Czech B et al. (2008) Dual effects of miR156-targeted *SPL* genes and CYP78A5/KLUH on plastochron length and organ size in *Arabidopsis thaliana*. Plant Cell 20: 1231–1243.

Wang J-W, Czech B and Weigel D (2009) miR156-regulated SPL transcription factors define an endogenous flowering pathway in *Arabidopsis thaliana*. Cell 138: 738–749.

Wang L, Peterson RB and Brutnell TP (2011) Regulatory mechanisms underlying C4 photosynthesis. New Phytol 190: 9–20.

Wang Z, Chen X, Wang J et al. (2007) Increasing maize seed weight by enhancing the cytoplasmic ADP-glucose pyrophosphorylase activity in transgenic maize plants. Plant Cell Tiss Org Cult 88: 83–92.

Wang ZM, Devos KM, Liu CJ et al. (1998) Construction of RFLP-based maps of foxtail millet, *Setaria italica* (L.) P. Beauv. Theor Appl Genet 96: 31–36.

Ward JMJ, Stromberg EL, Nowell DC et al. (1999) Gray leaf spot: A disease of global importance in maize production. Plant Dis 83: 884–895.

Warnick TA, Methé BA and Leschine SB (2002) *Clostridium phytofermentans* sp. nov., a cellulolytic mesophile from forest soil. Int J Syst Evol Microbiol 52: 1155–1160.

Weigel D and Glazebrook J (eds) (2002) *Arabidopsis*: A Laboratory Manual. Cold Spring Harbor Laboratory Press, Cold Spring Harbor, NY, USA, 343 pp.

Weigel D and Mott R (2009) The 1001 Genomes Project for *Arabidopsis thaliana*. Genome Biol 10: 107.

Welz HG and Geiger HH (2000) Genes for resistance to northern corn leaf blight in diverse maize populations. Plant Breed 119: 1–14.

West T, Dunphy-Guzman K, Sun A et al. (2009) Feasibility, economics, and environmental impact of producing 90 billion gallons of ethanol per year by 2030. US Department of Energy Publications, Lincoln, NE, USA, Paper 86, pp 1–30.

Westerbergh A, Doebley J (2004) Quantitative trait loci controlling phenotypes related to the perennial versus annual habit in wild relatives of maize. Theor Appl Genet 109: 1544–1553.

Wisser RJ, Kolkman JM, Patzoldt ME et al. (2011) Multivariate analysis of maize disease resistances suggests a pleiotropic genetic basis and implicates a GST gene. Proc Natl Acad Sci USA 108: 7339–7344.

Wisser RJ, Balint-Kurti PJ and Nelson RJ (2006) The genetic architecture of disease resistance in maize: A synthesis of published studies. Phytopathol 96: 120–129.

Wyman CE, Dale BE, Elander RT et al. (2005) Coordinated development of leading biomass pretreatment technologies. Bioresour Technol 96: 1959–1966.

Wyman CE (2007) What is (and is not) vital to advancing cellulosic ethanol. Trends Biotechnol 25: 153–157.

Yamaguchi-Shinozaki K and Shinozaki K (2005) Organization of cis-acting regulatory elements in osmotic- and cold-stress-responsive promoters. Trends Plant Sci 10: 88–94.

Yamamoto K, Sakamoto H and Momonoki YS (2011) Maize acetylcholinesterase is a positive regulator of heat tolerance in plants. J Plant Physiol 168: 1987–1992.

Yordem BK, Conte SS, Ma JF et al. (2011) *Brachypodium distachyon* as a new model system for understanding iron homeostasis in grasses: phylogenetic and expression analysis of Yellow Stripe-Like (YSL) transporters. Ann Bot 108: 821–833.

Yu J, Holland JB, McMullen MD et al. (2008) Genetic design and statistical power of nested association mapping in maize. Genetics 178: 539–551.

Yu TS, Kofler H, Häusler RE et al. (2001) The Arabidopsis *sex1* mutant is defective in the R1 protein, a general regulator of starch degradation in plants, and not in the chloroplast hexose transporter. Plant Cell 13: 1907–1918.

Yuan JS, Tiller KH, Al-Ahmad H et al. (2008) Plants to power: bioenergy to fuel the future. Trends Plant Sci 13: 421–429.

Zeeman SC, Kossmann J and Smith AM (2010) Starch: its metabolism, evolution, and biotechnological modification in plants. Annu Rev Plant Biol 61: 209–234.

Zhang F, Mackenzie AF and Smith DL (1993) Corn yield and shifts among corn quality constituents following application of different nitrogen fertilizer sources at several times during corn development. J Plant Nutr 16: 1317–1337.

Zhang N, Gibon Y, Gur A et al. (2010) Fine quantitative trait loci mapping of carbon and nitrogen metabolism enzyme activities and seedling biomass in the maize IBM mapping population. Plant Physiol 154: 1753–1765.

Zhang Q, Pettolino FA, Dhugga KS et al. (2011a) Cell wall modifications in maize pulvini in response to gravitational stress. Plant Physiol 156: 2155–2171.

Zhang S, Xu Y and Hanna MA (2011b) Pretreatment of corn stover with twin-screw extrusion followed by enzymatic saccharification. Appl Biochem Biotechnol 166: 458-469.

Zhang Y, Culhaoglu T, Pollet B et al. (2011c) Impact of lignin structure and cell wall reticulation on maize cell wall degradability. J Agri Food Chem 59: 10129–10135.

Zhong R, McCarthy RL, Lee C et al. (2011) Dissection of the transcriptional program regulating secondary wall biosynthesis during wood formation in poplar. Plant Physiol 157: 1452–1468.

Zhu X-G, Long SP and Ort DR (2010) Improving photosynthetic efficiency for greater yield. Annu Rev Plant Biol 61: 235–261.

Zwonitzer JC, Coles ND, Krakowsky MD et al. (2010) Mapping resistance quantitative trait loci for three foliar diseases in a maize recombinant inbred line population-evidence for multiple disease resistance? Phytopathol 100: 72–79.

CHAPTER 5

Molecular Genetics of Bioenergy Traits

Michael G. Muszynski[1], and *Marna D. Yandeau-Nelson[2]*

ABSTRACT

The long history of maize as a molecular genetic system positions this cereal species as a central crop for bioenergy utilization. Owing to the extensive list of molecular genetic tools and genomic technologies that have been and continue to be applied to deciphering the regulation of key biochemical, physiological and developmental pathways, we are now able to target select bioenergy traits for modulation, using state-of-the-art genetic engineering techniques. Examples of manipulating lignin accumulation, endosperm starch composition and even plant architecture are showing promise for optimizing maize for bioenergy production. Moreover, beyond its use for bioenergy production itself, maize is also a leading model molecular genetic system for which knowledge from maize can be applied to other species, from other grasses to algae, for bioenergy use improvement. Focused efforts to further illuminate the molecular control of bioenergy traits in maize portend a bright future for this important grain crop as part of our national energy supply.

Keywords: lignin, cellulose, plant architecture, forward genetics, reverse genetics, TILLING, Uniform *Mu*, *Ac/Ds*, RNAi, targeted mutagenesis

---

[1]Department of Genetics, Development and Cell Biology, Iowa State University, Ames, IA 50011, USA.
[2]Department of Biochemistry, Biophysics and Molecular Biology, Iowa State University, Ames, IA 50011, USA.
*Corresponding author: mgmuszyn@iastate.edu

Introduction

The history of maize as a genetic system began more than 100 years ago with its use as a model to study cytogenetics and to link gene function to a chromosomal locus (Rhoades 1984). Since that time, maize has matured into a model system with a sequenced genome and well equipped arsenal of genetic, molecular and genomic tools. In this chapter, we describe the molecular genetic tools that are currently used to identify genes, characterize gene function and connect functions to pathways that underlie bioenergy traits. We also outline some strategies being employed to manipulate development, biochemistry and physiology that can lead to improvement in bioenergy traits in maize and related grass species.

The Road Ahead: Pathways and Components Underlying Bioenergy Traits

A number of biochemical, physiological and developmental pathways influence grain and biomass composition and yield. Since grain and biomass (i.e., stover) are the raw materials used for bioenergy production, many studies have been devoted to understanding the molecular determinants affecting various grain and biomass traits. Significant efforts have been applied to understand the regulation of (1) the biosynthesis of the major constituents of the cell wall, namely, lignin and cellulose; (2) carbohydrate biosynthesis, storage and turnover in the maize endosperm; and (3) vegetative growth affecting plant architecture and biomass. In the following sections, we summarize the current progress in our understanding of the molecular mechanisms affecting each of these bioenergy traits in maize.

The Cell Wall—Lignin

Cellulose and lignin are the two most abundant biopolymers in nature, with lignin accounting for about 30% of the terrestrial carbon fixed in the biosphere every year (Boerjan et al. 2003). Lignin is a major component of secondary plant cell walls and is crucial for the structural stiffness and strength of the cell wall. Because of its direct impacts on crop use in agriculture and industry, the biochemistry of lignin biosynthesis has been intensively studied for more than a century. Over the last decade there have been rapid advances in our understanding of the enzymatic pathways leading to lignin biosynthesis and our ability to manipulate these pathways through genetic engineering (Li et al. 2008). Lignin polymers are primarily composed of three types of hydroxycinnamyl alcohol monomers or monolignols; these are *p*-coumaryl, coniferyl and sinapyl alcohols.

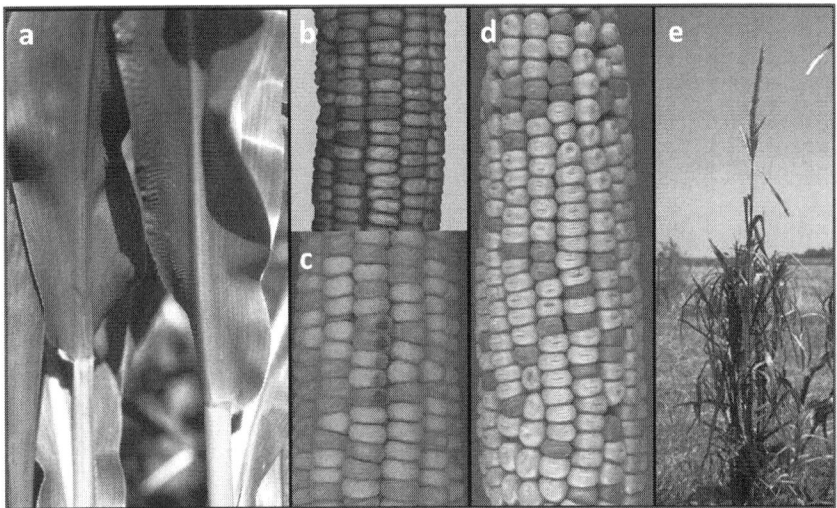

Figure 1 Phenotypes of maize mutants important for bioenergy traits. (a) The *bm1* mutant (left) affects lignin accumulation and shows pigment depostion in the leaf midrib and parts of the sheath comapred to a wild type sibling (right); (b) segregation of *ae* kernels on an otherwise homozygous *wx1* ear showing their interaction on starch composition (translucent kernels); (c) segregation of opaque appearing *wx1* kernels on a test cross ear, cut surfaces show differential iodine staining; (d) selfed ear segregating *du1* glassy appearing kernels; (e) heterozygous *Cg1* plant displaying narrow leaves and tiller proliferation. Images in a and b courtesy of Cold Spring Harbor Laboratory Press (Mutants of Maize 1997). All images are avaiable on the MaizeGDB website.

Color image of this figure appears in the color plate section at the end of the book.

When these monolignols are polymerized, they form *p*-hydroxyphenyl (H), guaiacyl (G) and syringyl (S) lignin. Lignins found in grasses, like maize, incorporate relatively equivalent amounts of G and S units but more H units than dicots (Boerjan et al. 2003). Monolignol biosynthesis occurs via the phenylpropanoid pathway, starting with the deamination of phenylalanine to form cinnamic acid followed by a series of ring hydroxylations, *O*-methylations and side-chain modifications.

Detailed reviews are available describing the core biosynthetic pathway of the lignin polymer and the phenotypic consequences of disruption of specific steps in the pathway (Boerjan et al. 2003; Bonawitz and Chapple 2010). Here we focus on what is known about this pathway in maize. Current research indicates that about 10 enzymes participate in monolignol biosynthesis and mutations in some of these enzymes lead to reduced lignin concentration or altered lignin composition. Recessive mutations affecting lignin content and composition in maize were first identified by the appearance of a characteristic reddish-brown pigment accumulating in the mid-vein or midrib of adult phase leaves. Because of this diagnostic pigment accumulation, these mutants were named *brown mid-rib (bm)* mutants (Jorgenson 1931; Kuc and Nelson 1964). Similar *bm* mutants have been identified in other grasses and several have been shown to have increased digestibility when used for forage or silage and a few were shown to significantly increase the conversion rate of biomass to biofuel by increasing the accessibility of cell wall polymers to hydrolytic enzymes during biofuel production (Frontini et al. 2005; Dien et al. 2009). Six *brown midrib* loci (*bm1* to *bm6*) have been genetically identified and studied in maize (Sattler et al. 2010; Ali et al. 2010a). All six mutants have the characteristic reddish-brown pigment accumulating in the midrib and the pith of the stalk. Each *bm* mutant also shows either a reduction in lignin accumulation or alteration in the composition of lignin. Some *bm* mutants have been reported to have reduced biomass yield, reduced stalk strength and increased susceptibility to disease to different degrees (Sattler et al. 2010). Whether these negative pleiotropic effects are directly caused by the *bm* mutation or more significantly influenced by genetic background is still an open question.

The *bm1* mutation maps to the short arm of chromosome 5 (GRMZM5G844562) and affects the accumulation of cinnamyl alcohol dehydrogenase2 (*ZmCAD2*) that catalyzes the last reduction of hydroxyl-cinnamaldehydes to their corresponding alcohols (Mansell et al. 1974). Initially the *bm1* mutation was thought to control the expression of the *ZmCAD2* gene, as expression profiling of near-isogenic *bm1* mutants and wild type sibling lines indicated five CAD genes and several other genes were differentially expressed (Guillaumie et al. 2007). However, *bm1* was later shown to encode the *ZmCAD2* gene using reverse genetics. Two independent *Mutator (Mu)* transposon insertion mutations in *ZmCAD2* were identified, which resulted in a brown midrib phenotype and also failed to complement the *bm1-ref* allele (Cao 2007). Sequencing *ZmCAD2* in the *bm1-ref* mutant identified a 2-bp insertion in exon 2 leading to a frame shift mutation. Taken together, these results prove *bm1* encodes *ZmCAD2* (Cao 2007). The *bm3* mutation maps to the short arm of chromosome 4 (AC196475.3_FG004) and encodes caffeic acid *O*-methyltransferase (COMT), which forms the precursor to S-lignin (Vignols et al. 1995). As expected, *bm3* mutants have greatly reduced S-lignin composition with only minor reductions in H- and G-lignin (Marita et al. 2003). Surprisingly, a novel lignin monomer, 5-hydroxy guaiacyl, was found at significant levels in the *bm3* mutant. A direct regulator of COMT expression was identified as the R2R3-MYB transcription factor *ZmMYB31* (Fornalé et al. 2010). *ZmMYB31* was shown to function as a negative regulator of the *ZmF5H* gene, which is involved in lignin biosynthesis. Overexpression of *ZmMYB31* in *Arabidopsis* indicated this negative regulator reduced lignin content without altering lignin composition (Fornalé et al. 2010). The *bm2* mutant maps to the bottom of the long arm of chromosome 1 within a 1.3 Mb region. Lignin analysis of *bm2* mutants indicates this mutation alters the tissue-specific patterns of lignin deposition but the underlying gene has not yet been cloned (Vermerris and Boon 2000). The *bm4* mutation maps to the bottom of the long arm of chromosome 9 within a 152-kb region based on flanking markers. This locus has also not been cloned and *bm4* mutants have only modest changes in lignin composition (Marita et al. 2003; Guillaumie et al. 2007). Little is known regarding the *bm5* and *bm6* loci as they have only recently been identified as non-allelic to the other four *bm* loci and have yet to be mapped (Ali et al. 2010a). Preliminary tests of biomass conversion efficiencies of stover from isogenic *bm1* and *bm3* lines showed significant improvement over wild type controls indicating targeted modifications to lignin content and lignin subunit composition holds promise to enhance maize as a bioenergy crop (Vermerris et al. 2007).

The Cell Wall—Cellulose

Cellulose is the most common biopolymer on earth and is the main structural component of the primary cell wall of green plants. Therefore, this unbranched polymer of β-linked glucose residues arranged in linear chains is a primary feedstock for bioenergy production. Chapter 2 in this book discusses the structure and composition of cellulose in detail and thus in this chapter we describe what is known about the genes controlling cellulose biosynthesis in maize. Evidence from bacterial and plant species supports the idea that the catalytic subunit of the cellulose synthase complex is encoded by a family of *cellulose synthaseA* (*CeSA*) genes, which function as β-glycosyltransferases (Delmer 1999). Using sequence homology searches of maize expressed sequence tag (EST) libraries, 12 maize *CeSA* genes (*ZmCesA1–12*) were identified (Holland et al. 2000; Appenzeller et al. 2004). Expression and phylogenetic analyses suggests that *ZmCesA1–9* are involved in primary cell wall formation while *ZmCesA10–12* are involved in secondary cell wall formation (Appenzeller et al. 2004). Consistent with this idea, the *ZmCesA10–12* genes are highly expressed in stalk tissue, which is rich in secondary cell walls. Proteins having high amino acid similarity to CesA proteins, known as the cellulose synthase-like (CSL) proteins, have been found in all plant species examined. The *CSL* genes exist as a superfamily, which is divided into several subfamilies. To date, maize has 41 *CSL* genes residing in subfamilies *ZmCSLA* (11), *ZmCSLC* (12), *ZmCSLD* (5), *ZmCSLE* (4), *ZmCSLF* (7), *ZmCSLH* (1) and *ZmCSLG* (1). Whether all these are functional or some are nonfunctional pseudogenes has not been fully determined (van Erp and Walton 2009).

Transcript profiling has helped determine the expression patterns of genes regulating lignin and cellulose accumulation and also to identify novel genes affecting cell wall biogenesis. A maize cell wall database was constructed, named MAIZEWALL, that allows for the query of organ-specific expression patterns of cell wall-related genes in maize (Guillaumie et al. 2007). Over 700 maize cell wall-related genes populating MAIZEWALL were identified by homology and keyword searches of existing databases for known cell wall regulating genes from other species. A macroarray, constructed using the 3' UTR of about 650 of these cell wall related genes, was used to determine expression patterns in roots, leaves and young stems. In addition, potential preferential routes for lignin biosynthesis in different organs were identified through an in depth transcriptome analysis of the gene families encoding enzymes of the lignin biosynthetic pathway. Transcript analysis of developing organs (e.g., elongating and nonelongating maize hypocotyls) or between cell wall mutants and their wild type sibs using different microarray platforms has also provided new insights into the dynamics of cell wall biogenesis (Bosch et al. 2011; Tamasloukht et al. 2011). With the decrease in cost and rapid increase in sequencing depth, open-ended RNA transcriptome sequencing platforms (e.g., Illumina RNA-Seq) are predicted to identify the interacting molecular networks underlying cell wall biology. A more detailed discussion of various "omics" platforms is found in Chapter 7 of this volume.

Endosperm Starch

Genes affecting the cell walls of maize are one target for improved bioenergy traits but genes affecting carbohydrate biosynthesis, storage and turnover in the maize endosperm offer additional targets for improving maize as a bioenergy crop. A limiting step in using starch for bioethanol production is that this complex carbohydrate must be digested into simpler monosaccharides by amylases and glucoamylases prior to fermentation by microbes or fungi to produce bioethanol. This step requires extended digestion times at high temperatures to produce fermentable sugars. Starches that can be digested at lower temperatures or at a faster rate would enhance bioethanol production. Our understanding of the genetics and biochemistry of starch synthesis in crops, particularly in maize, has greatly improved in the past decade. For a detailed review the reader is directed to a recent review by Keeling and Myers (2010). Below, we focus on a few examples where the composition of starch in the maize endosperm has been manipulated that may lead to improved bioethanol production.

Maize endosperm starch is composed of both amylose and amylopectin and the ratio is controlled by the action of several genes. The wildtype maize *waxy* (*wx*) gene encodes granule-bound starch synthase I (*GBSSI*) and when nonfunctional results in endosperm starch having a higher amylopectin:amylose ratio (Nelson and Rines 1962). The *dull1* (*du1*) gene encodes starch synthase II (*SSII*) and *du1* mutants have altered starch composition and branching properties (Gao et al. 1998). The *amylose extender* (*ae*) gene encodes starch branching enzyme IIb (*SBEIIb*) and *ae* mutants have higher amylose content and less highly branched amylopectin (Stinard et al. 1993; Kim et al. 1998). These starch composition mutants have been tested singly and in combination to assess if they are more easily hydrolyzed prior to fermentation to produce ethanol (Adams et al. 2011). The high amylopectin *wx* mutant was shown to have the highest ethanol yields and the lowest proportion of residual starch after fermentation using two commercial amylase/glucoamylase enzyme blends. The *ae* mutant has higher amylose content and forms digestion-resistant starch granules that resulted in low ethanol yields and higher residual starch post-fermentation (Adams et al. 2011). Thus, manipulation of starch composition towards more amylopectin may increase the efficiency of starch hydrolysis in the production of ethanol. Transcriptional analysis of double mutants lacking both *wx* and *ae* functions were shown to favor the flux of excess carbohydrates into glycolysis, the pentose phosphate pathway, and cell wall biosynthesis, but not toward the synthesis of alternative starch storage compounds (Li et al. 2010). Therefore, comprehensive analysis of starch composition mutants that increase or decrease ethanol yields may provide insights into which chemical or structural properties to modify in endosperm starch synthesis for enhanced bioethanol production.

As an alternative to testing known starch synthesis mutants, high throughput screens for novel mutants with altered starch digestibility have been done (Groth et al. 2008). Using an ethyl methanesulfonate (EMS) mutagenized maize population, almost 500 M_3 families were tested using a novel single-kernel starch digestibility assay. Forty three mutant lines were identified that segregated single kernels that had more rapid digestion of gelatinized starch and higher yields of fermentable sugars. These included mutants exhibiting dominant, semi-dominant and recessive gene action. Twenty-two additional mutants were found that digested uncooked starch more rapidly, suggesting there are distinct classes of mutants with rapidly digestible starch and that starch digestibility is to a large extent under genetic control (Groth et al. 2008). Given this analysis was performed on only 1/10th of the total EMS population available (from the Maize TILLING Project, see below) the authors estimate there may be more than 200 genes in the maize genome affecting starch digestibility, which offers numerous targets to test for improved bioethanol production.

Plant Architecture

Another target to modulate for improved bioenergy traits is plant architecture. Since cellulosic biofuel production comes primarily from vegetative tissues, manipulation of developmental regulators that increase vegetative growth may lead to a higher production of plant biomass in maize. Numerous genetic studies have identified key regulators of processes such as vegetative meristem activity, flowering time, cell elongation, photosynthetic efficiency, and secondary wall biosynthesis, which affect plant biomass production by altering plant architecture (Jakob et al. 2009; Demura and Ye 2010). Important biomass determinants including plant height, flowering time, tiller number per plant and stem thickness are regulated by complex genetic networks. However, a recent study indicates modulating the expression of a single microRNA gene has the potential to alter all these traits for improved bioenergy yield. Plants switch from a juvenile to an adult phase of development prior to flowering (Poethig 2010). In maize and other plants, juvenile phase tissues possess traits making them more amenable to conversion to fermentable sugars and thus are more desirable for bioenergy conversion. The maize *Corngrass1* (*Cg1*) gene encodes an unusual tandem microRNA belonging to the miR156 class of microRNAs (Chuck et al. 2007). In the dominant *Cg1* maize mutant, prolonged expression of miR156 promotes juvenile tissue identity, delays flowering and increases tiller number. Overexpression of the maize *Cg1* microRNA in several plant species, including the biofuel feedstock switchgrass (*Panicum virgatum*), prevented flowering, thereby locking the plants into a persistent vegetative state, and increased tiller number thus increasing biomass accumulation (Chuck et al. 2011). *Cg1* overexpression in switchgrass caused reduced lignin accumulation and increased starch accumulation in stems, leading to increased digestibility and glucose release. Altogether, plant biomass, lignin accumulation and starch content were altered by overexpressing a single regulatory gene leading to significant enhancements in bioenergy production. Altering plant architecture has the potential to also modify plant biochemical composition, thereby collectively improving several bioenergy traits.

An alternative to altering plant architecture by transgene overexpression is to exploit maize germplasm diversity. For example, tropical maize is photoperiod sensitive and requires short days to flower and set seed (Colasanti and Muszynski 2009). When tropical lines are grown at temperate latitudes, flowering is delayed resulting in increased vegetative growth and biomass accumulation. Tropical maize grown in temperate environments typically do not produce grain and thus photoassimilates accumulate in the stem as high amounts of extractable sucrose, glucose and fructose. Thus tropical maize has the potential to be used both as a source of sugar, similar to how sweet sorghum or sugar cane are used as feedstocks, and as a lignocellulosic bioenergy feedstock (White et al. 2011). Altering plant architecture either by transgenesis or the novel utilization of maize germplasm diversity presents new ways to enhance maize as a bioenergy crop and represents an underutilized strategy that may hold great promise.

Put it in Drive: Gene Isolation using Forward Genetics

Much of our understanding regarding the molecular pathways underlying bioenergy traits in maize and other species stems from the isolation and characterization of individual genes. Gene isolation in maize is typically performed using either forward or reverse genetic strategies. In this section, we describe forward genetic gene isolation methods and provide examples of screens applied to isolating genes relevant to bioenergy traits.

Forward genetic screens rely on identifying a mutant phenotype of interest and then using map-based cloning to molecularly isolate the underlying gene. Any developmental, biochemical or physiological process can be targeted for mutagenesis as long as the desired mutant phenotype can be "detected", either visibly, chemically or by some alternate methodology. Maize is amenable to both chemical and transposable element (TE) mutagenesis. Detailed methods exist for EMS mutagenesis and screening in maize, which often has a higher forward mutation rate than mutagenesis with TEs (Neuffer 1994; Candela and Hake 2008). EMS mutagenesis is also favored for mutagenesis within a specific inbred background or for performing suppressor/enhancer screens for genes that interact within a pathway of interest (Candela and Hake 2008). Many EMS induced mutants with visible phenotypes in defined inbred lines can be queried through the Maize Genetics and Genomics Database (MaizeGDB, http://www.maizegdb.org/) and can be obtained through the Maize Genetics Cooperation Stock Center (MGCSC, http://maizecoop.cropsci.uiuc.edu/). Although primarily constructed for use in reverse genetic screens (described in the following section), the maize TILLING (targeting induced local lesions in genomes) project has developed two large EMS mutagenized populations in the W22 and B73 inbred backgrounds that are also useful for forward genetic screens (Till et al. 2004; Weil and Monde 2007). As described above for endosperm starch, this population was screened with a single-kernel starch digestibility assay and mutations resulting in more rapid starch digestion were identified (Groth et al. 2008).

Once mutants are identified and their inheritance confirmed, the underlying gene is isolated by map-based, also known as positional, cloning techniques. Positional cloning has long been a powerful method to map and clone genes in species with small genomes. With the sequencing of the maize genome and a profusion of genetic maps and markers available, positional cloning is now feasible and commonplace in maize (Bortiri et al. 2006). In short, a polymorphic testcross or F_2 mapping population is made in which the new mutant segregates and DNA isolated from wild type and mutant individuals is assayed for cosegregation of markers linked to the mutant phenotype. A recombination interval is defined where the mutated locus resides and reiterative mapping of larger-sized populations with more closely linked markers refines the recombination interval to include a few or only one candidate gene. Maize genes have been cloned using populations as small as a few hundred individuals or have required populations as large as several thousand individuals (Bortiri et al. 2006). The population size often depends on the amount of repetitive sequence and the gene density in the chromosomal region where the mutated gene is located. Genes located near telomeres reside in gene rich regions and genes located near centromeres reside in gene poor regions, making positional cloning of genes near centromeres more laborious. Once a candidate gene has been identified, the gene is sequenced from both the progenitor inbred and the mutant line and the causative DNA lesion is determined. Having several independently derived mutant alleles is essential to proving the correct gene sequence has been identified.

Forward screens can also be performed using TE-mutagenized populations and several techniques exist for cloning genes "tagged" by TEs (Frey et al. 1998; Brutnell 2002; Settles 2009). These populations make use of endogenous maize transposable element systems and were developed largely to be used in reverse genetic screens but have also been successfully exploited in forward genetic screens (McCarty and Meeley 2009). The most widely used TE populations make use of either the *Mutator* or *Activator/ Dissociaton (Ac/Ds)* family of elements (McCarty et al. 2005; Vollbrecht et al. 2010). These populations are described in more detail in the following section on reverse genetic tools. One *Mutator* TE population, *UniformMu*, was used in a forward screen using NIR spectroscopy of mature leaf blade tissue to identify cell wall mutants (Vermerris et al. 2007; Penning et al. 2009). Out of 2,200 F_2 families screened, about 40 maize mutants with altered spectrotypes were identified but otherwise have no visible mutant phenotypes (details about these mutants can be searched at http://cellwall. genomics.purdue.edu). It is expected that a subset of these mutants will have enhanced biomass conversion properties due to alterations in the composition of their cell walls (Carpita and McCann 2008).

Put it into Reverse: Reverse Genetic Tools for Maize Crop Improvement

The second method used to isolate a gene with a defined function is reverse genetics. Reverse genetic studies focus on inducing a mutation and an associated mutant phenotype in a gene of unknown function. The maize research community has committed much time and effort to developing a suite of nontransgenic reverse genetic resources, including different transposon-tagged mutant and point mutant collections. These collections have been generated via the combination of traditional mutagenesis techniques coupled with high throughput mutation detection. Importantly, many of these reverse genetic resources have been built in the same progenitor background (i.e., inbred line W22), allowing for direct comparison of mutants identified from the different collections. Each of the well-developed resources described below can be or already is being employed for the purpose of studying bioenergy related traits.

TILLING

Unlike transposon-tagged alleles, which often cause loss-of-function phenotypes, point mutations induced by chemical mutagens can confer allelic series of both nonsense (i.e., introduction of stop codon or splice site defect) or missense (i.e., nonsynonymous amino acid changes) mutations within a gene of interest. Nonsense mutations often lead to loss-of-function phenotypes whereas missense mutations can lead to various changes in function. The TILLING strategy of point mutation induction followed by high-throughput discovery of induced mutations has been applied to numerous plant systems, including *Arabidopsis*, rice, soybean, pea, barley and maize (reviewed in Tadele et al. 2010).

The maize TILLING resource was generated by mutagenesis of pollen from either the inbred B73 or W22 with EMS, which induces G/C to A/T transitions. After mutagenesis, DNA is collected and pooled in 96-well format from M_1 plants. Computational tools such as CODDLe (codons optimized to detect deleterious lesions; McCallum et al. 2000) are available to predict regions within target genes that have the highest likelihood of yielding mutations with observable phenotypes (e.g., induction of stop codons, disruption of splice junctions, or induction of non-conservative missense mutations). PCR amplification of these identified regions of target genes is followed by heat denaturing and then reannealing to form heteroduplexes of mutant and wild type sequences. Incubation of PCR products with CEL I, an S1-family nuclease that recognizes and cleaves at mismatched base pairs, generates cleaved products that are separated either by HPLC or in sequence analyzers (Till et al. 2004a, b, 2006). Of 576 reported mutations identified by the Maize TILLING project, 45% are non-silent mutations and 14% are missense mutations that are predicted to be damaging to protein function (Weil 2009).

With the growing popularity of next-generation sequencing methods (discussed below), point mutation detection in maize is now moving toward resequencing of target genes using different sequencing technologies (e.g., Illumina/Solexa, Roche 454, ABI SOLiD (Weil and Monde 2007; Weil 2009)). These NGS technologies allow for either deep sequencing of numerous mutagenized plants across a few target genes or sequencing of fewer mutagenized plants at a large number of target loci in a single sequencing run. Multidimensional pooling coupled with TILLING by targeted sequencing followed by rigorous mutation calling has already been demonstrated in both wheat and rice (Missirian et al. 2011; Tsai et al. 2011).

TILLING has been used to identify several *brown mid-rib* (*bmr*) mutants in an EMS-mutagenized sorghum population. The leaf blades of 1.8% of resulting mutant families exhibited a brown midrib phenotype, which is associated with altered lignin composition within the cell wall. TILLING of these mutants revealed detrimental missense mutations within the *caffeic acid O-methyltransferase* (*COMT*) gene (Xin et al. 2008), which is involved in the biosynthesis of monolignol. Forage from similar *bmr* mutants was shown to be more digestible and led to increased milk production in cattle (Aydin et al. 1999; Oliver et al. 2004). The reduced content and increased digestibility of lignin in stalks of *bmr* mutants in sorghum make these genes promising targets in the generation of bioenergy feedstocks.

Mutants identified by TILLING of EMS-mutagenized populations can provide both loss-of-function nonsense mutants as well as missense mutants that confer either gain- or change-of-function phenotypes. Such mutants could be very useful in combination with the current set of *bm* knockout mutants in maize (Vignols et al. 1995; Morrow et al. 1997).

Mutator *Transposon Insertion Populations*

Transposon insertion populations have been widely used to identify insertions within genes of interest. The *Mutator* transposon system is widely used for reverse genetics studies due to *Mu's* highly mutagenic nature (Robertson 1978; Alleman and Freeling 1986; Settles et al. 2004) and its ability to germinally insert into genic regions (Hanley et al. 2000) unlinked to the original insertion site (Lisch et al. 1995). Both the Trait Utility System for Corn (TUSC) developed by Pioneer Hi-Bred International in 1995 (Bensen et al. 1995; Meeley and Briggs 1995) and the publically accessible Maize Targeted Mutagenesis (MTM) database (http://mtm.cshl.edu/)

developed at Cold Spring Harbor (May et al. 2003) utilize PCR-based screens of traditionally bred corn populations containing genome-wide *Mutator* insertions to identify *Mu* insertions within genes of interest. By silencing the activity of *Mu* (Slotkin et al. 2003), MTM populations control transposition and reduce false positives caused by somatic transposition events. Both the TUSC and MTM populations derive from numerous inbred lines, making backcrossing to a shared background necessary before comparing a series of mutant alleles. To that end, the *UniformMu* insertion population (http://www.maizegdb.org/documentation/uniformmu/) recently developed at the University of Florida utilizes a color-converted W22 inbred, into which active *Mutator* was introgressed seven back-cross generations to generate a *Mu*-insertion population within a homogenous genetic background. Advantages of this population include the homogenous genetic background, moderate copy number (~ 57 insertions/genome) and high mutation rate (e.g., the frequency of new seed mutants averaged ~ 7% per generation) (McCarty et al. 2005).

High-throughput screening methods have been developed to efficiently screen for germinal *Mu* insertions within these *Mu*-active populations. The TAIL (thermal asymmetric interlaced)-PCR method previously developed for screening of T-DNA insertions in *Arabidopsis* was modified to the MuTAIL procedure, specific for screening complex pools of *Mu* insertions (Settles et al. 2004). Libraries of cloned MuTAIL PCR products have been sequenced and the resulting *Mu*-flanking sequence tags (FSTs) can be easily screened via BLAST to identify insertions of interest within the *UniformMu* population (McCarty et al. 2005). The FSTs generated by this method are primarily (e.g., 89% in one study) associated with stable, germinal *Mu* insertions (Settles et al. 2007), making MuTAIL a very efficient method for identifying stable insertions in genes of interest. With the advent of next generation sequencing technologies (see discussion below), a new Illumina-based FST sequencing approach allows for both the identification of FSTs in genes of interest and the cataloging of insertions unlinked to those genes of interest (Williams-Carrier et al. 2010). The sequencing depth and multiplexing ability associated with next generation sequencing technologies maximizes insertion detection.

A *Mu* insertion was identified within the *cinnamoyl-CoA reductase1* (*ccr1*) gene, which catalyzes the formation of monolignols and is considered to be a key enzyme in determining lignin content and quality. The mutant exhibited reduced expression of *ccr1* transcripts, a slight reduction in lignin content, significant changes in lignin structure and increased cell wall digestibility with no adverse effects on plant growth or morphology (Tamasloukht et al. 2011). On a much larger scale, reverse genetic screening of the *UniformMu* population resulted in the identification of 72 insertions within 63 cell wall genes (Penning et al. 2009) including *COMT*, *cinnamoyl-*

CoA reductase (*ccr*) and *cinnamyl aldehyde dehydrogenase* (*cad*), which is a vast increase in the number of available mutants for cell wall-related genes. For example, insertions have been identified in the *cellulose synthase-like* gene family, including a *CsID5* insertion mutant with blocked root hair elongation (Penning et al. 2009). Characterization of these cell wall-related mutants will increase our understanding of cell wall composition and structure and potentially impact biomass yield and quality, both important factors in the use of maize for bioenergy.

Activator/Dissociation *Transposon Insertion Populations*

The *Ac/Ds* transposon system provides another routinely used tagging tool in maize. The *Ac/Ds* and *Mutator* systems can be viewed as complementary approaches, since *Mutator* tends to transpose to genetically unlinked sites, whereas *Ac* and *Ds*, tend to jump to linked sites (Dooner and Belachew 1989; Kermicle et al. 1989; Weil et al. 1992), many of which are <10 cM from the donor locus. The *Ac/Ds* tagging system (http://plantgdb.org/prj/AcDsTagging/index.php) therefore provides opportunities to identify an allelic series at a gene of interest as well as a mode for regional mutagenesis of nearby genes.

Activator, the autonomous element of the *Ac/Ds* system, has been used as a genome-wide gene tagging tool for regional mutagenesis (Brutnell and Conrad 2003; Kolkman et al. 2005). A collection of *Ac*-containing W22 lines has been developed in which there is a single genetically and physically positioned *Ac* element in each line and the collection consists of insertions on each chromosome. Transposition events can be identified via several inverse PCR methods, which can detect transposed *Ac* elements up to 8-kb from the donor locus (Kolkman et al. 2005). This regional mutagenesis system was successfully used at the *pink scutellum1* locus, which encodes a lycopene-β-cyclase, where 17 *Ac*-insertion alleles were identified and 19 excision alleles were generated. Excision products are useful because they can generate stable knock-out alleles as well as allelic series with varying levels of mutant phenotype (Bai et al. 2007).

Focus has recently moved from *Ac*-based to *Ds*-based transposon tagging because, unlike the mutable *Ac*, the nonautonomous *Ds* element provides stable insertions that can be easily mobilized by the autonomous *Ac* element and then be segregated away by genetic crossing. Over 1500 unique sequence-indexed *Ds* insertion lines, each containing 1–2 elements per line, have been generated in a uniform W22 background and >80% of these insertions have been physically mapped to the B73 genome sequence (Ahern et al. 2009; Vollbrecht et al. 2010), primarily in gene-rich areas of the genome (Vollbrecht et al. 2010). Each of these lines was derived from a single *Ds* insertion allele of the *R1* gene (*r1-sc:m3*) on chromosome 10. In addition, the lines contain *Ac-im*, which is an immobilized source of transposase that can mobilize *Ds* elements but cannot transpose itself due to a 10-bp deletion in the left terminal inverted repeat. *Ac-im* catalyzes germinal excisions, as evidenced by the recovery of transposed *Ds* insertions from the *r1-sc:m3* allele in 4.5% of progeny (Conrad and Brutnell 2005).

Regional mutagenesis experiments can be conducted by first using web-based tools (http://www.plantgdb.org/prj/AcDsTagging/) to identify several appropriate *Ds*-donor lines that contain *Ds* elements near the gene of interest followed by either crossing to a reference allele or screening for transposed insertions within the gene of interest using PCR (Ahern et al. 2009).

The advent of these recent *Ac/Ds* and *Mutator* tagging programs along with the TILLING system provides the researcher several complementary methods for reverse genetics experiments. Because the W22 background is used in many of these programs, one can easily compare mutant alleles obtained from the different systems without the need for extensive backcrossing. In addition, the introgression of these mutant alleles into this uniform background allows for the detection of subtle phenotypes (e.g., cell wall mutants).

To the next generation: Potential applications of Sequencing and Genotyping Technologies for Dissecting Bioenergy Traits

Sequence-based Technologies

The emergence of non-Sanger based massively parallel sequencing methods, so-called "next-generation sequencing (NGS)" technologies, has greatly decreased the costs and increased both the depth and rapidity of sequencing efforts. NGS platforms include Roche 454, Illumina's Solexa Genome Analyzer, Applied Biosystems' SOLiD and Life Technologies' Ion Torrent (for reviews of these technologies and their biochemistries, see Mardis 2008; Metzker 2010). The NGS platforms and relevant NGS bioinformatics tools (for review, see Varshney et al. 2009) have been widely used for genome resequencing efforts, differential gene expression analyses, genome-wide marker discovery (i.e., for genetic mapping or marker-assisted selection) and genotyping (Varshney et al. 2009; Brautigam and Gowik 2010). These NGS technologies can increase the speed of marker development—for example, single nucleotide polymorphism (SNP)- and simple sequence repeat (SSR)-based markers—and marker saturation (Davey et al. 2011) for gene or quantitative trait loci (QTL) mapping of traits (e.g., bioenergy traits). Recently, a physical map of the *Arabidopsis* genome was generated using Illumina-derived short reads of restriction digested 2-D bacterial artificial chromosome (BAC) pools. This whole genome profiling technique was simulated on maize short read data, and shown to also be applicable for physical map generation of more complex genomes (van Oeveren et al. 2011).

Although NGS can be very useful for marker development, whole genome resequencing can often not identify polymorphisms within genes of interest. Targeted resequencing of specific genes or even chromosomal intervals can greatly improve the sequencing depth and ability to uncover SNPs to be used as high-density genetic markers in regions of interest. Two methods that reduce genome complexity prior to NGS sequencing have recently been employed in maize: Complexity reduction of polymorphic sequences (CRoPS) and repeat subtraction-mediated sequence capture (RSSC). CRoPS involve NGS of amplified fragment length polymorphism (AFLP) fragment libraries from two or more genetically diverse genomes that are enriched for low complexity and presumably genic sequences due to digestion with methylation-sensitive enzymes (van Orsouw et al. 2007). The CRoPS strategy was recently applied to two proprietary maize lines (i.e., no reference sequence is needed) and ~ 1,100 SNPs were identified and mapped in a F_2 mapping population (Mammadov et al. 2010). Similar to CRoPS, RSSC also limits the resequencing of repetitive sequences and, in addition, selects target regions for resequencing. RSSC was recently used to sequence contiguous chromosomal intervals as well as targeted (dispersed) genes in maize inbreds B73 and Mo17 (Fu et al. 2010). Briefly, a sequencing library was hybridized to a repeat array and nonhybridizing sequences (i.e., low copy fraction) were subsequently hybridized to a capture array containing target sequences. Captured sequences were then eluted and sequenced via 454 NGS technology. Such complexity-reduction approaches to NGS will be useful in generating high-density markers to map causative bioenergy-related QTLs to high resolution. In addition, RSSC will allow for the sequence characterization of novel alleles in diverse nonreference genomes.

High-throughput Genotyping

Several genotyping technologies are now available for simultaneous high-throughput screening of hundreds, or even thousands, of markers (for review, see Appleby et al. 2009). Illumina offers high-throughput genotyping technologies that allow for medium (GoldenGate) or high density (Infinium) multiplexing. The GoldenGate microbead-based array technology is based on allelic discrimination directly on genomic DNA, followed by the generation of allele-specific PCR products that are captured by specific microbeads. The allele is then detected by primer-specific fluorescent signals. The Infinium II platform is based on single base extension from immobilized SNP-specific primers hybridized to amplified genomic DNA and it can assay ~ 500,000 SNPs per array, which makes it ideal for genome-wide genotyping projects (Fan et al. 2006). In contrast to Illumina's fluorescence-based detection methods, Sequenom's iPlex MassARRAY technology (Appleby et al. 2009) utilizes single base extension followed by mass detection of extended primer products using matrix-assisted laser desorption/ionization time of flight (MALDI-TOF) mass spectrometry.

These genotyping arrays are especially useful for high-density genetic mapping, genome-wide association studies and marker-assisted selection. For example, a GoldenGate assay comprising 1536 SNPs was used to genotype and generate an integrated linkage map from 5000 recombinant inbred lines that comprise the nested association mapping (NAM) population in maize (McMullen et al. 2009). This same panel of SNPs has been used to genotype two other recombinant inbred mapping populations and a diverse panel of 154 inbreds (Yan et al. 2010) as well a survey of 770 inbreds that represent global germplasm diversity (Lu et al. 2009). Diversity in tropical and semitropical lines has also been assessed using this high throughput genotyping technology (Yan et al. 2009). The quantitative nature of Sequenom-based SNP assays was exploited to genetically map and assign complementation groups to 28 independent EMS-induced phenotypic mutants via bulk segregant analysis (Liu et al. 2010). Infinium technology has recently been used to generate an array of ~ 50,000 SNPs (covering 17,520 genes and 16,168 intergenic regions evenly distributed across the genome), which yielded two highly saturated genetic maps from two intermated recombinant inbred populations (Ganal et al. 2011). Combined, these new sequencing and genotyping technologies can be applied to the uncovering and fine dissection of genes involved in bioenergy traits in maize.

Utility of Transgenics to Study and Modify Bioenergy-related Traits

To genetically engineer bioenergy traits, novel genes or genes with altered function can be inserted into maize via transformation of a transgene construct. Transgenesis can be used to add a new function by expressing a gene from another species or a gene engineered to have an altered activity. Modifying the expression level or temporal-spatial expression pattern of an endogenous gene can be used to increase or decrease gene activity. And new methods exist to make targeted modifications to gene activities. In this section we outline a few examples of using transformation to engineer improved bioenergy traits in maize and describe some transgenic technologies available for bioenergy trait modification.

Heterologous Genes

Genetic engineering approaches are being used to create feedstocks that will be more amenable to biofuel production by facilitating conversion of biomass components into fermentable substrates for microbes. Engineered cell wall-degrading enzymes from bacteria or fungi can be inserted into the bioenergy crop genome directly to save costs and increase hydrolysis efficiency (Xin et al. 2011). One example is to identify and engineer novel starch degrading enzymes that are effective at lower temperatures to improve hydrolysis efficiency. A recombinant α-amylase gene from the fungus *Rhizomucor* was constructed using the starch binding domain (SBD) from *A. niger* glucoamylase attached through a glycosylated linker to the core amylase catalytic domain (Tawil et al. 2010). This recombinant α-amylase was shown to be very efficient at hydrolyzing raw maize starch granules at low temperature (32°C) to release only glucose. A bacterial α-amylase from *Anoxybacillus flavothermus,* also engineered to contain the *A. niger* SBD, was found to be very efficient in hydrolysis of concentrated raw maize starch granules (Tawil et al. 2012). Whether either of these modified enzymes will show commercial utility has yet to be tested. More intriguing is an example of an engineered thermostable α-amylase gene, called *amy797E;* a chimeric gene derived from gene shuffling of three wild type α-amylase genes from the archael order Thermococcales (Eichler 2001). The chimeric protein was selected for its increased thermostability and activity at the high temperatures required for starch hydrolysis in dry-grind ethanol production. The expression of *amy797E* is specifically targeted to the grain endosperm by the zein promoter where the protein is retained in the endoplasmic reticulum of endosperm cells. Syngenta used this thermostable α-amylase to produce Event 3272 high-amylase maize, which is in early development. It is intended to simplify the dry mill production of ethanol by eliminating the need for the addition of bacterial-produced amylase during liquification (Wolt and Karaman 2007). Introduction of as little as 3% high-amylase maize to the total maize processed was sufficient to replace exogenous α-amylase for bench-scale ethanol production (Singh et al. 2006).

RNA Interference (RNAi)

RNA interference (RNAi) is a naturally occurring gene silencing phenomenon in plants that involves sequence-specific RNA degradation (for thorough reviews, see Baulcombe 2004; Watanabe 2011). Briefly, double-stranded RNA is cleaved by the Dicer enzyme into short interfering RNAs (siRNA). Mediated by the RNA-induced silencing complex (RISC) that contains the RNase H Argonaute enzyme, the siRNAs induce the degradation of mRNAs with shared sequence identity.

RNAi is often associated with the introduction of transgenes and, based on this phenomenon, has been successfully co-opted as technology for generating genetic knockout and knockdown mutants. RNAi technology can provide advantages over insertion mutagenesis techniques (e.g., *Ac/Ds*, *Mutator*, etc.) due to the dominant nature of RNAi-induced mutations, the ability to knockdown expression of genes for which knockouts are lethal, and the ability to simultaneously reduce the expression of closely related genes (e.g., gene families). RNAi has been successfully used not only for gene function analyses but also in crop improvement (e.g., pathogen resistance, nutritional enhancement, allergen reduction) (for review, see Ali et al. 2010b).

RNAi technology has been used to both knockdown individual gene targets in maize (Houmard et al. 2007; Casati and Walbot 2008; Frizzi et al. 2010; Wu et al. 2010; Wu and Messing 2011) as well as in a larger-scale functional genomic analysis of 130 chromatin-related genes in maize (McGinnis et al. 2007). This large scale RNAi knockdown experiment yielded a broad spectrum of results, including numerous lines for which targeted mRNA expression was drastically reduced, examples of transgene silencing (i.e., no suppression of target mRNAs), variable expression across generations, and lines for which secondary targets were also silenced (McGinnis et al. 2007). Indeed, there have been several reports of RNAi-induced silencing of unintended targets in plants (i.e., off-target silencing) (for review, see Senthil-Kumar and Mysore 2011). This broad spectrum of success and failure underscores both the usefulness of RNAi as well as the variability that still exists for this technology. RNAi technology has also been developed for forward mutagenesis for a targeted population of transcripts. Petsch et al. (2010) report forward mutagenesis of *Arabidopsis* using a library of transitive RNAi constructs from cDNAs derived from mesophyll cells from *Arabidopsis* leaves, from which numerous mutants defective in photosynthesis were identified.

RNAi and antisense techniques have been used to study the variability in cell wall traits in response to varying the levels of expression of genes involved in lignin biosynthesis, including *COMT*, *CAD* and *CCR*. Antisense down regulation of COMT, encoded by *brown midrib3 (bm3)*, within sclerenchyma cells resulted in a transgenic line that exhibited the brown midrib phenotype and a 70–85% reduction in COMT activity (Piquemal et al. 2002). This knockdown line showed reduced lignin content and alterations in lignin composition, but to a lesser extent than the *bm3* knockout mutant. In addition, the line showed increased forage digestibility at levels very similar to those of the knockout, but did not exhibit the associated detrimental effects (e.g., reduced standability) observed in some *bmr* mutants (Cherney et al. 1991). Similarly, knocked down expression of *CCR* in maize results in decreased lignin content but also in increased cellulose (Sticklen 2009), which has practical applications for biomass processing for biofuel production (Sticklen 2008). RNAi down regulation of expression has also been performed on *bm1*, which encodes CAD, the enzyme that catalyzes the last reduction of hydroxyl-cinnamaldehydes to their corresponding alcohols (Mansell et al. 1974) prior to polymerization of monolignols. The *bm1* knockdown mutant exhibits altered cell wall structure, increased cellulose accumulation, higher digestibility, higher biomass and increased ethanol production from that biomass (Fornale et al. 2012).

RNAi could prove to be a powerful technology for generating maize with better forage crop qualities and with greater potential for cellulosic ethanol production. The down regulation of only a handful of genes involved in lignin biosynthesis has already established that small modifications of lignin content and cell wall structure can increase digestibility while preserving structural integrity of the plant, which can be greatly weakened in gene knockouts. The use of tissue specific and inducible RNAi constructs could further reduce potential negative impacts on the overall growth and development of the plant (for review, see Mansoor et al. 2006).

Targeted Mutagenesis

Whereas candidate genes for bioenergy traits in maize can be knocked out or knocked down by numerous methods reviewed above, altering specific genes *in vivo* via homologous recombination (i.e., gene targeting) has proven much more challenging in plants. Gene targeting not only allows for the study of gene function *in vivo*, but also allows for trait or crop improvement via addition, editing or stacking of genes as well as stacking of independent traits.

Recently, gene targeting strategies have become viable based on the bioengineering of several classes of nucleases that induce sequence specific double-strand breaks (DSBs). Desired modifications to the target sequence occur during subsequent DNA repair. Site specific genomic engineering using the techniques described below (zinc finger nucleases, homing endonucleases and transcription activator-like effector nucleases) could be particularly useful in engineering maize for biorenewable applications by modifying relevant loci (e.g., *bm* genes) as well as stacking combinations of biorenewable traits into specific genotypes.

Zinc Finger Nucleases (ZFN)

The bioengineered zinc finger nuclease (ZFN) is a fusion product of the DNA binding domain from the zinc finger protein (ZFP) class of eukaryotic transcription factors and the nuclease domain of the *Fok*I restriction enzyme (for review, see Urnov et al. 2010). For targeted mutagenesis, three or more Zn fingers, which each recognizing a specific DNA nucleotide triplet, are strung together (i.e., "modular assembly") to recognize a target DNA sequence (9-bp or longer), at which a targeted double-strand break (DSB) is catalyzed by the *Fok*I domain (Townsend et al. 2009). Constructs containing the engineered ZFN and a donor repair template are co-transformed. The ZFN-mediated DSBs stimulate DNA repair by the endogenous homologous recombination machinery at nearby sites via homologous recombination with a donor template that contains the intended mutations for the target sequence (Wright et al. 2005; Cai et al. 2009; Townsend et al. 2009). In addition to targeting specific changes via donor templates (i.e., genome editing), in the absence of a repair template, ZFN mediated DSBs at endogenous loci are repaired via nonhomologous end joining, which can generate targeted knockout mutants (Townsend et al. 2009).

Precise ZFN mediated gene addition has recently been demonstrated in maize (Shukla et al. 2009). A ZFN mediated DSB in the *Ipk1* gene, which encodes an enzyme that catalyzes the last step in phytate biosynthesis in seed, is repaired by a template that knocks out *Ipk1* function and concomitantly introduces a gene that confers herbicide resistance. The reduction of phytate and the introduction of herbicide resistance in a single step illustrate the possibility for trait stacking via ZFN mediated genome editing.

Transcription Activator-like Effector Nucleases (TALENs)

While ZFNs have been shown to successfully mediate gene editing in plants, the engineering of ZFNs to recognize specific DNA motifs can be laborious and technically challenging (for review, see Urnov et al. 2010). For example, some ZFs can interact with one another or with nucleotides outside of a specific nucleotide triplet (i.e., off-target recognition). The difficulties associated with engineering DNA binding specificities for ZFNs are not issues for transcription activator-like effector nucleases (TALENs). TALENs, therefore show extreme promise as mediators of genome editing and gene stacking in plants.

TALE proteins are found in plant pathogens from the genus *Xanthomonas* and are secreted into plant hosts, where they activate gene expression via modular binding of a domain of tandem 33–35 amino acid repeats to effector-specific DNA sequences (for review, see Bogdanove and

Voytas 2011). TALE proteins recognize specific DNA sequences via a pair of hypervariable amino acids residing in each tandem repeat, such that each repeat binds a specific nucleotide (Boch et al. 2009; Moscou and Bogdanove 2009) and TALEs can be synthesized to have specific and novel binding specificities (Boch et al. 2009). Because TALE proteins recognize DNA in a modular fashion, these proteins have been exploited for site specific induction of DSBs by coupling the TALE to the catalytic domain of the *Fok*I nuclease (for review, see Bogdanove and Voytas 2011), which has also been employed in ZFNs (see above). The utility of these TALE nucleases (TALENs) to induce targeted DSBs in plants has been demonstrated in transient assays in tobacco (Mahfouz et al. 2011) and *Arabidopsis* (Christian et al. 2010; Cermak et al. 2011). Repair of TALEN induced DSBs with user supplied repair templates has not yet been demonstrated.

Homing Endonucleases

Homing endonucleases are encoded within introns/inteins of microbes and facilitate self-splicing (for review, see Stoddard 2011). Because homing endonucleases (e.g., I-CreI) have very specific DNA recognition and cleavage specificity, they have been co-opted for gene targeting in plants. The LAGLIDADG family of endonucleases has been engineered to have altered DNA recognition specificities (for review, see Stoddard et al. 2007). In addition, endonucleases have been engineered to induce single strand nicks in the place of DSBs, which are more specific for repair via homologous recombination with a supplied template (i.e., gene replacement) as compared to mutagenic repair (Metzger et al. 2011). The utility of homing endonucleases for targeted mutagenesis and editing in maize has already been demonstrated. For example, transgenic maize harboring a defective herbicide resistance reporter gene and an I-*Sce*I recognition site was repaired via I-*Sce*I homing endonuclease that was cotransformed with a repair template (D'halluin et al. 2008). In a separate demonstration, the I-CreI homing endonuclease was used without a repair template to generate short insertion/deletion mutations viaDSBs in the promoter region of the *liguleless1* (*lg1*) gene. These DSBs were repaired via non-homologous end joining (NHEJ) to produce heritable mutations at a 3% frequency in this endogenous locus (Gao et al. 2010).

Future Prospects

Maize is an excellent target crop for bioenergy applications due to our understar ling of maize genetics, biochemistry and physiology. With the availability of numerous genetic resources, rich sources of genetic maps, phenotypically diverse germplasm and the sequenced genome, the tools are in place to improve maize as a bioenergy source. Because of this richness in available resources, maize also offers itself as an excellent model for pathway discovery that has the potential to be applied to other biological systems. For example, the cuticle covering maize silks is composed almost entirely (> 90%) of long-chain simple hydrocarbons (i.e., alkanes and alkenes), which account for ~ 2% of the dry weight of silks. These surface hydrocarbons closely resemble crude oil, and therefore the pathway(s) involved in their synthesis and regulation could have practical applications in the arena of biorenewable chemicals. However, the mechanism(s) of their synthesis from very long-chain fatty acid (VLCFA) precursors and the genes involved in the synthesis to hydrocarbons are currently unknown. Fine scale metabolite analyses from the cuticle of maize silks from the inbred B73 suggest that surface hydrocarbons are biosynthesized via three related pathways involving reduction of a VLCFA to the corresponding aldehyde and subsequent decarbonylation to yield a homologous series of odd-numbered linear alkanes and alkenes, ranging between 19 and 33 carbon atoms in length (Perera et al. 2010). Because hydrocarbon accumulation varies 5-fold between silks from inbreds B73 and Mo17 (Perera et al. 2010), the Syn4 and Syn10 intermated B73xMo17 (IBM) mapping populations were used to identify numerous high confidence QTLs that are associated with the biochemical steps in surface hydrocarbon production as well as several QTLs that affect the final abundance of individual constituents (N Lauter, BJ Nikolau and MD Yandeau-Nelson, personal communication). These quantitative experiments are being coupled with RNA seq transcriptome profiling from silks that hyper- or hypo-accumulate hydrocarbons to identify candidate genes at each locus. From understanding the enzymatic and regulatory factors associated with hydrocarbon production in maize, one can apply this knowledge to other systems (e.g., algae) that have the capacity to produce large amounts of biofuel with a smaller environmental footprint (Schenk et al. 2008; Wijffels and Barbosa 2010). While maize has many important potential roles as a bioenergy/biomass crop, specific genes/pathways in maize could be key players in bioenergy production in other bioengineered organisms. For example, expandable platforms are being built for the production of biorenewable chemicals using biocatalysts from many organisms (Nikolau et al. 2008). Exploring maize as a model system for studying bioenergy traits to apply to other organisms could be advantageous to avoid the difficulty of harvesting target molecules from a crop and also avoids the "fuel vs. food" debate, which is a major concern for many crop systems.

References

Adams J, Teunissen P, Robson G et al. (2012) Scanning electron microscopy and fermentation studies on selected known maize starch mutants using STARGEN™ enzyme blends. BioEnergy Res 5: 330–340.

Ahern KR, Deewatthanawong P, Schares J et al. (2009) Regional mutagenesis using *Dissociation* in maize. Methods 49: 248–254.

Ali F, Scott P, Bakht J et al. (2010a) Identification of novel *brown midrib* genes in maize by tests of allelism. Plant Breed 129: 724–726.

Ali N, Datta SK and Datta K (2010b) RNA interference in designing transgenic crops. GM Crops 1: 207–213.

Alleman M and Freeling M (1986) The *Mu*-transposable elements of maize—Evidence for transposition and copy number regulation during development. Genetics 112: 107–119.

Appenzeller L, Doblin M, Barreiro R et al. (2004) Cellulose synthesis in maize: isolation and expression analysis of the cellulose synthase (*CesA*) gene family. Cellulose 11: 287–299.

Appleby N, Edwards D and Batley J (2009) New technologies for ultra-high throughput genotyping in plants. In: Somers DJ, Langridge P, Gustafson JP (eds) Plant Genomics: Methods and Protocols, vol 513. Humana Press, New York, USA, pp 19–39.

Aydin G, Grant RJ and O'Rear J (1999) *Brown midrib* sorghum in diets for lactating dairy cows. J Dairy Sci 82: 2127–2135.

Bai L, Singh M, Pitt L et al. (2007) Generating novel allelic variation through *Activator* insertional mutagenesis in maize. Genetics 175: 981–992.

Baulcombe D (2004) RNA silencing in plants. Nature 431: 356–363.

Bensen RJ, Johal GS, Crane VC et al. (1995) Cloning and characterization of the maize *an1* gene. Plant Cell 7: 75–84.

Boch J, Scholze H, Schornack S et al. (2009) Breaking the code of DNA binding specificity of TAL-Type III effectors. Science 326: 1509–1512.

Boerjan W, Ralph J and Baucher M (2003) Lignin biosynthesis. Annl Rev Plant Biol 54: 519–546.

Bogdanove AJ and Voytas DF (2011) TAL effectors: customizable proteins for DNA targeting. Science 333: 1843–1846.

Bonawitz ND and Chapple C (2010) The genetics of lignin biosynthesis: Connecting genotype to phenotype. Ann Rev of Genet 44: 337–363.

Bortiri E, Jackson D and Hake S (2006) Advances in maize genomics: the emergence of positional cloning. Curr Opin in Plant Biol 9: 164–171.

Bosch M, Mayer C-D, Cookson A et al. (2011) Identification of genes involved in cell wall biogenesis in grasses by differential gene expression profiling of elongating and non-elongating maize internodes. J Exp Bot 62: 3545–3561.

Brautigam A and Gowik U (2010) What can next generation sequencing do for you? Next generation sequencing as a valuable tool in plant research. Plant Biol 12: 831–841.

Brutnell TP (2002) Transposon tagging in maize. Funct Integr Genom 2: 4–12.

Brutnell TP and Conrad LJ (2003) Transposon tagging using *Activator* (*Ac*) in maize. Meth Mol Biol 236: 157–176.

Cai CQ, Doyon Y, Ainley WM et al. (2009) Targeted transgene integration in plant cells using designed zinc finger nucleases. Plant Mol Biol 69: 699–709.

Candela H and Hake S (2008) The art and design of genetic screens: maize. Nat Rev Genet 9: 192–203.

Cao J (2007) Genetic dissection of the *rf2a*-mediated fertility restoration pathway in maize. PhD Dissertation (Publication No AAT 3289403) Iowa State University, Ames, USA.

Carpita NC and McCann MC (2008) Maize and sorghum: genetic resources for bioenergy grasses. Trends Plant Sci 13: 415–420.

Casati P and Walbot V (2008) Maize lines expressing RNAi to chromatin remodeling factors are similarly hypersensitive to UV-B radiation but exhibit distinct transcriptome responses. Epigenetics 3: 216–229.

Cermak T, Doyle EL, Christian M et al. (2011) Efficient design and assembly of custom TALEN and other TAL effector-based constructs for DNA targeting. Nucl Acids Res 39: e82.

Cherney JH, Cherney DJR, Akin DE et al. (1991) Potential of brown-midrib, low-lignin mutants for improving forage quality. In: Sparks DL (ed) Advances in Agronomy, vol 46. Academic Press, Inc., San Diego, USA, pp 157–198.

Christian M, Cermak T, Doyle EL et al. (2010) Targeting DNA double-strand breaks with TAL effector nucleases. Genetics 186: 757–761.

Chuck G, Cigan AM, Saeteurn K et al. (2007) The heterochronic maize mutant *Corngrass1* results from overexpression of a tandem microRNA. Nat Genet 39: 544–549.

Chuck GS, Tobias C, Sun L et al. (2011) Overexpression of the maize *Corngrass1* microRNA prevents flowering, improves digestibility, and increases starch content of switchgrass. Proc Nat Acad Sci 108: 17550–17555.

Colasanti J and Muszynski M (2009) The Maize Floral Transition. In: Bennetzen JL, Hake SC (eds) Handbook of Maize: Its Biology, Springer, New York, USA, pp 41–55.

Conrad LJ and Brutnell TP (2005) *Ac-Immobilized*, a stable source of *Activator* transposase that mediates sporophytic and gametophytic excision of *Dissociation* elements in maize. Genetics 171: 1999–2012.

D'halluin K, Vanderstraeten C, Stals E et al. (2008) Homologous recombination: a basis for targeted genome optimization in crop species such as maize. Plant Biotech J 6: 93–102.

Davey JW, Hohenlohe PA, Etter PD et al. (2011) Genome-wide genetic marker discovery and genotyping using next-generation sequencing. Nat Rev Genet 12: 499–510.

Delmer DP (1999) Cellulose biosynthesis: Exciting times for a difficult field of study. Annl Rev Plant Phys Plant Mol Biol 50: 245–276.

Demura T and Ye Z-H (2010) Regulation of plant biomass production. Curr Opin Plant Biol 13: 298–303.

Dien B, Sarath G, Pedersen J et al. (2009) Improved sugar conversion and ethanol yield for forage sorghum (*Sorghum bicolor* L. Moench) lines with reduced lignin contents. BioEnergy Res 2: 153–164.

Dooner HK and Belachew A (1989) Transposition pattern of the maize element *Ac* from the *bz-m2(Ac)* allele. Genetics 122: 447–457.

Eichler J (2001) Biotechnological uses of archaeal extremozymes. Biotech Adv 19: 261–278.

Fan JB, Gunderson KL, Bibikova M et al. (2006) Illumina universal bead arrays. Method Enzymol 410: 57–73.

Fornalé S, Capellades M, Encina A et al. (2012) Altered lignin biosynthesis improves cellulosic bioethanol production in transgenic maize plants down-regulated for cinnamyl alcohol dehydrogenase. Mol Plant 5: 817–830.

Fornalé S, Shi X, Chai C et al. (2010) *ZmMYB31* directly represses maize lignin genes and redirects the phenylpropanoid metabolic flux. Plant J 64: 633–644.

Frey M, Stettner C and Gierl A (1998) A general method for gene isolation in tagging approaches: amplification of insertion mutagenised sites (AIMS). Plant J 13: 717–721.

Frizzi A, Caldo RA, Morrell JA et al. (2010) Compositional and transcriptional analyses of reduced zein kernels derived from the *opaque2* mutation and RNAi suppression. Plant Mol Biol 73: 569–585.

Frontini M, Soutoglou E, Argentini M et al. (2005) TAF9b (formerly TAF9L) is a bona fide TAF that has unique and overlapping roles with TAF9. Mol Cell Biol 25: 4638–4649.

Fu Y, Springer NM, Gerhardt DJ et al. (2010) Repeat subtraction-mediated sequence capture froma complex genome. Plant J 62: 898–909.

Ganal MW, Durstewitz G, Polley A et al. (2011) A large maize (*Zea mays* L.) SNP genotyping array: Development and germplasm genotyping, and genetic mapping to compare with the B73 reference genome. PLoS One 6: e28334.

Gao HR, Smith J, Yang MZ et al. (2010) Heritable targeted mutagenesis in maize using a designed endonuclease. Plant J 61: 176–187.

Gao M, Wanat J, Stinard PS et al. (1998) Characterization of *dull1*, a maize gene coding for a novel starch synthase. Plant Cell 10: 399–412.

Groth D, Santini J, Hamaker B et al. (2008) High-throughput screening of EMS mutagenized maize for altered starch digestibility. BioEnergy Res 1: 118–135.

Guillaumie S, Pichon M, Martinant J-P et al. (2007a) Differential expression of phenylpropanoid and related genes in brown-midrib *bm1*, *bm2*, *bm3* and *bm4* young near-isogenic maize plants. Planta 226: 235–250.

Guillaumie S, San-Clemente H, Deswarte C et al. (2007b) MAIZEWALL. Database and developmental gene expression profiling of cell wall biosynthesis and assembly in maize. Plant Physiol 143: 339–363.

Hanley S, Edwards D, Stevenson D et al. (2000) Identification of transposon-tagged genes by the random sequencing of *Mutator*-tagged DNA fragments from *Zea mays*. Plant J 23: 557–566.

Holland N, Holland D, Helentjaris T et al. (2000) A comparative analysis of the plant cellulose synthase (*CesA*) gene family. Plant Physiol 123: 1313–1324.

Houmard NM, Mainville JL, Bonin CP et al. (2007) High-lysine corn generated by endosperm-specific suppression of lysine catabolism using RNAi. Plant Biotechnol J 5: 605–614.

Jakob K, Zhou F and Paterson A (2009) Genetic improvement of C4 grasses as cellulosic biofuel feedstocks. *In vitro* Cell Dev Biol Plant 45: 291–305.

Jorgenson LR (1931) Brown midrib in maize and its linkage relations. Agron J 23: 549–557.

Keeling PL and Myers AM (2010) Biochemistry and genetics of starch synthesis. Annual Rev Food Sci Technol 1: 271–303.

Kermicle JL, Alleman M and Dellaporta SL (1989) Sequential mutagenesis of a maize gene, using the transposable element *Dissociation*. Genome 31: 712–716.

Kim K-N, Fisher DK, Gao M et al. (1998) Molecular cloning and characterization of the *amylose-extender* gene encoding starch branching enzyme IIB in maize. Plant Mol Biol 38: 945–956.

Kolkman JM, Conrad LJ, Farmer PR et al. (2005) Distribution of *Activator* (*Ac*) throughout the maize genome for use in regional mutagenesis. Genetics 169: 981–995.

Kuc J and Nelson OE (1964) The abnormal lignins produced by the *brown-midrib* mutants of maize: I. The *brown-midrib-1* mutant. Arch Biochem Biophy 105: 103–113.

Li X, Chen GH, Zhang WY et al. (2010) Genome-wide transcriptional analysis of maize endosperm in response to *ae wx* double mutations. J Genet Genom 37: 749–762.

Li X, Weng J-K and Chapple C (2008) Improvement of biomass through lignin modification. Plant J 54: 569–581.

Lisch D, Chomet P and Freeling M (1995) Genetic characterization of the *Mutator* system in maize: Behavior and regulation of *Mu* transposons in a minimal line. Genetics 139: 1777–1796.

Liu S, Chen HD, Makarevitch I et al. (2010) High-throughput genetic mapping of mutants via quantitative single nucleotide polymorphism typing. Genetics 184: 19–26.

Lu YL, Yan JB, Guimaraes CT et al. (2009) Molecular characterization of global maize breeding germplasm based on genome-wide single nucleotide polymorphisms. Theor Appl Genet 120: 93–115.

Mahfouz MM, Li LX, Shamimuzzaman M et al. (2011) *De novo*-engineered transcription activator-like effector (TALE) hybrid nuclease with novel DNA binding specificity creates double-strand breaks. Proc Nat Acad Sci 108: 2623–2628.

Mammadov JA, Chen W, Ren RH et al. (2010) Development of highly polymorphic SNP markers from the complexity reduced portion of maize [*Zea mays* L.] genome for use in marker-assisted breeding. Theor Appl Genet 121: 577–588.

Mansell RL, Gross GG, Stockigt J et al. (1974) Purification and properties of cinnamyl alcohol-dehydrogenase from higher-plants involved in lignin biosynthesis. Phytochemistry 13: 2427–2435.

Mansoor S, Amin I, Hussain M et al. (2006) Engineering novel traits in plants through RNA interference. Trends Plant Sci 11: 559–565.

Mardis ER (2008) Next-generation DNA sequencing methods. Annu Rev Genom Hum Genet 9: 387–402.

Marita JM, Vermerris W, Ralph J et al. (2003) Variations in the cell wall composition of maize *brown midrib* mutants. J Agri Food Chem 51: 1313–1321.

May BP, Liu H, Vollbrecht E et al. (2003) Maize-targeted mutagenesis: A knockout resource for maize. Proc Nat Acad Sci 100: 11541–11546.

McCallum CM, Comai L, Greene EA et al. (2000) Targeting induced local lesions IN genomes (TILLING) for plant functional genomics. Plant Physiol 123: 439–442.

McCarty DR, Mark Settles A, Suzuki M et al. (2005a) Steady-state transposon mutagenesis in inbred maize. Plant J 44: 52–61.

McCarty DR and Meeley RB (2009) Transposon resources for forward and reverse genetics in maize. In: Bennetzen JL and Hake S (eds) Handbook of Maize: Its Biology. Springer New York, USA, pp 561–584.

McGinnis K, Murphy N, Carlson AR et al. (2007) Assessing the efficiency of RNA interference for maize functional genomics. Plant Physiol 143: 1441–1451.

McMullen MD, Kresovich S, Villeda HS et al. (2009) Genetic properties of the maize nested association mapping population. Science 325: 737–740.

Meeley RB and Briggs SP (1995) Reverse genetics for maize. Maize Genet Coop Newsl 69: 67–82.

Metzger MJ, McConnell-Smith A, Stoddard BL et al. (2011) Single-strand nicks induce homologous recombination with less toxicity than double-strand breaks using an AAV vector template. Nucl Acids Res 39: 926–935.

Metzker ML (2010) Sequencing technologies—the next generation. Nat Rev Genet 11: 31–46.

Missirian V, Comai L and Filkov V (2011) Statistical mutation calling from sequenced overlapping DNA pools in TILLING experiments. BMC Bioinformatics 12: 287.

Morrow SL, Mascia P, Self KA et al. (1997) Molecular characterization of a *brown midrib3* deletion mutation in maize. Mol Breed 3: 351–357.

Moscou MJ and Bogdanove AJ (2009) A simple cipher governs DNA recognition by TAL effectors. Science 326: 1501.

Nelson OE and Rines HW (1962) The enzymatic deficiency in the *waxy* mutant of maize. Biochem Biophys Res Commun 9: 297–300.

Neuffer MG (1994) Mutagenesis. In: Freeling M and Walbot V (eds) The Maize Handbook. Springer-Verlag, New York, USA; Berlin, Heidelberg, Germany, pp 212–219.

Nikolau BJ, Perera MA, Brachova L et al. (2008) Platform biochemicals for a biorenewable chemical industry. Plant J 54: 536–545.

Oliver AL, Grant RJ, Pedersen JF et al. (2004) Comparison of *brown midrib-6* and -18 forage sorghum with conventional sorghum and corn silage in diets of lactating dairy cows. J Dairy Sci 87: 637–644.

Penning BW, Hunter CT, Tayengwa R et al. (2009) Genetic resources for maize cell wall biology. Plant Physiol 151: 1703–1728.

Perera MADN, Qin WM, Yandeau-Nelson M et al. (2010) Biological origins of normal-chain hydrocarbons: a pathway model based on cuticular wax analyses of maize silks. Plant J 64: 618–632.

Petsch KA, Ma C, Scanlon MJ et al. (2010) Targeted forward mutagenesis by transitive RNAi. Plant J 61: 873–882.

Piquemal J, Chamayou S, Nadaud I et al. (2002) Down-regulation of caffeic acid O-methyltransferase in maize revisited using a transgenic approach. Plant Physiol 130: 1675–1685.

Poethig RS (2010) The past, present, and future of vegetative phase change. Plant Physiol 154: 541–544.

Rhoades MM (1984) The early years of maize genetics. Ann Rev Genet 18: 1–30.

Robertson DS (1978) Characterization of a mutator system in maize. Mutat Res 51: 21–28.

Sattler SE, Funnell-Harris DL and Pedersen JF (2010) *Brown midrib* mutations and their importance to the utilization of maize, sorghum, and pearl millet lignocellulosic tissues. Plant Sci 178: 229–238.

Schenk PM, Thomas-Hall SR, Stephens E et al. (2008) Second generation biofuels: high-efficiency microalgae for biodiesel production. Bioenergy Res 1: 20–43.

Senthil-Kumar M and Mysore KS (2011) Caveat of RNAi in plants: the off-target effect. Methods Mol Biol 744: 13–25.

Settles AM (2009) Transposon tagging and reverse genetics. In: Kriz AL, Larkins BA (eds) Molecular genetic approaches to maize improvement, vol 63. Springer, Berlin, Heidelberg, Germany, pp 143–159.

Settles AM, Holding DR, Tan BC et al. (2007) Sequence-indexed mutations in maize using the UniformMu transposon-tagging population. BMC Genomics 8: 166.

Settles AM, Latshaw S and McCarty DR (2004) Molecular analysis of high-copy insertion sites in maize. Nucl Acids Res 32: e54.

Shukla VK, Doyon Y, Miller JC et al. (2009) Precise genome modification in the crop species *Zea mays* using zinc-finger nucleases. Nature 459: 437–441.

Singh V, Batie CJ, Aux GW et al. (2006) Dry-grind processing of corn with endogenous liquefaction enzymes. Cereal Chem J 83: 317–320.

Slotkin RK, Freeling M and Lisch D (2003) *Mu* killer causes the heritable inactivation of the *Mutator* family of transposable elements in *Zea mays*. Genetics 165: 781–797.

Sticklen MB, inventor; Michigan State Univ, assignee. Altering regulation of maize lignin biosynthesis enzymes via RNAi technology. United States patent application, US WO/2008/069964. 2008 Dec 06.

Sticklen MB (2009) Expediting the biofuels agenda via genetic manipulations of cellulosic bioenergy crops. Biofuel Bioprod Bior 3: 448–455.

Stinard PS, Robertson DS and Schnable PS (1993) genetic isolation, cloning, and analysis of a mutator-induced, dominant antimorph of the maize *amylose extender1* locus. Plant Cell 5: 1555–1566.

Stoddard BL (2011) Homing endonucleases: from microbial genetic invaders to reagents for targeted DNA modification. Structure 19: 7–15.

Stoddard BL, Monnat RJ and Scharenberg AM (2007) Advances in engineering homing endonucleases for gene targeting: ten years after structures. In: Bertolotti R and Ozawa K (eds) Progress in gene therapy: Autologous and cancer stem cell gene therapy. World Scientific eBooks, Singapore, pp 135–167.

Tadele Z, Chikelu MBA and Till BJ (2010) TILLING for mutations in model plants and crops. In: Jain SM, Brar DS (eds) Molecular techniques in crop improvement, 2nd edn. Springer, New York, USA, pp 307–332.

Tamasloukht B, Lam MSJWQ, Martinez Y et al. (2011) Characterization of a *cinnamoyl-CoA reductase 1* (*CCR1*) mutant in maize: effects on lignification, fibre development, and global gene expression. J Exp Bot 62: 3837–3848.

Tawil G, Viksø-Nielsen A, Rolland-Sabaté A et al. (2012) Hydrolysis of concentrated raw starch: A new very efficient α-amylase from *Anoxybacillus flavothermus*. Carbohyd Polym 87: 46–52.

Tawil G, Viksø-Nielsen A, Rolland-Sabaté A et al. (2010) In depth study of a new highly efficient raw starch hydrolyzing α-amylase from *Rhizomucor* sp. Biomacromolecules 12: 34–42.

Till BJ, Reynolds S, Weil C et al. (2004a) Discovery of induced point mutations in maize genes by TILLING. BMC Plant Biology 4: 12.

Till BJ, Burtner C, Comai L et al. (2004b) Mismatch cleavage by single-strand specific nucleases. Nucleic Acids Res 32: 2632–2641.

Till BJ, Zerr T, Comai L et al. (2006) A protocol for TILLING and Ecotilling in plants and animals. Nat Protoc 1: 2465–2477.

Townsend JA, Wright DA, Winfrey RJ et al. (2009) High-frequency modification of plant genes using engineered zinc-finger nucleases. Nature 459: 442–445.

Tsai H, Howell T, Nitcher R et al. (2011) Discovery of rare mutations in populations: TILLING by sequencing. Plant Physiol 156: 1257–1268.

Urnov FD, Rebar EJ, Holmes MC et al. (2010) Genome editing with engineered zinc finger nucleases. Nat Rev Genet 11: 636–646.

van Erp H and Walton J (2009) Regulation of the *cellulose synthase-like* gene family by light in the maize mesocotyl. Planta 229: 885–897.

van Oeveren J, de Ruiter M, Jesse T et al. (2011) Sequence-based physical mapping of complex genomes by whole genome profiling. Genome Res 21: 618–625.

van Orsouw NJ, Hogers RCJ, Janssen A et al. (2007) Complexity reduction of polymorphic sequences (CRoPS (TM)): A novel approach for large-scale polymorphism discovery in complex genomes. PLoS One 2: e1172.

Varshney RK, Nayak SN, May GD et al. (2009) Next-generation sequencing technologies and their implications for crop genetics and breeding. Trends Biotechnol 27: 522–530.

Vermerris W and Boon JJ (2000) Tissue-specific patterns of lignification are disturbed in the *brown midrib2* mutant of maize (*Zea mays* L.). J Agri Food Chem 49: 721–728.

Vermerris W, Saballos A, Ejeta G et al. (2007) Molecular breeding to enhance ethanol production from corn and sorghum stover. Crop Sci 47: S-142-S-153.

Vignols F, Rigau J, Torres MA et al. (1995) The *brown midrib3* (*bm3*) mutation in maize occurs in the gene encoding caffeic acid O-methyltransferase. Plant Cell 7: 407–416.

Vollbrecht E, Duvick J, Schares JP et al. (2010) Genome-wide distribution of transposed dissociation elements in maize. Plant Cell 22: 1667–1685.

Watanabe Y (2011) Overview of plant RNAi. Meth Mol Biol 744: 1–11.

Weil CF (2009) Tilling in Grass Species. Plant Physiol 149: 158–164.

Weil CF, Marillonnet S, Burr B et al. (1992) Changes in state of the *wx-m5* allele of maize are due to intragenic transposition of *Ds*. Genetics 130: 175–185.

Weil CF and Monde RA (2007) Getting the point—Mutations in maize. Crop Sci 47: S60–S67.

White WG, Moose SP, Weil CF et al. (2011) Tropical maize: Exploiting maize genetic diversity to develop a novel annual crop for lignocellulosic biomass and sugar production routes to cellulosic ethanol. In: Buckeridge MS and Goldman GH (eds) Springer New York, USA, pp 167–179.

Wijffels RH and Barbosa MJ (2010) An outlook on microalgal biofuels. Science 329: 796–799.

Williams-Carrier R, Stiffler N, Belcher S et al. (2010) Use of Illumina sequencing to identify transposon insertions underlying mutant phenotypes in high-copy *Mutator* lines of maize. Plant J 63: 167–177.

Wolt JD and Karaman S (2007) Estimated environmental loads of alpha-amylase from transgenic high-amylase maize. Biomass Bioenergy 31: 831–835.

Wright DA, Townsend JA, Winfrey RJ et al. (2005) High-frequency homologous recombination in plants mediated by zinc-finger nucleases. Plant J 44: 693–705.

Wu Y and Messing J (2011) Novel genetic selection system for quantitative trait loci of quality protein maize. Genetics 188: 1019–1022.

Wu YR, Holding DR and Messing J (2010) gamma-zeins are essential for endosperm modification in quality protein maize. Proc Nat Acad Sci 107: 12810–12815.

Xin Z, Watanabe N and Lam E (2011) Improving efficiency of cellulosic fermentation via genetic engineering to create "smart plants" for biofuel production routes to cellulosic ethanol. In: Buckeridge MS, Goldman GH (eds). Springer New York, USA, pp 181–197.

Xin ZG, Wang ML, Barkley NA et al. (2008) Applying genotyping (TILLING) and phenotyping analyses to elucidate gene function in a chemically induced sorghum mutant population. BMC Plant Biol 8: 103.

Yan JB, Shah T, Warburton ML et al. (2009) Genetic characterization and linkage disequilibrium estimation of a global maize collection using snp markers. PLoS One 4: e8451.

Yan JB, Yang XH, Shah T et al. (2010) High-throughput SNP genotyping with the GoldenGate assay in maize. Mol Breed 25: 441–451.

CHAPTER 6

Molecular Breeding for Bioenergy Traits

Brandon Jeffrey and Thomas Lübberstedt*

ABSTRACT

Maize is currently being used to produce starch ethanol. Two of the processes being investigated to produce lignocellulosic fuels from maize are biochemical conversion to ethanol and thermochemical conversion to bio-oil. Here we review some of the research that has investigated the genetic variance that could potentially be employed in improving the conversion of maize into lignocellulosic biofuels. In addition, mapping methods used to study bioenergy-related traits are summarized.

Keywords: QTL mapping, cell wall traits, ethanol, bio-oil

Introduction

First generation biofuels, such as ethanol, produced from starch or sugar, are a mature industry in the US, where starch from maize has driven the industry. Limited availability of corn starch can increase prices of food and feed for livestock, particularly in years with poor growing conditions like the dry summer of 2012. Next-generation biofuels produced from lignocellulosic materials can be produced through two routes: biochemical and thermochemical. Current research indicates that the optimal plant ideotype for these conversion platforms differs significantly. The stalks

Department of Agronomy, Iowa State University, Ames, IA 50011, USA.
*Corresponding author: thomasl@iastate.edu

and leaves left behind after harvest of maize, known as stover, have been identified as a promising lignocellulosic feedstock for next generation biofuels.

Biochemical Conversion

There are four steps in the traditional ethanol production process: pretreatment, enzymatic hydrolysis, fermentation, and post-processing procedures. The goal of pretreatment is to increase the access to cellulose and other polysaccharides for the hydrolysis step (Wyman et al. 2005). After pretreatment, the hydrolysis step uses enzymes to break down cellulose and hemicelluloses into free sugars. Large amounts of enzymes are required in this step in order to achieve the efficiency necessary in a high volume process (Houghton et al. 2006). The amount and cost of enzymes is the greatest barrier to producing inexpensive ethanol. After hydrolysis, free sugars are converted into ethanol via microbes or synthetic catalysts. Additional steps, such as purification, separation, etc., are then required to meet fuel standards for ethanol.

The cell wall, by nature, is resistant to degradation. Lignin has been shown to have a significant negative effect on ethanol yield with only a modest rise in lignin content (Chen and Dixon 2007). This is due to the binding of lignin to cellulose, inhibiting the accessibility of enzymes to the cell-wall polysaccharides. The genetic variation available in maize has been explored with regard to ethanol production (Lorenz et al. 2009, 2010; Lewis et al. 2010). A large amount of variation for ethanol production traits including stover yield and composition, cob yield and composition, and theoretical ethanol potential was found among 49 maize cultivars with correlations between traits being mostly significant. Among the 49 cultivars, 25 silage hybrids were analyzed for general and specific combining abilities and significant differences were found for all traits analyzed (Lorenz et al. 2009). The results of this study suggest that the amount of biofuels produced from the lignocellulosic constituents of maize could be increased using existing genetic variation.

In 2010, Lewis et al. assessed the feasibility of incorporating stover traits for ethanol production into an existing breeding program for yield traits. They measured the amount of glucose in dry stover, the amount of glucose released from the stover after pretreatment and saccharification, and lignin content as the stover quality traits. The correlations among the stover quality traits and yield and other agronomic traits were neutral or positive. The authors did not directly measure stover yield, but, through estimates, suggest that stover-quality traits may be a better target for breeding than stover yield for ethanol production.

Thermochemical Conversion

Thermal decomposition of biomass produces gases, liquids (termed bio-oil), and solids (char or biochar). The yields of these three products depend on the conversion technology (which heavily influence process parameters) and biomass used. Five parameters are known to play a role in the final bio-oil quality and quantity: the heating rate achieved by the reactor, the final temperature of the reactor, the alkali metal content of the biomass, the organic composition of the biomass, and the particle size of the biomass (Fahmi et al. 2007). The amounts of lignin, cellulose, and hemicellulose within the biomass are known to play a significant role in the outcome of the bio-oil (Butler et al. 2011). In fast pyrolysis, cellulose is degraded primarily into sugars and water (as well as some aldehydes and carboxylic acid) that contribute to the liquid bio-oil product, while hemicellulose produces more solids (char) and acids than cellulose (Shen et al. 2010). Hemicellulose also contributes a large portion to the gaseous products (Oasmaa et al. 2010). Lignin degrades into high molecular weight oligomers (Bridgwater and Czernik 2004; Fahmi et al. 2008) and into phenolic compounds (Amen-Chen et al. 2001). While a higher lignin content gives bio-oil a beneficial higher heating value, it also lowers the quality of the bio-oil from a long term stability standpoint (Fahmi et al. 2008; Nowakowski et al. 2010).

The Genetics of the Maize Cell Wall

Vermerris et al. (2007) identified a number of different genetic methods for improving maize for biofuel conversion: utilization of current mutants, development of new mutants, evaluation of transgenic approaches, and employment of plant breeding programs. While plant cell walls have consistent materials, the amounts and arrangements of cellulose, hemicellulose, lignin, pectin, and protein can differ significantly between cell types. Many factors have been suggested as playing a role in the biosynthesis of the cell wall, but there is still little understanding regarding how the expression of genes are regulated for primary cell wall formation (Zhong and Ye 2007). While some of the genes for cell wall construction are known, a large number of genes are either unknown, or their role is uncertain (see Chapter 5 of this book). Because the cell wall plays a central role in the production of lignocellulosic biofuels, more understanding regarding cell wall biosynthesis and the genes and regulators underlying its biosynthesis is needed. Mapping studies will play a critical role in understanding the quantitative inheritance of biofuels related traits in maize.

Mapping Methods

Quantitative Trait Loci (QTL) Mapping

Population Types for QTL Mapping

The population types available in maize include (but are not limited to) backcross (BC), F_2, recombinant inbred lines (RILs), advanced intercross lines (AILs), doubled haploid lines (DHLs), and nested association mapping (NAM) (Table 1).

Backcross populations are developed by crossing an F_1 with one of the parental lines. This population can be produced quickly, but will produce a relatively low resolution map. Backcross genotypes cannot be proliferated unless they can be reproduced asexually. As such this population is limited with respect to the accumulation of large amounts of information as compared to populations that can be continually multiplied sexually (Burr and Burr 1991). BC populations will only contain two genotypes

Table 1 Population types for QTL mapping.

Population	Created By...	Advantages	Disadvantages
F_2	Selfing an F_1	Quick and easy to create	Few recombination events means low level of precision
Backcross (BC)	Crossing F_1 to a parental line	Quick and easy to create	Few recombination events means low level of precision
Recombinant inbred lines (RILs)	Selfing of F_1 and successive generations	High levels of recombination, can be continually reproduced	Many rounds of mating means a long time to produce
Advanced intercross lines (AILs)	Random mating of an F_2 population that resulted from a cross of inbred parents	High levels of recombination required for fine-mapping	Many rounds of mating means a long time to produce
Doubled haploid lines (DHLs)	Chromosome doubling of a haploid	One step creation of a line that is homozygous at every locus. Good for investigating additive effects, linkage effects, and additive epistasis	Haploids are created at a low frequency, DHLs difficult and expensive to create. Expression of undesirable recessive traits and mutants
Nested association mapping population (NAM) (maize)	25 families of diverse maize lines crossed to B73. These lines then bred to create 200 or more NILs per family	High allele diversity and statistical power. Very high mapping resolution. Combines linkage (QTL) and association analysis	Time consuming and expensive to create due to diverse founder lines, many rounds of mating and genotyping

at any given locus and therefore cannot be used to analyze additive and dominance effects. However, backcrossing is useful to improve several target traits or to introduce new traits to existing populations. A donor is crossed to the existing material (the recurrent parent) to improve the target trait(s). Additional generations lead to backcross inbred lines (BIL) and will use the recurrent parent such that only the target traits remain of the donor parent (Xu 2010).

F_2 populations are created by the selfing of an F_1. Like BC populations, F_2 populations can be produced quickly, but will have relatively low genetic resolution due to only one generation of effective recombination. Very large populations are needed to achieve a high map resolution. F_2 populations are more complex to analyze than BC due to the presence of three possible genotypes at a locus, which allows the possibility of investigating additive and dominance effects.

RILs are produced from repeated selfing of individuals starting from F_2 individuals until (almost) homozygosity is achieved. Due to the homozygosity, RILs can be reproduced indefinitely for evaluation in multiple experiments. Because numerous generations are required to achieve homozygosity, more recombination events occur during the production of RILs compared to BC or F_2. This creates a more accurate and higher resolution genetic map, increasing the chance of finding recombinants between linked loci (Xu 2010). RILs will not be useful for traits that have small amounts of genetic variation in the parental lines used to create the RILs (Burr et al. 1988). The main disadvantage of RILs is the time required to create them (Burr and Burr 1991).

AILs were introduced by Darvasi and Soller (1995). AILs are produced by randomly and sequentially intercrossing offspring of F_1, with the next generations (F_3, F_4, F_5, etc.) being created by randomly intercrossing the previous generation, with founding parents being two inbred lines. The probability of a recombination event between any two loci is enhanced. AILs show a five-fold reduction in the size of a confidence interval estimating QTL positions in comparison to an F_2 population in an F_{10} AIL population. This is due to the large number of generations substantially increasing the cumulative number of recombination events. A single AIL can be more effective than a large number of RILs for fine-mapping, but RILs can be preferred in cases where environmental variance needs to be reduced in order to evaluate a trait that has QTL with low heritability (Darvasi and Soller 1995).

DHLs are produced by chromosome doubling of haploids through *in vitro* or *in vivo* methods. DH lines can be difficult to produce, but lines that are entirely homozygous and homogeneous are produced in one step (Xu 2010). This is a distinct advantage in evaluation of environmental effects, as DH lines can be identically reproduced as many times as needed across multiple environments, multiple studies, etc. As a result of their genetic makeup, there is no dominance or dominance related epistatic effects to be evaluated in DH lines. This allows better analysis of additive, additive related epistatic, and linkage effects (Xu 2010). While DH lines offer many benefits, they do have several disadvantages. Haploids can be difficult and expensive to obtain in large numbers and also eliminate potentially interesting lethal mutants in the haploid phase. DH lines may also suffer from reduced genetic diversity (Xu 2010). Since DH lines have only undergone one round of recombination, the genetic resolution is lower as compared to RIL populations (Burr and Burr 1991).

The NAM population was developed to make use of the best features from linkage (QTL) and association mapping (McMullen et al. 2009). The NAM population consists of 25 families of diverse maize lines, each containing more than 200 near-isogenic lines (NILs). About 136,000 recombination events were observed in this population. This means that there are three recombination events per gene on average and allows for much higher resolution mapping. The NAM population provides high statistical power, high allele diversity and short range of linkage disequilibrium that allow for very high resolution mapping. Through single nucleotide polymorphism (SNP) information from the founding lines, nucleotide polymorphisms can be tested more directly than identity-by-descent (Yu et al. 2008).

QTL Mapping methods

Quantitative traits differ substantially from qualitative traits. Qualitative traits are usually controlled by one, or a few, gene(s) that has a distinguishable effect on the target phenotype. Quantitative traits are generally controlled by multiple genes that each has a small effect on the target phenotype. In addition, environmental effects, as well as genotype x environment interactions, can also play a large role when evaluating quantitative traits. While qualitative traits can be grouped into classes and often studied as segregating classes, quantitative traits require the application of proper statistical methods based on trait distributions. Because the factors underlying quantitative traits can be much more difficult to elucidate, a variety of mapping methods have been developed for a diversity of population structures. Since the introduction of molecular markers in the 1980s, it has become possible to determine the location of a QTL through linkage (single marker, simple interval, composite interval, and multiple interval methods) or association analysis in a more efficient manner. In addition, the contribution of individual QTL to the phenotype can be established. Table 2 summarizes the mapping methods covered in this chapter.

Table 2 Mapping methods.

Mapping Method	Advantages	Disadvantages
Single marker analysis	Quick	Cannot differentiate size of QTL from distance between marker and QTL
Simple interval mapping (SIM)	Can estimate both position and effect of QTL	Linked QTL often cannot be separated leading to ghost QTL or missing QTL
Composite interval mapping (CIM)	Better control over linked QTL than SIM	Closely linked QTL can be missed
Multiple interval mapping (MIM)	Able to separate linked QTL. Finds multiple QTLs and can analyze epistatis. Can estimate genetic value, genetic variance, and heritability	Higher computational burden. Selection of best QTL model is challenging
Association mapping	Higher resolution than linkage (QTL) analysis, high genetic diversity, no need for a breeding population	Population structure in natural populations can be difficult to model

Single marker analysis

For each marker genotype and QTL genotype combination, a genotypic frequency can be calculated based on the recombination rate and population type. For calculating the sample mean and variance, we assume that the values of the QTL are normally distributed with homogenous variances over the different QTL genotypes. Testing for marker-trait associations can be carried out by comparing the sample means for each marker across genotype classes by ANOVA, or by regression (Xu 2010). This is the easiest and simplest method of QTL detection, but single marker analysis cannot determine the size of QTL effects or the distance between marker and QTL. Both estimates are confounded since analysis occurs only at individual marker positions (Lander and Botstein 1989). In single marker analysis, we assume that QTL trait values and variances are normally distributed (Xu 2010). If a QTL is not located at a marker locus, significantly more progeny will be required as the variance explained by the marker will decrease in relation to the recombination frequency.

Simple Interval Mapping (SIM)

In proposing interval mapping, Lander and Botstein (1989) addressed several shortcomings of single marker analysis. By using maximum likelihood (rather than linear regression), both a phenotypic value and a logarithm (base 10) of odds (LOD) score can be calculated for a QTL at any location on the genetic map. A QTL is found when a LOD value is higher than a predetermined critical value (values between 2 and 3 are often used). SIM uses a likelihood ratio test at every position within the single marker interval to test for a putative QTL. Both single marker analysis and SIM are methods used for locating a single QTL. Haley and Knott (1992) proposed a regression model for interval mapping and found little difference in results when compared with maximum likelihood. Closely linked QTLs are difficult to separate by SIM, which can lead either to the discovery of false QTLs or the failure in discovery of true QTLs. Using interval mapping with regression analysis, Haley and Knott (1992) had trouble separating QTLs that were as far as 20 cM apart. SIM has a higher statistical power than single marker analysis for QTL detection and, therefore, requires fewer progeny (Haley and Knott 1992; Lander and Botstein 1992). In SIM, we assume no interference and that the three possible QTL genotypes follow normal distributions. As a result, the effect of QTLs on the desired trait is a combination of these three normal distributions for the given marker locus (Xu 2010).

Composite Interval Mapping (CIM)

Composite interval mapping expands on SIM and single marker analysis by allowing the detection of multiple QTLs. Single marker analysis and SIM can show false ("ghost") QTLs in cases where multiple QTLs are linked and in coupling phase on the same chromosome. Markers between these linked QTLs may show, inaccurately, the highest phenotypic score (Xu 2010). Simple interval mapping can also give less accurate results for unlinked QTLs. CIM uses other markers, outside the interval being tested, as cofactors to control the genetic background. While scanning a particular marker interval for presence of a QTL, CIM eliminates the effects of other QTLs by using multiple regression analysis (Jansen 1993; Zeng 1993). For these reasons, CIM is more precise than SIM and single marker analysis (Zeng 1993). While CIM improves upon SIM in identifying QTLs, closely linked QTLs with opposite effects can contribute to missing QTLs. This occurs because CIM is unable to simultaneously consider, and remove the variation associated with, multiple QTLs that have already been found in the search for other QTLs. As such, linked QTLs with opposite effects on the phenotype can cancel out each other (Kao et al. 1999). For example, CIM was unable to find two QTLs in radiata pine due to one QTL 61 cM away from the left marker in the 3rd interval of linkage group 1 having an effect of 81.05 and a second QTL at the left marker of the 4th interval in linkage group 1 having an effect of –92.99. These positions are 11.8 cM apart (Kao et al. 1999). Multiple interval mapping was used to distinguish these QTLs.

Multiple Interval Mapping (MIM)

Multiple interval mapping was proposed by Kao et al. (1999) to apply SIM and CIM to a multiple QTL model and incurs a much heavier computational burden. Whereas SIM and CIM use one interval at a time to find a QTL, MIM uses multiple intervals concurrently to find multiple putative QTLs. MIM not only discriminates among separate linked QTLs, but also allows for the search and analysis of epistatic QTLs as well as the estimation of genotypic effects, the estimation of genotypic variance components, and the heritability of individual traits. MIM obtains better accuracy and power for QTL mapping, but identifying the best QTL model becomes a more complicated task (Kao et al. 1999). Because genotypic data at QTL is not directly observed (marker data is), maximum likelihood estimation of QTL position and effects is used to infer the distribution of the genotype of QTL. If there are a large number of QTLs, these estimates can quickly become very difficult to manage. Kao and Zeng (1997) developed formulas to handle this problem that assume no crossing-over interference, which means independence between flanking marker genotypes. To search for QTL to fit into the model, model selection methods are used as it is not possible to consider all model possibilities. Kao et al. (1999) discuss several of these selection methods.

Association Mapping

Association mapping, also referred to as linkage disequilibrium (LD) or gametic-phase disequilibrium (GPD) mapping, is a high resolution method. Association mapping is dependent on how the LD was created and the distance over which LD decays. Population subgroups, selection, drift, recombination and mating design are just some of the factors that can impact LD (Gaut and Long 2003). Because there are many factors that can affect LD, it can vary greatly between species, within species, and from locus to locus (Flint-Garcia et al. 2003). Association mapping makes use of populations with uncharacterized ancestry, or natural populations. This methodology has several advantages over linkage analysis: higher allele number and diversity, higher resolution, and no need to establish a breeding population (Buckler and Thornsberry 2002).

By definition, LD is the nonrandom association of alleles between separate loci. Because of complete linkage between markers and QTL alleles in F_1, large numbers of meioses are required to reduce associations between distant markers and QTL alleles in order to achieve high resolution mapping. Population types with a high number of generations (AILs, NILs, etc.) are one way to address this problem and to reduce LD. However, these designed crosses do not reduce disequilibrium as much as random mating (Jannink and Walsh 2002).

Applications of Mapping for Selection

Marker-Assisted Selection

For application of MAS, plant breeders use associations (linkage) between the desired traits and molecular markers. The first step is to establish linkage of relevant genes or QTL with markers. This is a relatively simple process for Mendelian inherited traits, as classical breeding procedures, such as backcrossing, can be used to introduce the alleles. For quantitatively inherited traits, the process is more complicated due to the need to manipulate many areas of the genome because many genes, often dispersed throughout the genome, with small effects must be incorporated. Also, quantitative traits confer additional difficulty in MAS due to epistatic and genotype by environment interactions complicating QTL discovery (Ribaut and Hoisington 1998). In addition, QTL mapping is required to analyze the association between markers and traits. After establishing these associations, breeders can introduce the desired alleles by F_2 improvement, gene stacking, or marker-assisted recurrent selection procedures to develop improved genotypes (Bernardo 2010).

DNA markers enjoy several advantages over phenotypic assays. DNA markers are often more accurate than phenotyping as they can be scored without ambiguity and are not muddied by phenotyping characteristics like trait heritability, environmental factors, and quantitative trait considerations (Xu and Crouch 2008; Xu 2010); better time efficiency as DNA markers can be evaluated at an earlier stage in development than phenotypes, and a potential reduction in cost as genotyping can be less expensive than phenotyping (Xu 2010). The markers chosen should be (1) 2 cM or less from the target gene, (2) polymorphic between genotypes that contain and lack the target gene, and (3) cost effective (Mohler and Singrün 2004).

According to Xu (2003), MAS requires (1) good and well characterized markers, (2) high-density molecular maps, (3) well characterized marker-trait associations, (4) efficient genotyping systems, and (5) effective data analysis. Limitations of MAS include the need to perform QTL analyses initially and the limited transferability of QTL from one mapping population to other breeding populations (Bernardo 2008). If the application of MAS requires performance of QTL experiments within the desired breeding population first (i.e., QTL analysis has not previously been performed in the breeding population for the desired traits), it delays progress in the breeding process. In addition, the pitfalls of QTL mapping must also be considered. Beavis (1998) points out that QTLs are actually just statistical constructs and are thus subject to evaluation of type I errors, power, precision, and accuracy. Type I errors lead to the finding of "ghost" QTL and a lack of statistical power leads to missing true QTL. QTL methods based on the null hypothesis of no QTL present, such as SIM, use this incorrect assumption and are subject to biased estimates of genetic effects and statistical significance. CIM produces more accurate and precise estimates than SIM, but including too many cofactors reduces power. In addition to the method used, the population used can also have a great impact. In small population sizes, detected QTLs are upwardly biased because the effects of QTL that are not found are lumped into the effects of those detected (Beavis 1998). This is known as "The Beavis Effect." As a result of these shortcomings of QTL analysis, selection based on marker information alone may not increase efficiency of breeding for quantitative traits compared to classical selection procedures.

Genomic Selection

Rather than using marker-trait associations that require linkage to be assessed between markers and traits, genomic selection, also called genome-wide selection (GWS) uses all available markers. MAS is limited by the amount of variance that is accounted for by known QTL (Meuwissen et al. 2001). In GWS, genetic effects are estimated for each individual marker in training populations, which are characterized for the traits of interest, while using high density and low cost marker systems. Based on the estimates obtained in the training population, "genomic estimated breeding values" (GEBVs) are determined for individuals in breeding populations, which are not evaluated in field trials. Selection is then based on these GEBVs. By employing all available markers, GWS can reduce the chance of missing QTL or finding "ghost" QTL (Guo et al. 2012). While simulations for genomic selection have generally concluded that this analysis can be more successful in increasing trait means, it does not provide any understanding of the underlying genes (Bernardo and Yu 2007; Guo et al. 2012). Genome-wide selection uses a large number of markers in lieu of a phenotype and thus avoids some of the problems associated with QTL analysis, such as the failure to incorrectly estimate the number and effect of genes, and their epistatic effects, for a trait of interest.

Mapping Studies

Biochemical Conversion of Lignocellulosic Materials

Lewis et al. (2010) used a testcross population comprised of 223 recombinant inbreds from the intermated B73 x Mo17 (IBM) population and the parental inbreds testcrossed to a Monsanto inbred to evaluate the feasibility of incorporating stover quality traits for ethanol production into a breeding program already in place for grain yield and agronomic traits (Lee et al. 2002). The authors measured the amount of cell wall glucose in dry stover, the amount of glucose released from the stover following thermochemical pretreatment and enzymatic saccharification, and the amount of cell wall lignin to evaluate the quality of corn stover in regards to ethanol conversion. A stover quality selection index based on these three traits was calculated, its foundation resting on high glucose, high glucose release, and low lignin content. Agronomic data was also collected and included: grain yield, grain moisture, root lodging, stalk lodging, and plant height. A yield selection index was calculated according to Bernardo (1991).

Narrow sense heritabilities (h^2) were 0.57, 0.63, 0.68, and 0.67 for glucose, glucose release, lignin, and stover quality index, respectively. Grain yield and yield index were 0.50 and 0.57, respectively. Glucose, glucose release, and stover quality indices were positively correlated (both phenotypically and genotypically) both to grain yield and yield index, while lignin was negatively correlated (both genotypically and phenotypically) to the same parameters. As a result of their analyses, the authors conclude that stover quality traits for ethanol production could be incorporated into an existing breeding program without detrimental effects to grain yield or other agronomic traits.

Lorenzana et al. (2010) used CIM to find 152 QTLs for glucose release and stover cell wall composition traits. The population used was the same as used by Lewis et al. (2010). This study found between four and ten QTLs for each of the 11 traits measuring cell wall composition and glucose conversion. Phenotypic variance (R^2) values ranged from 3% to 12%, and QTLs were found on all the ten maize chromosomes. The authors also mapped QTLs for stover cell wall components on a cell wall basis. Between five and eight QTLs were found for each of the nine traits. R^2 values ranged from 3% to 27%, and QTLs were found on all the ten maize chromosomes. The majority of all QTLs found in this study had small genetic effects.

The desired traits for ethanol conversion are high glucose content and release, and low lignin content. Positive correlations among traits suggest that selecting for glucose content may lead to higher levels of xylose and cell wall concentration, which may be advantageous as it has been suggested that future technology could allow higher ethanol yield through utilization of xylose and arabinose (Sun and Cheng 2002). Klason lignin and p-coumarate esters were negatively correlated on a dry matter basis –0.64 and -0.61, respectively, on a cell wall basis -0.64 and -0.65, respectively with glucose release. As a result, glucose release could be improved by selecting for low Klason lignin and p-coumarate esters levels. Due to the genetic variance and heritability found for the cell wall composition and glucose release traits, the authors conclude that these traits could be improved through selection.

Massman et al. (2012 e-published) used the above-mentioned study to evaluate trait gains using marker-assisted recurrent selection (MARS) against trait gains using genome-wide selection (GWS) for three selection cycles. A set of 287 SNPs markers were genotyped for two multiple trait selection indices: stover index and yield + stover index. For MARS, 58 SNP markers were significant ($P = 0.10$) for stover index, and 59 SNP markers were significant for yield + stover index. Four different selection programs were used: GWS for stover index, MARS for stover index, GWS for stover + yield index, and MARS for stover + yield index. After the three selection cycles, the population produced by GWS achieved a 14% higher stover index using both wet chemistry and NIRS measurement results, a 33% higher yield + stover index (for wet chemistry methods and 50% for NIRS) than the population produced by MARS. Evaluation on the yield index alone failed to discriminate a significant difference among selection procedures. However, the yield index in the GWS population was significantly higher for selection cycles two and three as compared to cycle 1, but not when measured against the population produced by MARS. Analyzing individual traits, as opposed to an index, expressed gains that were not significant. These results suggest that GWS is able to produce better trait improvement than MARS for trait indices.

Cell Wall and Forage Mapping Studies

Since few mapping studies directly relating to ethanol conversion have been completed, mapping studies for forage quality and cell wall related traits are reviewed as it has been shown that these traits impact ethanol conversion (Dien et al. 2006).

Using CIM, Méchin et al. (2001) evaluated a set of 100 silage maize RILs for QTL among 11 agronomic, cell wall digestibility and lignification traits (measured for whole above-ground plant). RILs were analyzed on a *per se* and testcross (with F252) basis. Narrow sense heritabilities ranged from 0.49 to 0.70 in the RILs and from 0.12 to 0.58 in the testcross population. Twenty-eight QTLs were found over the 11 traits in the testcross population. For each trait (except for *in vitro* dry matter digestibility), at least one QTL was detected, with a maximum of seven QTLs for *in vitro* cell wall digestibility. R^2 values ranged from 5.7% to 20.2%. In the *per se* population, 20 QTLs were found over the 11 traits. No QTLs were found for neutral detergent fiber and dry matter yield and four QTLs were found for *in vitro* digestibility of nonstarch and nonsoluble carbohydrate parts with R^2 values ranging between 6.5% and 15.3%. Most of the QTLs found for cell wall digestibility (but not agronomic traits) were consistent across the two populations and suggest that breeders could evaluate cell wall digestibility in early generations.

Cardinal et al. (2003) used RILs to analyze QTLs for neutral detergent fiber (NDF), acid detergent fiber (ADF) and acid detergent lignin (ADL) in leaf-sheath and stalk tissues. The RILs were developed from B73 and B52 due to their differences in ADF, NDF, and lignin values, and CIM was used for QTL mapping. The plot-basis broad sense heritabilities for the three traits ranged from 0.51 to 0.63 for leaf-sheath, and from 0.71 to 0.78 for stalk. The entry-mean broad sense heritabilities were much higher and ranged between 0.87 and 0.96 for both tissue types. All of the traits were positively correlated among both tissue types. Between eight and 12 QTLs were found for each of the three traits for each tissue type. The QTLs for each trait explained between 45% and 65% of the phenotypic variance. The high heritabilities and relatively large cumulative R^2 per trait suggest that selection should be effective. There were clusters of QTLs found that affected traits in one type of tissue and not the other.

Krakowsky et al. (2005, 2006) evaluated QTLs for ADF, NDF, and ADL in maize stalks (2005) and leaf sheaths (2006) in RILs created with crosses of B73 and De811. Broad sense heritability was 0.92 for NDF, 0.92 for ADF, and 0.74 for ADL in the stalks. Narrow sense heritabilities for the leaf sheath tissue were 0.94, 0.93, and 0.67, respectively. Fifteen QTLs were found for NDF in the leaf sheath that explained 58% of the phenotypic variance, and 16 QTLs were found for NDF in the stalk that explained 71% of the phenotypic variance. Thirteen QTLs were found for ADF in the leaf sheath that explained 54% of the phenotypic variance, and 18 QTLs were found for ADF in the stalk that explained 70% of the phenotypic variance. Fourteen QTLs were found for ADL in the leaf sheath that explained 51% of the phenotypic variance, and 10 QTLs were found for ADL in the stalk that explained 50% of the phenotypic variance. The authors compared their study with that of Cardinal et al. (2003) and found only a small portion of QTLs in common.

Truntzler et al. (2010) conducted a meta-analysis of QTLs for digestibility and cell wall composition traits across 14 mapping studies using 11 mapping populations. Fifty-nine QTLs for digestibility traits and 150 QTLs for cell wall composition traits were included in the analysis. Of these QTLs, 26 associated with digestibility and 42 associated with cell wall composition were found to be metaQTLs. Many of the metaQTLs were found in relatively few of the initial studies, which suggest that additional QTL studies using more diverse lines could reveal additional QTLs for digestibility and cell wall component traits. Several regions found in this study are promising, for further research and for use in a breeding program, as these regions contained metaQTLs that encompassed digestibility traits, cell wall component traits, and candidate genes.

While there is a set of putative pathways for lignin synthesis, some steps have yet to be confirmed (Barrière et al. 2007). Current data suggest that in excess of 37 genes could be part of maize monolignol synthesis (Guillaumie et al. 2007). Using acid detergent lignin and neutral detergent fiber determination methods, 58 QTLs have been found in maize for lignification. These QTLs were found at 43 locations throughout the genome with most of them lacking validated candidate genes (Barrière et al. 2007). The QTLs were spread over the maize genome, except for chromosome 7, which contained only two QTLs with low R^2 values. Unexpectedly, no QTLs were found near the *bm2* and *bm4* mutations, and only one was found near *bm1*. QTL could be expected to be near these loci as *brown midrib* (*bm*) mutants impact some aspect of lignin synthesis and show altered cell wall composition (Barrière and Argillier 1993; Barrière et al. 2007). The *bm3* gene infers a large negative impact on lignin content and unexpectedly colocated with only two QTLs. Very little QTL data is available regarding hydroxyphenyl, guaiacyl, and syringyl lignin units (Barrière et al. 2007).

Thermochemical Mapping Studies

No mapping studies investigating the thermochemical conversion of maize have been completed. Mapping cell wall related traits could provide insight, but how the cell wall impacts thermochemical conversion is not fully understood. The authors of this chapter are currently working on a QTL mapping study using the intermated B73 x Mo17 (IBM) population to investigate individual bio-oil compounds as measured by pyrolysis-gas chromatography-mass spectrometry (publication in preparation).

Conclusion

All of the data available suggest that maize could be significantly improved as a lignocellulosic feedstock for ethanol conversion. Because ethanol conversion traits have not been selected for in traditional maize breeding programs, gains will likely be very significant and the amount of ethanol produced from maize stover increased substantially. In addition, ethanol conversion traits have been found to have either positive or nonsignificant correlation with agronomic traits such as grain yield, suggesting that ethanol conversion traits could be selected for simultaneously with agronomic traits.

Acknowledgements

I want to thank the Plant Sciences Institute at the Iowa State University (Ames, IA, USA) and the Symbi GK-12 NSF program for providing my funding. I also want to thank the people at the Center for Sustainable and Environmental Technologies at Iowa State University for greatly contributing to my educational training.

References

Barrière Y and Argillier O (1993) Brown-midrib genes of maize: a review. Agronomie 13: 865–876.

Barrière Y, Riboulet C, Méchin V et al. (2007) Genetics and Genomics of Lignification in Grass Cell Walls Based on Maize as Model Species. In: Teixeira da Silva JA (ed) Genes, Genomes, and Genomics, Global Science Books, London UK, pp. 134–151.

Beavis W (1998) QTL analyses: Power, precision, and accuracy. In: Paterson AH (ed) Molecular Dissection of Complex Traits. CRC Press, Boca Raton, FL, USA, pp 145–162.

Bernardo R (1991) Retrospective index weights used in multiple trait selection in a maize breeding program. Crop Sci 31: 1174–1179.

Bernardo R and Yu J (2007) Prospects for genomewide selection for quantitative traits in maize. Crop Sci 47: 1082–1090.

Bernardo R (2008) Molecular markers and selection for complex traits in plants: Learning from the last 20 Years. Crop Scie 48: 1649–1664.

Buckler ES and Thornsberry JM (2002) Plant molecular diversity and applications to genomics. Curr Opin Plant Biol 5: 107–111.

Burr B, Burr FA, Thompson KH et al. (1988) Gene mapping with recombinant inbreds in maize. Genetics 118: 519–526.

Burr B and Burr FA (1991) Recombinant inbreds for molecular mapping in maize: theoretical and practical considerations. Trends Genet 7: 55–60.

Butler E, Devlin G, Meier D et al. (2011) A review of recent laboratory research and commercial developments in fast pyrolysis and upgrading. Renew Sustain Energy Rev 15: 4171–4186.

Cardinal AJ, Lee M and Moore KJ (2003) Genetic mapping and analysis of quantitative trait loci affecting fiber and lignin content in maize. Theor Appl Genet 106: 866–874.

Czernik S and Bridgwater AV (2004) Overview of applications of biomass fast pyrolysis oil. Energy Fuels 18: 590–598.

Darvasi A and Soller M (1995) Advanced intercross lines, an experimental population for fine genetic mapping. Genetics 141: 1199–1207.

Dien BS, Jung HJG, Vogel KP et al. (2006) Chemical composition and response to dilute-acid pretreatment and enzymatic saccharification of alfalfa, reed canarygrass, and switchgrass. Biomass Bioenergy 30: 880–891.

Fahmi R, Bridgwater AV, Donnison I et al. (2008) The effect of lignin and inorganic species in biomass on pyrolysis oil yields, quality and stability. Fuel 87: 1230–1240.

Flint-Garcia SA, Thornsberry JM and Buckler(IV) ES (2003) Structure of linkage disequilibrium in plants. Annu Rev Plant Biol 54: 357–374.

Gaut BS and Long AD (2003) The lowdown on linkage disequilibrium. Plant Cell 15: 1502–1506.

Guillaumie S, San-Clemente H, Deswarte C et al. (2007) MAIZEWALL. Database and developmental gene expression profiling of cell wall biosynthesis and assembly in maize. Plant Physiol 143: 339–363.

Guo Z, Tucker DM, Lu J et al. (2012) Evaluation of genome-wide selection efficiency in maize nested association mapping populations. Theor Appl Genet 124: 261–275.

Haley CS and Knott SA (1992) A simple regression method for mapping quantitative trait loci in line crosses using flanking markers. Heredity 69: 315–324.

Hästbacka J, de la Chapelle A, Kaitila I et al. (1992) Linkage disequilibrium mapping in isolated founder populations: diastrophic dysplasia in Finland. Nat Genet 2: 204–211.

Jannink JL and Walsh B (2002) Association mapping in plant populations. In: Kang MS (ed) Quantitative Genetics, Genomics and Plant Breeding. CAB International, Wallingford, UK, pp 59–68.

Jansen RC (1993) Interval mapping of multiple quantitative trait loci. Genetics 135: 205–211.

Kao CH and Zeng ZB (1997) General formulas for obtaining the MLEs and the asymptotic variance-covariance matrix in mapping quantitative trait loci when using the EM algorithm. Biometrics 53: 359–371.

Kao CH, Zeng ZB and Teasdale RD (1999) Multiple interval mapping for quantitative trait loci. Genetics 152: 1203–1216.

Krakowsky MD, Lee M and Coors JG (2005) Quantitative trait loci for cell-wall components in recombinant inbred lines of maize (*Zea mays* L.) I: stalk tissue. TheorAppl Genet 111: 337–346.

Krakowsky MD, Lee M and Coors JG (2006) Quantitative trait loci for cell-wall components in recombinant inbred lines of maize (*Zea mays* L.) II: leaf sheath tissue. Theor Appl Genet 111: 337–346.

Lander ES and Botstein D (1989) Mapping Mendelian factors underlying quantitative traits using RFLP linkage maps. Genetics 121: 185–199.

Lee M, Sharopova N, Beavis WD et al. (2002) Expanding the genetic map of maize with the intermated B73 × Mo17 (IBM) population. Plant Mol Biol 48: 453–461.

Lewis MF, Lorenzana RE, Jung HJG et al. (2010) Potential for simultaneous improvement of corn grain yield and stover quality for cellulosic ethanol. Crop Sci 50: 516–523.

Lorenz AJ, Coors G, de Leon N et al. (2009) Characterization, genetic variation, and combining ability of maize traits relevant to the production of cellulosic ethanol. Crop Sci 49: 85–98.

Lorenz AJ, Coors G, Hansey CN et al. (2010) Genetic Analysis of Cell Wall Traits Relevant to Cellulosic Ethanol Production in Maize. Crop Sci 50: 842–852.

Lorenzana RE, Lewis MF, Jung HJG et al. (2010) Quantitative Trait Loci and Trait Correlations for Maize Stover Cell Wall Composition and Glucose Release for Cellulosic Ethanol. Crop Sci 50: 541–555.

Massman JM, Jung HJG and Bernardo R (2012) Genomewide Selection versus Marker-Assisted Recurrent Selection to Improve Grain Yield and Stover-Quality Traits for Cellulosic Ethanol in Maize. Crop Sci: Published ahead of print 21 Aug 2012; doi: 10.2135/cropsci2012.02.0112.

McMullen MD, Kresovich S, Villeda HS et al. (2009) Genetic Properties of the Maize Nested Association Mapping Population. Science 325: 737–740.

Méchin V, Argillier O, Hébert Y et al. (2001) Genetic Analysis and QTL Mapping of Cell Wall Digestibility and Lignification in Silage Maize. Crop Sci 41: 690–697.

Meuwissen THE, Hayes BJ and Goddard ME (2001) Prediction of Total Genetic Value Using Genome-Wide Dense Marker Maps. Genetics 157: 1819–1829.

Mohler V and Singrün C (2004) General considerations: marker-assisted selection. In: Lörz H and Wenzel G (eds) Biotechnology in Agriculture and Forestry, Vol 55 Molecular Marker Systems in Plant Breeding and Crop Improvement. Springer-Verlag, Berlin, Germany, pp 305–317.

Nowakowski DJ, Bridgwater AV, Elliott DC et al. (2010) Lignin fast pyrolysis: Results from an international collaboration. J Analyt Appl Pyrolysis 88: 53–72.

Oasmaa A, Solantausta Y, Arpiainen V et al. (2010) Fast Pyrolysis Bio-Oils from Wood and Agricultural Residues. Energy Fuels 24: 1380–1388.

Ribaut JM and Hoisington D (1998) Marker-assisted selection: new tools and strategies. Trends Plant Sci 3: 236–239.

Shen DK, Gu S and Bridgwater AV (2010) The thermal performance of the polysaccharides extracted from hardwood: Cellulose and hemicelluloses. Carbohyd Polym 82: 39–45.

Sun Y and Cheng J (2002) Hydrolysis of lignocellulosic materials for ethanol production: A review. Bioresour Technol 83: 1–11.

Truntzler M, Barrière Y, Sawkins MC et al. (2010) Meta-analysis of QTL involved in silage quality of maize and comparison with the position of candidate genes. Theore Appl Genet 121: 1465–1482.

Vermerris W, Saballos A, Ejeta G et al. (2007) Molecular Breeding to Enhance Ethanol Production from Corn and Sorghum Stover. Crop Sci 47: S142–S153.

Xu Y (2003) Developing marker-assisted selection strategies for breeding hybrid rice. Plant Breed Rev 23: 73–174.

Xu Y and Crouch JH (2008) Marker-Assisted Selection in Plant Breeding: From Publications to Practice. Crop Sci 48: 391–407.

Xu Y (2010) Marker-assisted selection: Theory. In: Xu Y (ed) Molecular Plant Breeding. CAB International, Wallingford, UK, pp 286–336.

Xu Y (2010) Molecular dissection of complex traits: Theory. In: Xu Y (ed) Molecular Plant Breeding. CAB International, Wallingford, UK, pp 195–249.

Xu Y (2010) Populations in genetics and breeding. In: Xu Y (ed) Molecular Plant Breeding. CAB International, Wallingford, UK, pp 113–151.

Yu J, Holland JB, McMullen MD et al. (2008) Genetic design and statistical power of nested association mapping in maize. Genetics 178: 539–551.

Zeng ZB (1993) Theoretical basis for separation of multiple linked gene effects in mapping of quantitative trait loci. Proc Natl Acad Sci USA 90: 10972–10976.

Zeng ZB (1994) Precision mapping of quantitative trait loci. Genetics 136: 1457–1468.

CHAPTER 7

Genomics Resources

Eugenia Barros[1], and E. Jane Morris[2]*

ABSTRACT

The fastest growing use of maize is for the production of fuel ethanol using the enzymatic conversion of corn starch to glucose and then to ethanol as well by converting the cellulosic (non-food) parts of maize to ethanol. However for the production of ethanol from maize to be economically viable, improvements need to be made not only in maize yield and agronomic performance, but also in properties such as grain quality with respect to starch conversion, lower lignin content and modified cell wall structure. This chapter highlights the new advances in functional genomics and related technologies that have the potential to make maize more amenable to ethanol production.

Keywords: starch, ethanol, lignin, biomass, proteomics, metabolomics, functional genomics

Introduction

Maize is considered as one of the future biofuel crops, with the potential to contribute to the first generation biofuels by the fermentation of corn starch into ethanol and to the second generation biofuels by the chemical/enzymatic hydrolysis of lignocellulosic materials into ethanol. Starch is

[1]CSIR Biosciences, Meiring Naude Road, Brummeria, Pretoria 0001, South Africa.
[2]Department of Biochemistry, University of Pretoria and African Centre for Gene Technologies, PO Box 75011, Lynnwood Ridge 0040, South Africa.
*Corresponding author: ebarros@csir.co.za

the main constituent of the maize kernel and is composed of two glucose polymers, amylose and amylopectin. Glucose is produced after starch hydrolysis following gelatinization, liquefaction and saccharification. Lignin, on the other hand, is one of the most important components of the plant cell walls and the one that limits the enzymatic degradability and digestibility of cellulose and hemicellulose in the biomass into fermentable sugars (Torney et al. 2007).

Functional genomics technologies seek to understand the relationship between the maize genome and its phenotype, the properties and function of all the maize genes and the gene products and to interpret the natural variation that exists in maize at the gene, protein and metabolite levels during plant development and in the different parts of the plant. Recent advances in functional genomics and other inter-related "omics" technologies have provided crucial information for the optimization of maize as a bioenergy crop. This chapter reviews the molecular data that has been generated using these technologies and highlights the sources that are available to plant breeders to screen for the best maize germplasm using marker-assisted selection and to genetic engineers to manipulate the maize plant with the genes coding for traits that will improve bioethanol production.

Structural and Functional Genomics for Improved Starch and Biomass

Various methodologies have been used to understand starch metabolism and to find strategies that can alter the synthesis of starch to make it more amenable for conversion into bioethanol. Work has also been reported on the lignin biosynthetic pathway to look for ways to reduce the lignin content of maize and/or facilitate lignin breakdown for the easier conversion of biomass into bioethanol.

Lignin is the second most abundant organic compound in plants and the one that hinders the conversion of plant biomass into fermentable sugars. One hundred and two candidate expression sequence tags (ESTs) have been identified in maize and their function linked to cell-wall digestibility using suppression subtractive hybridization (SSH) combined with microarray-based expression profiling (Shi et al. 2007). The maize microarrays used in this study and in similar studies were the cDNA unigene microarrays produced by the laboratory of Prof. P. Schnable (Iowa University, USA), that contain 11,827 maize ESTs clustered into 9,841 unigenes, accounting for 20% of about 50,000 maize genes (http://www.plantgenomics.iastate.edu/maizechip/). Guillaumie et al. (2007) developed a comprehensive maize cell wall macroarray containing gene-specific tags (GSTs) generated from a wide diversity of maize lignified cell types. In addition a detailed and user friendly maize cell wall database, MAIZEWALL, was constructed containing a complete bioinformatic analysis of each gene and gene expression data. This database can be accessed from the home page http://www.polebio. scsv.ups-tlse.fr/MAIZEWALL (Guillaumie et al. 2007) and is a valuable source for future candidate molecular markers to allow plant breeders to select the best maize germplasm with either lower lignin content or cell walls easier to degrade. The MAIZEWALL website has links to other cell wall websites and contains: 1) gene sequence and gene differential expression data; 2) protein sequence information on maize as well as other species including rice, poplar and papaya.

Fifteen genes associated with the quantitative trait loci (QTL) for lignin content and cell-wall digestibility were identified by Thomas et al. (2010) and these genes can be further characterized to be potentially converted into molecular markers. Bosch et al. (2011) identified differentially expressed genes involved in cell wall biogenesis by comparing the transcriptome profiles of actively elongating internodes of maize stalks with internodes that had stopped growing, using a 70-mer maize oligonucleotide array. The identified genes included those coding for enzymes involved in carbohydrate metabolism such as glycosyl transferases (GTs), glycosyl hydrolases (GHs) and xyloglucan endotransglucosylases/hydrolases (XTHs). Genes associated with lignin biosynthesis were identified, as well as cell wall and plasma membrane proteins that are involved in cell-wall related processes, such as leucine-rich proteins (LRPs) and genes coding for arabinogalactan proteins (AGPs).

The exploitation of cell wall modifying enzymes such as cellulases has been shown to increase biomass by over expression in a number of plant species. The increased biomass is due to cleavage of cross-links between cell wall sugars, loosening the cell wall and allowing unregulated cellular expansion. There is scope for using a variety of cell wall modifying enzymes identified by proteomics either in combination or by targeting them to specific tissue types to increase plant biomass (Ito et al. 2010). Proteomics techniques have also been used to study the effect of inhibitors of cellulose biosynthesis on maize cell wall structure and plasticity (Melida et al. 2010). In a proteomics study of cell wall proteins associated with maize root apical cell elongation (which also requires a loosening of the cell wall), it was noteworthy that a high proportion of identified proteins were associated with carbohydrate metabolism (Zhu et al. 2006).

The sequencing of the maize genome provided the foundation for the blueprint of many genes including those that are key to the understanding of the metabolic and regulatory pathways involved in starch metabolism and in biomass breakdown/degradation (Schnable et al. 2009). By combining the information supplied by the genome sequence with the available DNA microarrays Sekhon et al. (2011) embarked on a comprehensive gene expression study of the genes for enzymes involved in lignin biosynthetic pathways across developmental stages and plant organs. In the process they generated a maize transcriptome atlas that is a valuable source for gene discovery and for functional genomics with application in biofuel research in maize. For example, when planning the manipulation of a gene involved in a biochemical pathway, it is crucial to select the exact developmental stage of the plant, i.e., if one wants to modify a gene involved in a particular biochemical pathway, it is essential to know the expression profile of that gene at the exact developmental stage of the plant in order for it to make the biggest impact.

The gene for cinnamoyl-CoA reductase (CCR1) that catalyzes the lignin-specific branch of monolignol biosynthesis has been confirmed to be the key enzyme controlling the quantity and the structure of lignin in maize. By studying the expression profiles of cDNAs of two CCR1 mutants (*ZmCCR1* and *ZmCCR2*) Tamasloukht et al. (2011) showed a downregulation of these genes that resulted in a decrease in lignin content without negative effects to plant growth. Another way to decrease the lignin content is to keep the plant at a juvenile state; Chuck et al. (2011) overexpressed the maize mutant, *Corngrass1* (Gg1), in maize and produced plants with more branches and leaves that had a juvenile morphology. These mutants can be utilized to produce better biofuel crops.

A genetically modified variety of maize that expresses α-amylase, one of the enzymes responsible for starch breakdown, has been approved by the US Department of Agriculture (Waltz 2011). This maize will contribute to a more efficient ethanol production.

Functional Genomics Towards Improved Agronomic Performance for Biofuel Production

One of the factors that make maize attractive as a biofuel crop is its highly efficient C_4 photosynthetic pathway. Crops using C_4 photosynthesis on average out-yield and are more water and nitrogen use efficient than C_3 crops (Hibberd and Covshoff 2010). Recent functional genomics studies have explored gene regulation and protein accumulation in the key mesophyll and bundle sheath cells involved in C_4 photosynthesis, opening possibilities for future manipulation of the photosynthetic pathway at the cellular level including the manipulation of starch biosynthesis (Friso et al. 2010; Majeran et al. 2010). The use of high-resolution and high-sensitivity mass spectrometry to enable proteomic analysis at the cellular and subcellular level shows great promise for the future.

The integration of functional genomics studies at all levels (transcriptomics, proteomics and metabolomics) offers new possibilities to understand and manipulate systemic changes in the plant's response to the external environment. This is demonstrated in a study of the response of maize to UV-B irradiation (Casati et al. 2011). Tolerance to high UV-B levels is essential if maize yields are to be maintained in the face of ozone depletion and climate change.

Application of maize as a biofuel crop would be enhanced if it were able to grow on marginal land. In this regard tolerance to salt, heavy metals and other abiotic stressors are important. Genomic, transcriptomic and proteomic approaches have been used to study salt tolerance in crops other than maize. The response of the maize proteome to salt stress was studied by Zorb et al. (2010) using 2-D gel electrophoresis (2-DE). Bai et al. (2011) investigated the role of nitric oxide and related G-protein at the proteomic level in protecting against salt stress. Such studies need to be complemented by an understanding of alterations at the metabolite level. Gavaghan et al. (2011) used ^{1}H NMR metabolomics profiling of maize, complemented with multivariate data analysis, to identify possible biomarkers for osmotic adjustment in high saline environments.

Less attention has been given to studying heavy metal tolerance in maize at the genomics level. Using 2-DE, Requejo and Tena (2005) identified the upregulation of a number of oxidative stress regulated proteins in response to arsenic exposure and subsequently a number of QTLs for arsenic accumulation have been identified and mapped (Ding et al. 2011). Maron et al. (2008) performed a detailed temporal analysis of root gene expression under aluminium stress using microarrays with Al-tolerant and Al-sensitive maize genotypes, and subsequently mapped identified ESTs to genes coding for multidrug and toxic compound extrusion (MATE) family of transporters (Maron et al. 2010).

One of the most complex forms of abiotic stress is drought. Drought tolerance is a key success factor for maize if it is to achieve widespread use as a biofuel crop, yet it involves physiological and morphological adaptations in the plant, which are not yet well understood at the molecular level. At the transcriptional level the expression of genes associated with drought stress has been studied in developing maize kernels (Luo et al. 2010) with the aim of identifying drought-responsive genes that could be used in germplasm assessment. Other studies at the proteomic level have used 2-DE to examine responses to abscisic acid and drought (Hu et al. 2011) as well as to characterize small heat shock proteins associated with combined drought and heat stress (Hu et al. 2010). Zhu et al. (2007) also used 2-DE to investigate the cell wall proteins involved in maize primary root elongation under water stress, which are associated with an increase in longitudinal cell wall extensibility in the apical region.

As mentioned by Luo et al. (2010) there is a linkage between drought stress and susceptibility to biotic stress. Identification of stress-responsive genes requires a multidisciplinary approach that includes genome-wide genetic and physical chromosome mapping, isolation and sequencing of important genes, microarray analysis, and proteomic analyses (Afroz et al. 2011). Chivasa et al. (2005) studied changes in the maize extracellular matrix in response to pathogen elicitors at the proteomics level using 2-DE. More recently Pechanova et al. (2011) also used 2-DE to examine the role of the maize cob (rachis) in resistance to the fungus *Aspergillus flavus*. Although maize lines with resistance to this fungus are available, the introgression of resistance from these lines into elite commercial cultivars has not been successful due to the lack of selectable markers; it is to be hoped that studies of this nature will aid the search for suitable markers. Lanubile et al. (2010) studied the defense response of maize to *Fusarium verticillioides* infection at the RNA level using oligonucleotide arrays; genetic variation linked to resistance to *Fusarium* ear rot exists although the resistance is not complete. Mohammadi et al. (2011) examined the response of two different maize varieties to infection by the fungus *Fusarium graminearum* using iTRAC (isobaric tagging for relative and absolute quantification) shotgun proteomics. Responses to virus infection have received less attention, but a recent paper by Li et al. (2011), also using 2-DE, showed that maize infected with rice black-streak dwarf virus showed significant impairment of morphology and metabolism, including effects on starch metabolism and on the composition and secretion of cell wall structural polymers. This clearly demonstrates the potential negative impact of disease on maize as a biofuel crop.

A number of studies at the proteomics level have examined characteristics of maize roots that contribute to optimal root structure and nutrient uptake for enhanced yield. These include the use of nanoLC-MS/MS to study the proteins of maize root mucilage, which contributes to water holding and ameliorates the effects of heavy metals (Ma et al. 2010). The technique has also been used to study root hairs, which significantly increase the water and nutrient uptake of plants by enlarging their root surface (Nestler et al. 2011). Li et al. (2007) used 2-DE to study the adaptations of maize roots to phosphorus deficiency.

Although significant advances have been made in recent years, metabolomics techniques are less developed than transcriptomics or proteomics, and their application to maize has been limited to date. However, a team of researchers from Monsanto (Skogerson et al. 2010) recently undertook detailed metabolic profiling of maize grain using GC–time of flight mass spectrometry (GC-TOF MS). Although it is early days, the authors highlight a potential role for metabolic profiling in assisting the process of selecting elite germplasm in biotechnology development, or marker-assisted breeding. Studies of this nature have the potential to be complemented by *in silico* studies such as that of Saha et al. (2011) who undertook a genome-scale metabolic reconstruction of maize metabolism. These authors point out that attempting to focus on a single pathway, such as a reduction in lignin content for improved biofuel production, without quantitatively assessing the system-wide implications, may lead to loss of plant viability and fitness.

Metabolomics techniques have also been used to investigate the phenomenon of heterosis in maize (i.e., the superior performance of maize hybrids over their inbred parents) (Römisch-Margl et al. 2010), complementing transcriptomic studies of heterosis (e.g., Guo et al. 2006; Meyer et al. 2007; Hoecker et al. 2008; Stupar et al. 2008; Paschold et al. 2010) and exploration of copy number variations (Beló et al. 2009). The authors showed that inbred and hybrid genotypes could be distinguished based on the quantification of only a few metabolites. A deeper understanding of the contribution of parent genotypes to maize hybrid vigor can contribute towards the optimal selection of inbred parents to improve maize as a bioenergy crop.

The challenges facing agriculture and the need for high yielding maize crops able to adapt to future climate changes has prompted the development of new approaches to tackle functional genomics and plant breeding. Phenomics is the new research area that comprises a set of technologies aiming at fast tracking the selection of the most appropriate germplasm for a specific agricultural application be it food security or biofuel production (Furbank and Tester 2011). Using the phenomics approach maize plants best suited for biofuel production and high yielding crops adapted to changing climatic conditions will be identified. Basically phenomics aims to bridge the gap between genomics, plant function, plant physiology and agricultural traits using a suite of technologies that range from noninvasive imaging, spectroscopy, image analysis, robotics and high-performance computing for field phenotyping; carbon isotope discrimination (CID), infrared thermography, chlorophyll fluorescence analysis and digital growth analysis for phenoptyping of abiotic stress tolerance and, nondestructive screening techniques for phenomic screening for biotic stress tolerance, among others.

Impact of New Technologies on Developing Genetic Resources

Genetic engineering of maize plants to both increase cell wall biomass and to make the cell walls more degradable to fermentable sugars, without compromising the life cycle of the maize plant, is one of the areas that would have a tremendous impact in the biofuel industry.

Table 1 Omics sources available for maize.

Omics description		Link	Reference
Genomics Sources	B73 Reference Genome Version 2	(http://gbrowse.maizegdb.org/cgi-bin/gbrowse/maize_v2)	Maize Genetics and Genomics Database
	Haplotype Map	http://www.sciencemag.org/cgi/content/full/326/5956/1112/DC1	Gore et al. 2009
	Cell wall genes	http://polebio.scsv.ups-tlse.fr/MAIZEWALL	Guillaumie et al. 2007
	Maize Genetics and Genomics Database	http://www.maizegdb.org	Guillaumie et al. 2007
	GreenPhylDB for plant comparative genomics	http://greenphyl.cirad.fr/v2/cgi-bin/index.cgi	Rouard et al. 2011
Transcriptomics Sources	Transcripts	http://maizesequence.org/index.html	Friso et al. 2010
	Maize Seed *in situ* Hybridization database (MASISH)	http://masish.uab.cat/masish/	Miquel et al. 2011
	MaizePLEX (part of PLEXdb) expression resources for maize	http://www.plexdb.org/plex.php?database=Corn	Dash et al. 2011
	Maize microarray annotation database	http://maizearrayannot.bi.up.ac.za/	Coetzer et al. 2011
Proteomics Sources	ZmGI maize proteome database	http://compbio.dfci.harvard.edu/index.html	Friso et al. 2010
	PRoteomicsID Entifications database (PRIDE)	http://www.ebi.ac.uk/pride/	Friso et al. 2010
	Plant Proteomics Database (PPDB) for maize and Arabidopsis	http://ppdb.tc.cornell.edu/	Sun et al. 2009
	Protein Families Involved in Transduction of Signalling in maize (ProFITS)	http://bioinfo.cau.edu.cn/ProFITS	Ling et al. 2010
Metabolomics Sources	Plant NMR metabolomic profiles (MeRy-B)	http://www.cbib.u-bordeaux2.fr/MERYB/index.php	Ferry-Dumazet et al. 2011
	Golmmetabolome database	http://gmd.mpimp-golm.mpg.de/Default.aspx	Kopka et al. 2005
	Maize metabolic pathways (MaizeCyc)	(http://www.gramene.org/pathway/maizecyc.html)	-
	KaPPA-View4 Classic	http://kpv.kazusa.or.jp/kpv4/	Sakurai et al. 2011
	Gene-to-gene and metabolite-to-metabolite coexpression analysis		

In order to obtain a better understanding of maize metabolic pathways and regulation of carbon flow, much research still needs to be done to get to the point where the outcomes could contribute to the design of better materials. In future it is likely that manipulation at the regulatory level will be more successful in altering multigenic traits than intervention at the level of single genes. This will demand a systems biology approach, integrating all aspects of structural and functional genomics and utilising sophisticated bioinformatics.

To facilitate the access to the technologies that are currently available in maize and that can be applied towards a better production of ethanol from maize a user friendly summary was compiled with links to the most up to date 'Omics' sources (Table 1).

References

Afroz A, Ali GM, Mir A et al. (2011) Application of proteomics to investigate stress-induced proteins for improvement in crop protection. Plant Cell Rep 30: 745–763.

Bai X, Yang L, Yang Y et al. (2011) Deciphering the protective role of Nitric Oxide against salt stress at the physiological and proteomic levels in maize. J Proteome Res 10: 4349–4364.

Beló A, Beatty MK, Hondred D et al. (2010) Allelic genome structural variations in maize detected by array comparative genome hybridization. Theor Appl Genet 120: 355–367.

Bosch M, Mayer C-D, Cookson A et al. (2011) Identification of genes involved in cell wall biogenesis in grasses by differential gene expression profiling of elongating and non-elongating maize internodes. J Exp Bot 62: 3545–3561.

Casati P, Campi M, Morrow A et al. (2011) Transcriptomic, proteomic and metabolomics analysis of UV-B signaling in maize. BMC Genomics 12: 321.

Chivasa S, Simon WJ, Yu XL et al. (2005) Pathogen elicitor-induced changes in the maize extracellular matrix proteome. Proteomics 5: 4894–4904.

Chuck GS, Tobias C, Sun L et al. (2011) Overexpression of the maize *Corngrass 1* microRNA prevents flowering, improves digestibility and increases starch content of switchgrass. Proc Natl Acad Sci USA 108: 17550–17555.

Coetzer N, Myburg AA and Berger DK (2011) Maize microarray annotation database. Plant Meth 7: 31 doi: 10.1186/1746-4811-7-31.

Dash S, Van Hemert J, Hong L et al. (2011) PLEXdb: gene expression resources for plants and plant pathogens. Nucl Acids Res doi: 10.1093/nar/gkr938.

Ding D, Li W, Song G et al. (2011) Identification of QTLs for arsenic accumulation in maize (*Zea mays* L.) using a RIL population. PLoS ONE 6(10): e25646. doi: 10.1371/journal.pone.0025646.

Ferry-Dumazet H, Gil L, Deborde C et al. (2011) MeRy-B: a web knowledgebase for the storage, visualization, analysis and annotation of plant NMR metabolomics profiles. BMC Plant Biol 11: 104.

Friso G, Majeran W, Huang M et al. (2010) Reconstruction of metabolic pathways, protein expression, and homeostasis machineries across maize bundle sheath and mesophyll chloroplasts: large-scale quantitative proteomics using the first maize genome assembly. Plant Physiol 152: 1219–1250.

Furbank RT and Tester M (2011) Phenomics—technologies to relieve the phenotyping bottleneck. Trends Plant Sci 16: 635–644.

Gavaghan CL, Li JV, Hadfield ST et al. (2011) Application of NMR-based metabolomics to the investigation of salt stress in maize (*Zea mays*). Phytochem Anal 22: 214–224.

Gore MA, Chia J-C, Elshire RJ et al. (2009) A first-generation haplotype map of maize. Science 236: 1115–1117.

Guillaumie S, San-Clemente H, Deswarte C et al. (2007) MAIZEWALL. Database and developmental gene expression profiling of cell wall biosynthesis and assembly in maize. Plant Physiol 143: 339–363.

Guo M, Rupe MA, Yang X et al. (2006) Genome-wide transcript analysis of maize hybrids: allelic additive gene expression and yield heterosis. Theor Appl Genet 113: 831–845.

Hibberd JM and Covshoff S (2010) The regulation of gene expression required for C_4 photosynthesis. Annu Rev Plant Biol 61: 181–201.

Hoecker N, Keller B, Muthreich N et al. (2008) Comparison of maize (*Zea mays* L.) F_1-hybrid and parental inbred line primary root transcriptomes suggests organ-specific patterns of nonadditive gene expression and conserved expression trends. Genetics 179: 1275–1283.

Hu X, Li Y, Li C et al. (2010) Characterization of small Heat Shock Proteins associated with maize tolerance to combined drought and heat stress. J Plant Growth Regul 29: 455–464.

Hu X, Lu M, Li C et al. (2011) Differential expression of proteins in maize roots in response to abscisic acid and drought. Acta Physiol Plant 33: 2437–2446.

Ito J, Petzold CP, Mukhopadhyay A et al.(2010) The role of proteomics in the development of cellulosic biofuels. Current Proteom 7: 121–134.

Kopka J, Schauer N, Krueger S et al.(2005) GMD@CSB.DB: the GolmMetabolome Database. Bioinformatics 21: 1635–1638.

Lanubile A, Pasini L andMarocco A (2010) Differential gene expression in kernels and silks of maize lines with contrasting levels of ear rot resistance after *Fusarium verticillioides* infection. J Plant Physiol 167(16): 1398–1406.

Li K, Xu C, Zhang J et al. (2007) Proteomic analysis of roots growth and metabolic changes under phosphorus deficit in maize (*Zea mays* L.) plants. Proteomics 7: 1501–1512.

Li K, Xu C and Zhang J (2011) Proteome profile of maize (*Zea mays* L.) leaf tissue at the flowering stage after long-term adjustment to rice black-streaked dwarf virus infection. Gene 485: 106–113.

Ling Y, Du Z, Zhang Z et al. (2010) ProFITS of maize: a database of protein families involved in the transduction of signalling in the maize genome. BMC Genomics 11: 580.

Luo M, Liu J, Lee RD et al. (2010) Monitoring the expression of maize genes in developing kernels under drought stress using oligo-microarray. J Integr Plant Biol 52: 1059–1074.

Ma W, Muthreich N, Liao C et al. (2010) The mucilage proteome of maize (*Zea mays* L.) primary roots. J Proteome Res 9. 2968–2976.

Majeran W, Friso G, Ponnala L et al. (2010) Structural and metabolic transitions of C4 leaf development and differentiation defined by microscopy and quantitative proteomics in maize. Plant Cell 22: 3509–3542.

Maron LG, Kirst M, Mao C et al. (2008) Transcriptional profiling of aluminum toxicity and tolerance responses in maize roots. New Phytol 179: 116–128.

Maron LG, Piñeros MA, Guimarães CT et al.(2010) Two functionally distinct members of the MATE (multi-drug and toxic compound extrusion) family of transporters potentially underlie two major aluminum tolerance QTLs in maize. Plant J 61: 728–740.

Mélida H, Encina A, Álvarez J et al. (2010) Unravelling the biochemical and molecular networks involved in maize cell habituation to the cellulose biosynthesis inhibitor dichlobenil. Mol Plant 3: 842–853.

Meyer S, Pospisil H and Scholten S (2007) Heterosis associated gene expression in maize embryos 6 days after fertilization exhibits additive, dominant and overdominant pattern. Plant Mol Biol 63: 381–391.

Miquel M, López-Ribera I, Ràmia M et al. (2011) MASISH: a database for gene expression in maize seeds. Bioinformatics 27: 435–436

Mohammadi M, Anoop V, Gleddie S et al. (2011) Proteomic profiling of two maize inbreds during early gibberella ear rot infection. Proteomics 11: 3675–3684.

Nestler J, Schutz W and Hochholdinger F (2011) Conserved and unique features of the maize (*Zea mays* L.) root hair proteome. J Proteome Res 10: 2525–2537.

Paschold A, Marcon C, Hoecker N et al. (2010) Molecular dissection of heterosis manifestation during early maize root development. Theor Appl Genet 120: 383–388.

Pechanova O, Pechan T, Williams WP et al. (2011) Proteomic analysis of the maize rachis: Potential roles of constitutive and induced proteins in resistance to *Aspergillus flavus* infection and aflatoxin accumulation. Proteomics 11: 114–127.

Requejo R and Tena M (2005) Proteome analysis of maize roots reveals that oxidative stress is a main contributing factor to plant arsenic toxicity. Phytochemistry 66: 1519–1528.

Römisch-Margl L, Spielbauer G, Schützenmeister A et al. (2010) Heterotic patterns of sugar and amino acid components in developing maize kernels. Theor Appl Genet 120: 369–380.

Rouard M, Guignon V, Aluome C et al. (2011) GreenPhylDB v2.0: comparative and functional genomics in plants. Nucl Acids Res 39: D1095–D1102.

Saha R, Suthers PF and Maranas CD (2011) *Zea mays* iRS1563: a comprehensive genome-scale metabolic reconstruction of maize metabolism. PLoS ONE 6(7): e21784. doi: 101371/journal.pne.0021784.

Sakurai N, Ara T, Ogata Y et al. (2011) KaPPA-View4: a metabolic pathway database for representation and analysis of correlation networks of gene co-expression and metabolite co-accumulation and omics data. Nucl Acids Res 39: D677–684

Schnable PS, Ware D, Fulton RS et al. (2009) The B73 maize genome: complexity, diversity and dynamics. Science 326: 1112–1115.

Sekhon RS, Lin H, Childs KL et al. (2011) Genome-wide atlas of transcription during maize development. Plant J 66: 553–563.

Shi C, Uzarowska A, Ouzunova M et al. (2007) BMC Genomics 8: 22.

Skogerson K, Harrigan GG, Reynolds TL et al. (2010) Impact of genetics and environment on the metabolite composition of maize grain. J Agri Food Chem 58: 3600–3610.

Stupar RM, Gardiner JM, Oldre AG et al. (2008) Gene expression analyses in maize inbreds and hybrids with varying levels of heterosis. BMC Plant Biol 8: 33.

Sun Q, Zybailov B, Majeran W et al. (2009) PPDB, the Plant Proteomics Database at Cornell. Nucl Acids Res 37: D969–D974.

Tamasloukht B, MS-JWQ Lam, Martinez Y et al. (2011) Characterization of a cinnamoyl-CoA reductase 1 (CCR1) mutant in maize: effects on lignifications, fibre development and global gene expression. J Exp Bot 62: 3837–3848.

Thomas JS, Guillaumie C Verdu et al. (2010) Mol Breed 25: 105–124.

Torney F, Moeller L, Scarpa A et al. (2007) Genetic engineering approaches to improve bioethanol production from maize. Curr Opin Biotechnol 18: 193–199.

Waltz E (2011) Amylase corn sparks worries. Nat Biotechnol 29: 294.

Zhu J, Chen S, Alvarez S et al. (2006) Cell wall proteome in the maize primary root elongation zone. I. Extraction and identification of water soluble and lightly ionically bound proteins. Plant Physiol 140: 311–325.

Zhu J, Alvarez S, Marsh EL et al. (2007) Cell wall proteome in the maize primary root elongation zone. II. Region-specific changes in water soluble and lightly ionically bound proteins under water deficit. Plant Physiol 145: 1533–1548.

Zorb C, Schmitt S and Mühling KH (2010) Proteomic changes in maize roots after short-term adjustment to saline growth conditions. Proteomics 10: 4441–4449.

CHAPTER 8

Genome Sequencing
Past and Present

W. Brad Barbazuk[1,2,]** *and Wenbin Mei*[1]

ABSTRACT

Maize nucleotide sequences have been routinely sampled for well over 10 years, and early efforts were successful in producing collections of EST and genome survey sequences that were of great use to the maize genetics and genomics community. Given its agricultural importance and its status as a model for plant genetics and cytogenetics, the maize genome was an obvious target for comprehensive DNA sequencing. Coordinated efforts to obtain a whole-genome sequence of maize inbred B73 began in 2006, and culminated in the publication of a whole-genome sequence in 2009 (Schnable et al. 2009). While the release of the reference genome sequence for maize enables a better understanding of maize genome structure and function, the collection of maize DNA sequence continues. Over the past five years, advances in sequencing technology have facilitated large-scale, high-throughput, sequence-based approaches to better understand genetics, genome structure and many biological phenomena such as genome evolution, gene expression, gene structure and the epigenetic interactions that augment expression. Continued use of Next Generation sequencing technologies in maize are expanding maize sequence collections to include tissue and genotype specific sets, which are identifying the genome mechanics behind

[1]Department of Biology, University of Florida, 220 Bartram Hall, PO Box 118525, Gainesville Fl 32611, USA.
[2]The University of Florida Genetics Institute, Cancer & Genetics Research Complex, 2033 Mowry Road, PO Box 103610, Gainesville, FL 32610, USA.
*Corresponding author: bbarbazuk@ufl.edu

its domestication, allowing a better understanding of its breeding history, and enabling efforts that will lead to improvements in yield and quality.

Keywords: Genomic sequencing, EST, methyl filtration, B73, comparative sequencing, high throughput

Introduction

Early studies suggested the existence of significant collinearity between grass genomes, which prompted the expectation that comparative genomics would enable the application of knowledge gained by sequencing the smaller rice genome to help in positional cloning of maize genes underlying important agronomic traits (Goff et al. 2002; Yu et al. 2002). However, substantial differences in gene organization observed between maize inbreds illustrated the surprisingly dynamic nature of the maize genome, underscoring the importance of understanding maize at the genome level (Fu and Dooner 2002). A whole-genome sequence of maize inbred B73 was published in 2009 and genomic techniques centered on DNA sequencing continue to address important questions related to maize gene regulation, epigenetics, genome structure, genome evolution and agronomic improvement (Schnable et al. 2009). This chapter focuses on the technologies, resources and outcomes, both past and present, relevant to sequencing the maize genome, and includes a brief discussion on the use of sequence data from related species for maize comparative genomics. However, our emphasis will be on the major findings of the whole-genome sequence of maize and the subsequent use of second generation, high-throughput sequencing to further sample the maize genome and transcriptome (Schnable et al. 2009). The rapid improvement in DNA sequence technology underpinning the available 'next-generation' sequencing (NGS) platforms is facilitating large-scale, high-throughput, sequence-based approaches to better understand the dynamic genome of maize.

Past Efforts

Maize genome sequencing began over 10 years ago with several reports of DNA sequence from specific genes or small genomic intervals published in the late 1990s to early 2000s. For example: ribosomal RNA genes (Messing et al. 1984); maize zein genes (Kridl et al. 1984; Kirihara et al. 1988; Song and Messing 2002); *sh2* and *a1* loci (Chen et al. 1997); *adh1* (Tikhonov et al. 1999); the *rp3* rust resistance locus (Webb et al. 2002); the *Rp1* resistance gene complex (Ramakrishna et al. 2002); the phytochrome A region (Morishige et al. 2002). However, large-scale sequencing of maize genomic DNA was not pursued in the public domain until early in the 21st century at which time maize genome sequence investigations mediated by reduced representation (see below) methods (Whitelaw et al. 2003), BAC-end sequencing (Messing et al. 2004) and random BAC clones (Haberer et al. 2005; Bruggmann et al. 2006) proceeded to a joint NSF, USDA and DOE supported collaborative genome sequencing effort that released a whole-genome sequence in 2009 (Schnable et al. 2009).

Prior to the jointly-funded effort to produce a maize whole-genome sequence most large-scale efforts utilized a reduced representation approach. Reduced representation simplifies the problem of sequencing a large complex genome by targeting only a portion of the genome, usually genic, and avoiding the bulk of the non-genic and repeat content. In addition to having a genome that is multiple-gigabases in size, the maize genome is a segmental allotetraploid that originated approximately 11 million years ago from progenitors that diverged 20.5 million years before (Gaut and Doebley 1997; Gaut et al. 2000; Gaut 2001). Subsequent to polyploidization, a major retrotransposon expansion occurred that has resulted in repetitive DNA accounting for upwards of 80% of the maize genome (Flavell et al. 1974; SanMiguel et al. 1998; Meyers et al. 2001; Palmer et al. 2003; Whitelaw et al. 2003); with the maize gene content widely dispersed throughout (Tikhonov et al. 1999; Fu and Dooner 2002; Song and Messing 2002; Ilic et al. 2003; Langham et al. 2004). Consequently, approaches that avoid the largely repetitive regions and preferentially sequence gene regions to characterize the maize genome are discussed below.

EST sequencing

The most widely applied reduced representation method is expression sequence tag (EST) sequencing. With this methodology, random cDNAs, derived from segments of individual mRNA molecules can be sequenced to obtain expressed nucletide information without requiring knowledge of the genome arrangement. Because cDNA is derived from mRNA, these sequences define the transcribed portion of the genome, much of which accounts for the protein-coding portion. In addition to identifying expressed regions, a comprehensive random cDNA sequencing project can also enumerate the relative amounts of each transcript present, thus providing expression information. The main drawback to cDNA sequencing as a method for gene discovery is that EST data shows uneven representation of genes due to expression biases, resulting in an incomplete collection of gene sequences, regardless of the size of the data set. In addition, untranscribed flanking regions that may contain the regulatory sequences are absent from

ESTs. Over 2 million EST sequences from a wide variety of maize tissues and genotypes have been collected and are available from Genbank.

Efforts at clustering and assembling maize ESTs to remove redundancy and provide as-long-as-possible reference transcript sets have been made and these collections are available from three main sources. The maize gene index (Quackenbush et al. 2000, 2001; http://compbio.dfci.harvard.edu/cgi-bin/tgi/gimain.pl?gudb=maize) hosted by the Computational Biology and Functional Genomics Laboratory at the Dana Farber Cancer Institute consists of 112,256 unigene clusters constructed from 1,698,645 input sequences. A collection of greater than 93 thousand unigene sequences is available from GenBank at the NCBI (http://www.ncbi.nlm.nih.gov/unigene?term=Zea%20mays%5BOrganism%5D), while a collection of 115,221 unigene contigs assembled from 1,293,797 input sequences is housed in PlantGDB (Dong et al. 2004; http://www.plantgdb.org/prj/ESTCluster/AssemblyRawData.php).

Of particular interest is a collection of full-length cDNA sequences produced by researchers at the University of Arizona and Stanford University (Soderlund et al. 2009). This collection, consisting of 27,455 full-length cDNA sequences collected from 27 B73 tissues and abiotic stress treatments, provides a resource describing the precise primary sequence of a significant fraction of maize transcripts and thus remains as a high value collection of maize gene associated sequences, which played an instrumental role during the annotation phase of the whole-genome sequence of maize (Schnable et al. 2009).

Methylation filtration

In general, plants exhibit differential methylation of gene and repetitive sequence. While repetitive DNA is heavily methylated at 5'-CG-3' and 5'-CNG-3' sequences, gene sequences and low-copy elements are relatively undermethylated (Bennetzen et al. 1994; Rabinowicz et al. 1999; Lippman et al. 2004). The differential methylation that exists in plant genomes can enable preferential recovery of gene-rich DNA fragments (hypomethylated) through techniques such as methylation filtration or digestion with methylation-sensitive restriction enzymes. Methylation-sensitive enzymes provide a simple and reproducible method to enrich for low-copy genomic DNA. In general, treatment of the maize genome with a methyl-sensitive restriction enzyme will result in cuts only at restriction sites that are not methylated. Because the majority of repeat sequences in maize are methylated, this tends to maintain large blocks of repeat sequences on long strands, while the hypomethylated, and presumably gene-rich, portion is substantially fragmented. Size selection followed by sequencing will enrich for gene-associated sequences. Emberton et al. 2005 demonstrated the use of methyl-sensitive enzymes in maize to create a hypomethylated partial restriction (HMPR) library that demonstrated a 6-7-fold enrichment in gene-associated sequences relative to a library made with nonmethylation-sensitive restriction enzymes. This approach was utilized by Gore et al. (2009b), who used HMPR to sample genomic sequences from two maize inbred lines and identify gene-enriched single nucleotide polymorphisms (SNPs) between them. Over 126,000 SNPs were detected at an estimated false discovery rate of ~ 15%, demonstrating utility for large-scale detection of gene-associated SNP markers that can support fine-mapping and association studies.

The obvious drawback to using methylation-sensitive enzymes is that restriction enzyme-cut sites are not randomly distributed across the genome, so not all hypomethylated gene-rich regions will be represented. Two reduced representation techniques that significantly enrich for gene-associated sequences without experiencing the same sequence content biases that plague restriction enzyme based methods are methylation filtration (Rabinowicz et al. 1999) and high C_0t selection (Yuan et al. 2003). Methylation filtration uses the endogenous restriction-modification system of *E. coli* McrBC to eliminate random genomic clones containing methylated (i.e., repetitive) DNA inserts (Rabinowicz et al. 1999), and enriching for clones that containing low-copy DNA fragments (Rabinowicz et al. 1999). In contrast, high C_0t selection relies on DNA renaturation that was typically used to characterize the repetitive and unique DNA components of complex genomes (Britten et al. 1974). This technique uses DNA renaturation kinetics to examine the fraction of low-copy DNA within a genome. The renaturation rate of a particular sequence is proportional to the copy number found in the genome. In a typical C_0t experiment, the genome is sheared, fully heat-denatured and then slowly cooled to allow renaturation. DNA sequences that are repetitive, and thus present in high-copy number, will renature faster than single-copy sequences. Repetitive sequences that renature quickly can be separated from single strand low-copy sequences by hydroxyapatite (HAP) chromatography. The single-stranded fraction is double-stranded *in vitro* and cloned into sequencing vectors (Yuan et al. 2003).

In maize, two large-scale methyl filtration experiments (Palmer et al. 2003; Whitelaw et al. 2003) and one high C_0t experiment (Whitelaw et al. 2003) produced ~ 500,000 and ~ 400,000 methyl-filtered and high C_0t reads, respectively. These sequences have been clustered by two independent efforts producing contigs designated as AZMs (Whitelaw et al. 2003; Chan et al. 2006) or MAGIs (Fu et al. 2005). The AZM sets are available at http://maize.jcvi.org/, the MAGIs at: http://magi.plantgenomics.iastate.edu/. The main benefit of this collection is that it provided the research community with a significant fraction of the maize gene space, well in advance of the commencement of the whole maize genome sequence project. Analysis suggested (Whitelaw et al. 2003; Springer et al. 2004; Chan et al. 2006) that the maize gene space represents approximately 550 Mb, and the current collection of methyl-filtered and high cot sequences represented on average 75% of the nucleotides of ~ 95% of the cDNAs discovered as of 2006 (Chan et al. 2006).

Comparative genomic sequencing

Early work using restriction fragment length polymorphism (RFLP) markers revealed well conserved synteny between the grass genomes (Moore et al. 1995; Gale and Devos 1998). Based on their analysis, Gale and Devos (1998) suggested that the linear organization of genes in nine grass genomes could be described in terms of "25 rice linkage blocks". While it remained to be seen to what extent gene collinearity would be preserved at the 'micro' level, the gross level synteny results suggested that "...elucidation of the organization of the economically important grasses with larger genomes, such as maize ($2n$ = 10, 4,500 Mb DNA), will, to a greater or lesser extent, be predicted from sequence analysis of smaller genomes such as rice, with only 400 Mb..." (Gale and Devos 1998) and that "Comparative genetics will provide the key to unlock the genomic secrets of crop plants with bigger genomes than *Homo sapiens*." (Gale and Devos 1998).

Understanding the genome of rice was important for continued improvement of this cereal, which is a primary staple of much of the world population. However, its modestly sized genome (400 Mbp), relative ease of transformation, genetic maps, germplasm collections, and the discovery of general genome collinearity elevated it to a model organism for cereal research (reviewed by Cooke et al. 2007). During 2002, two publications reported the genome sequence of rice: *Oryza sativa* ssp. *japonica* var. Nipponbarre (Goff et al. 2002) and *Oryza sativa* ssp. *indica* var. 93-11 (Yu et al. 2002). Genome sequences of additional cereals such as sorghum (Paterson et al. 2009) and *Brachypodium* (The International Brachypodium Initiative 2010) followed, increasing the potential for leveraging comparative genomics analysis in maize. Several loci have been compared between maize and rice and/or sorghum that reveal numerous exceptions to microcollinearity (Cooke et al. 2007); however, sequence alignments between maize and other cereal genomes reveal significant gross-level collinear stretches (Fig. 8.1) that can identify candidate regions in maize that may contain orthologous loci to those known to affect agronomic traits in other cereals. To facilitate these efforts, comparative maps between maize, rice, sorghum, barley, wheat and oat have been computed and have been made publicly available on Gramene, a comparative genome mapping database for grasses and a community resource for rice (Ware et al. 2002). An example of the utility of comparative genomics in maize is provided by the identification of candidate genes affecting flowering time (Chardon et al. 2004) and cell wall composition (Hazen et al. 2003).

The Genome Sequence of B73 Maize

While the reduced-representation sequence resources and availability of other cereal genome sequences were extremely useful for examining the gene content and organization of maize genome, the ultimate maize genetic reference is a whole-genome sequence. A whole-genome sequence provides the highest-resolution genetic map of an organism, representing every single feature (genes, regulatory regions, noncoding features, etc.) in their relative order and orientation with respect to one another. While genome science may not be able to decipher all of the important features of a genome when it is first available, a comprehensive whole-genome sequence contains all of the important information and provides a substrate for ongoing discovery. The whole-genome sequence of the maize inbred line B73 commenced in 2006, and was led by a consortium of investigators from Washington University in St. Louis, Iowa State University, The University of Arizona and Cold Spring Harbor Laboratories under a joint United States Department of Agriculture, Department of Energy and the US National Science Foundation (NSF) effort funded by NSF under the auspices of the National Plant Genome Initiative (NPGI). The effort culminated in a release of the primary sequence and annotation (Schnable et al. 2009) in addition to several companion reports detailing specific features of the genome sequence and analysis. A summary of the maize genome sequence and primary findings follows.

The sequencing strategy utilized was Sanger sequencing a minimum tiling path of BACs selected from the whole genome BAC fingerprint-based physical map (Wei et al. 2007, 2009). The BAC-fingerprint map framework was constructed from the fingerprints of over 292,000 agarose and 350,000 HICF-based fingerprints from BAC clones (Wei et al. 2007), derived from three deep coverage BAC libraries each constructed with a different restriction enzyme (Coe and Schaeffer 2005). Over 25,000 markers were placed upon the fingerprint map (Wei et al. 2007), allowing integration to the comprehensive maize genetic map (Coe and Schaeffer 2005) and providing a chromosomal context to the BAC map fingerprint contigs and genome sequences (see below). A collection of 16,868 BAC clones constituting a minimally redundant tiling path through the BAC fingerprint physical map were selected and these served as substrates for shotgun

Synteny dot plot Maize vs Rice

A

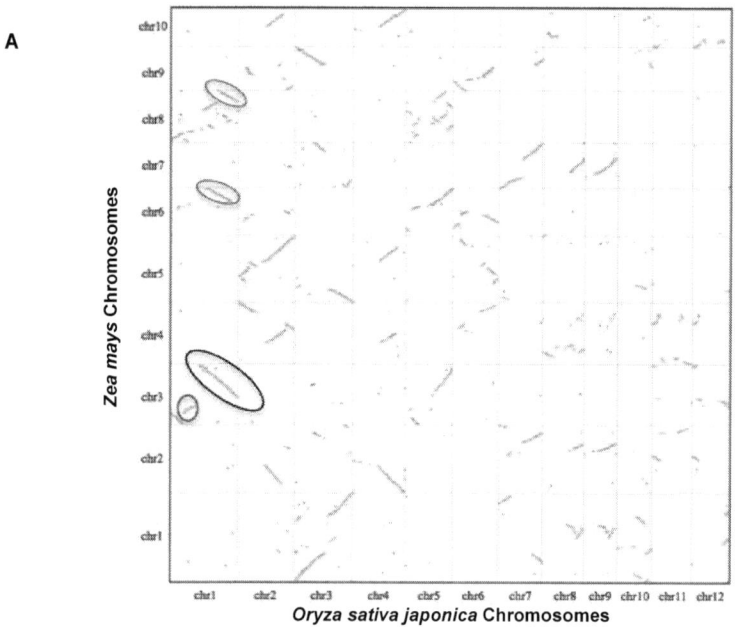

Synteny dot plot Maize vs Sorghum

B

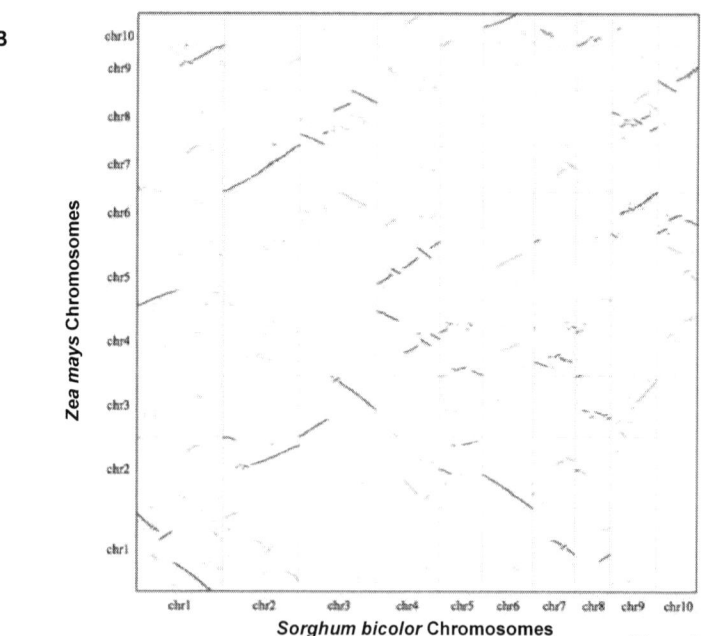

Figure 1 contd....

sequencing. Each BAC was sequenced to a depth of between 4–6×, and assembled independently of one another. Overlaps between neighboring clones were identified and consolidated, and unique (nonrepeat laden) regions of the primary assembly were subjected to automated and manual improvement (finishing) (Schnable et al. 2009). In addition, a genome-wide, high-resolution optical map was used to further refine the order of the sequence scaffolds (Zhou et al. 2009).

Annotation was carried out to add value to the assembly by identifying genes, repeat families, noncoding RNAs and other genome features. As eluded too by previous studies the maize genome was found to be highly repetitive (Meyers et al. 2001; Whitelaw et al. 2003). The maize genome is rich in transposable elements, which accounts for upwards of 85% of the genome. The vast majority of these are long terminal repeat retrotransposons (LTRs) accounting for 75% of the genome that are represented within 406 families. Additionally, approximately 8% of the genome sequence is defined by DNA transposons that derive from 855 families. Moreover, nearly 20,000 Helitrons representative of 8 families were also discovered. Helitrons are recently described transposable elements that do not generate terminal repeats and target-site duplications, instead using rolling circle amplification (Yang and Bennetzen 2009). The existence of a large collection of Helitrons within primarily gene-rich regions is particularly intriguing, because Helitrons have been implicated in gene movement and genome evolution, and these may have contributed significantly to the dynamic nature of the maize genome.

Figure 1 Dot plots illustrating colinearity between maize and rice (A), and maize and Sorghum (B). Note that regions of colinearity are recognized as line segments, for example, the extensive colinearity observed between chromosome 1 of rice and chromosome 3 of maize. The red oval represents a large portion of maize chromosome 3 that is colinear to rice chromosome 1, but the orthologous gene order is inverted relative to one another, as opposed to in the same orientation as depicted by an additional smaller stretch labeled marked with a red oval. Black ovals represent other portions of the maize genome (portions of maize chromosome 6 and maize chromosome 8) that exhibit colinearity to chromosome 1 of rice, and are indicative of duplicated regions in maize.

Plots were produced using the SynMap utility available as part of the CoGe Comparative genomics online analysis toolkit (http://genomevolution.org/CoGe/index.pl). Parameters used were:

Analysis Options:

Blast Algorithm: **Last**; Expected number of hits per sequence **4;** Maximum distance between two matches (-D): **20** genes), Minimum number of aligned pairs (-A, #57): **5** genes. Calculate syntenic CDS pairs and color dots: **Synonymous (ks)** substitution rates;

Display options:

Sort Chromosomes by: **Name**; Dotplot axis metric: **Genes**; Label Chromosomes; Skip Random/ Unknown Chromosomes.

Color image of this figure appears in the color plate section at the end of the book.

Gene discovery was accomplished by unifying the results of cross-species homology searches, results of alignments of available ESTs to the genor e sequence, addition of RNA-seq (see below, *High-Throughput Sequencing and Gene Expression Studies*) to identify transcribed regions, *ab initio* computational gene discovery methods, and conservation of gene order between maize, rice and sorghum. The results from the above analysis identified over 100,000 putative protein-coding gene loci (unfiltered set), that were reduced to a collection of 32,540 high-confidence (or filtered) gene loci by considering the above data sets simultaneously and extracting those loci that were supported by multiple pieces of evidence (Schnable et al. 2009). Subsequent analysis of the filtered set genes identified 11,892 maize gene families, of which 8,494 are also present within the genomes of *Arabidopsis*, rice and sorghum suggestive of a core set of gene families present in the flowering plants. A total of 10,571 such genes are conserved between maize, rice and sorghum suggestive of a core set of gene families in cereals, while 465 families are only found in maize and may represent maize-specific gene families (Schnable et al. 2009). That there exists a collection of 465 gene families that exist in maize but are not found in rice, sorghum or *Arabidopsis* is not that unexpected. Earlier work aimed at identifying maize transcripts associated with the shoot apical meristem identified several expressed sequences that had no similarity to any sequences present in GenBank at that time (Emrich et al. 2007). While absence of these families in *Arabidopsis*, rice and sorghum is intriguing, definitive identification of those families that are truly maize-specific requires a more comprehensive comparative analysis that samples species broadly throughout the major angiosperm clades.

Access to the maize genome sequence, reduced representation sequence collections, gene annotation, genetics and additional information is available from public portals; two of the most important will be mentioned here. The main project website for the B73 maize genome reference sequence is housed at the Cold Spring harbor Laboratory, and can be accessed at http://www.maizesequence.org. This site (Schnable et al. 2009) contains links to download sequence and annotation data, as well as provides access to an interactive genome browser, and tools to facilitate comparative genomics. A second public portal, Maize GDB (Lawrence et al. 2008; Harper et al. 2011; Schaeffer et al. 2011) provides access to maize community data collections such as sequences, maps, metabolic pathway info and quantitative trait loci (QTLs), etc.

Two additional maize genome sequence efforts were run in parallel to the B73 whole genome sequencing project: *Palomero Toluqueno*, which is highland popcorn from Mexico (Vielle-Calzada et al. 2009) and the *Zea mays* inbred, Mo17. The highland popcorn landrace was sequenced and compared to the B73 genome to identify sequence variability and help identify putative domestication loci (Vielle-Calzada et al. 2009). Maize inbred Mo17 was draft shotgun sequenced by the JGI/DOE using the Roche 454 and Illumina next-generation sequence technologies to provide an additional resource to study sequence variation in the maize genome. Information and links to the sequence data are available at http://genome. jgi.doe.gov/genome-projects/pages/project-status.jsf?projectId=401872 and http://www.phytozome.net/maize.php, respectively.

Next-Generation Sequencing Technologies

Although the whole-genome sequence reference for B73 maize is completed, and shotgun sequences of the maize inbred Mo17, in addition to the Palomero genome already exist, maize genome and transcriptome sequencing efforts continue within the maize community (Schnable et al. 2009; Vielle-Calzada et al. 2009). Much of the current maize sequencing activity involves the use of next-generation sequence methods to essentially "re-sequence" additional maize genotypes, to examine the methylation status of the maize genome to further delineate epigenetics, or to examine gene expression among tissues, various lines or within mutant backgrounds. Next-generation sequence technologies are relatively recent advances in sequencing chemistries and instrumentation that enable the acquisition of high volumes of DNA sequence at costs that are orders of magnitude lower than obtained with automated Sanger sequencing. While the benefits of inexpensive, high-throughput, DNA sequencing are numerous and obvious, the increased volume and decreased cost come at the expense of short read-lengths.

Five next-generation sequencing platforms that are available commercially and used routinely are: the Roche 454 (Holt and Jones 2008) the Illumina family consisting of the Genome Analyzer, the MiSeq and the HiSeq2000 (Bentley 2006; Mardis 2008), the Applied Biosystems SOLiD system (Holt and Jones 2008; Mardis 2008), the Ion Torrent PGM (Rothberg et al. 2011) and the Pacific Biosciences DNA sequencer (Eid et al. 2009; http://www.pacificbiosciences.com). The Pacific Biosciences (PacBio) machine is a single molecule sequencing system, while the remaining platforms clonally amplify template by PCR to produce enough substrate to provide sufficient signal for base determination. Roche-454, Ion Torrent, Illumina and PacBio systems all use a sequence-by-synthesis approach, while the ABI SOLiD sequences by ligation (Metzker 2010).

The Roche-454 platform utilizes iterative pyrosequencing. In a 454 reaction, beads containing clonally amplified template and bead-tethered pyrosequencing enzymes (polymerase, luciferase, sulfurylase) are deposited into the wells of a picotitre plate, and buffer containing one of the four nucleotides is passed across the plate. Base incorporation results in the emission of a light signal (catalyzed by sulphurylase and luciferase) that is detected by the apparatus. A single run of the 454 GS-FLX titanium and 454 GS-FLX+ instrument takes approximately 7 hours, and can produce over 1 million sequences with mean lengths of approximately 500 bp and 700 bp, respectively.

The Ion Torrent machine uses native dNTPs and a DNA polymerase for extension, but rather than detecting a florescent signal to signify base addition, the Ion Torrent system relies on a silicon chip to detect hydrogen ions released during base addition (Rothberg et al. 2011; Loman et al. 2012). Currently, Ion Torrent PGM instruments are capable of 1 Gbp of sequence with read-lengths of 200 bp on average, but lengths approaching 400 bp are expected by late 2012.

Unlike the 454 and Ion Torrent platforms, Illumina uses a glass plate to which templates are anchored and clonally amplified *in situ*, and base extension is accomplished with nucleotides labeled with four-color fluorescent reversible terminators (see Bentley 2006; Mardis 2008). This allows the simultaneous addition of all four labeled nucleotides across samples, while permitting only a single base extension each round. After extension and detection of the fluorescent signal, sequencing reagents are washed away, the fluorophore is cleaved, and the 3' hydroxyl is unblocked in preparation for the next round of extension. Because all four fluorescently tagged based are present the risk of mis-incorporation is low. Depending on the specific instrument, Illumina platforms can produce read-lengths of 36–150 bp with output capacities ranging from 6Gbp to 600Gbp in a single instrument run (http://www.illumina.com/systems/sequencing.ilmn).

Like Roche-454, Ion Torrent and Illumina, the Pacific Biosciences DNA sequencer follows a sequence-by-synthesis approach. However, rather than using fluorescently labeled terminators, the Pac-Bio system (Eid et al. 2009; http://www.pacificbiosciences.com) incorporates nucleotide tri-phosphates that are labeled on the gamma-phosphorous. The PacBio sequencing plate is called a SMRT cell, which is a small platform that consists of tens of thousands of tiny wells. These wells, termed zero-mode waveguides (ZMWs), serve double-duty as sequencing reaction and detection vessels. Each ZMW is illuminated from below, but the wavelength of the light is too large for it to pass easily through the wave guide, and instead can only illuminate a shallow zone at the bottom of the ZWM. A DNA template-polymerase complex is tethered to the bottom of the ZMW, and the phospholinked nucleotides are introduced to the ZMW chamber. Because all nucleotides are present, it is expected that all will fluoresce assuming the light illuminating the bottom of the ZMW was able to penetrate into the chamber. Because just a shallow zone of the ZMW is illuminated, only those phospholinked nucleotides that are within the shallow zone can fluoresce, and only those stationary long enough for their signal to be detected will be registered above background signals. When a base is incorporated, it is held in place for a brief time by the polymerase, and a brief pulse of light is given off by the fluorophore as a result of it absorbing attenuated light near the bottom of the ZMW chamber. This delay is sufficient to detect the fluorophore label above the background signals prior to phosphodiester bond formation and concomitant release of the fluorophore labeled pyrophosphate (Eid et al. 2009; http://www.pacificbiosciences.com). Another important difference in the PacBio platform is that only a single molecule of DNA is present in a ZMW—thus the PacBio system is a true single molecule sequencing system, which avoids some of the template errors that can occur as a result of clonal amplification. Additionally, the PacBio system has the potential to generate very long reads (well in excess of 1 Kb) and a very fast machine run time (~15 minutes). However, there is presently a higher error rate associated with PacBio sequence than with the other available NGS platforms.

The ABI-SOLiD systems are a departure from sequence-by-synthesis, and instead sequence by ligation (Metzker 2010). In its simplest form, a labeled probe will hybridize its complementary sequence adjacent to a primed sequence, and DNA ligase is used to join the fragments. Nonligated fragments are washed away, and a fluorescence detection system is used to determine the identity of the ligated probe (Metzker 2010). The ABI-SOLiD is a commercialized sequencing-by-ligation platform that is massively parallel, uses two-base encoded probes for improved accuracy, and can produce in excess of 60 Gb of DNA sequence/machine run.

The main difficulties presented by short read sequencing technology are data handling, which is a challenge due to the high volumes of data, and difficulty associated with the assembly of short reads with less than perfect base call accuracy. *De novo* assembly has been quite successful with 454 reads (Goldberg et al. 2006; Hofreuter et al. 2006; Hiller et al. 2007; Pearson et al. 2007; Smith et al. 2007) and longer read-lengths will only improve assembly. *De novo* assembly for the short-read platforms remains challenging, although several algorithms have been published (reviewed in Loman et al. 2012). Many of these assemblers can already build accurate contigs for bacterial genomes that are on the order of 10 kbp in length, but contig sizes produced by assembling complex eukaryotic genomes are significantly smaller. Specific discussion of assembly methods and available tools are largely beyond the scope of this chapter, and interested readers are directed to recent reviews detailing the algorithms (Pop 2009; Miller et al. 2010) and available assembly packages (Bao et al. 2011).

Short read-lengths may confound assembly programs, making *de novo* large genome assembly difficult, but assembly of short reads from cDNA is a less complex problem, which makes NGS sequence good for transcriptome characterization (Martin and Wang 2011). In addition, deep, inexpensive, short reads have demonstrated excellent utility in many resequencing applications (Bentley 2006) such as transcript profiling (Cullum et al. 2011; Jain 2012), polymorphism discovery (Barbazuk et al. 2007a, b; Van Tassell et al. 2008) mutation detection (Wurtzel et al. 2010; Audo et al. 2012; Jiang et al. 2012) and *in vivo* DNA binding site detection (Johnson et al. 2007; Cullum et al. 2011). The availability of a reference genome simply requires that each resequencing read be of sufficient length and quality to align uniquely to the genome. The availability of a maize reference genome sequence for the inbred B73 (Schnable et al. 2009), and a steady increase in the robustness, throughput, and efficiency of NGS sequencing systems, provide effective sequence-based platforms that have enabled further characterization of the maize genome and are providing raw material for future maize improvement. Several NGS based analyses are underway in maize and have produced many large maize NGS datasets that are available in public sequence repositories (Table 8.1).

Maize Genome Sequencing in the Post-Genome Era

One of the first, and most powerful applications of NGS resequencing is to identify variation either within, or among populations (Perez-Enciso and Ferretti 2010; Nielsen et al. 2011). Generating a high volume of short reads from one or more individuals and aligning these to a reference is an efficient method for discriminating sequence variants, making NGS resequencing excellent systems for SNP discovery. One of the first uses of NGS in maize was for the identification of SNPs that could be used as molecular markers. Because maize is highly polymorphic, sequence comparison between genotypes provides a vast resource for SNPs. Barbazuk et al. (2007b) identified over 36,000 SNPs between the Mo17 and B73 maize inbred lines mined from shoot apical meristem (SAM). Specific transcripts were sequenced with a Roche 454 sequencer (Emrich et al. 2007). A subset of these SNPs were used to genotype a set of 297 B73/Mo17 recombinant inbred lines (RIls; Fu et al. 2006) from the intermated B73 x Mo17 (IBM) population (Lee et al. 2002), identifying > 1,000 new SNP based genetic markers (Liu et al. 2010) that were placed on the maize genetic map and provided the foundation of a high-throughput genetic mapping platform that was successfully demonstrated on a set of 37 mutants (Liu et al. 2010).

Maize resequencing for SNP discovery: the maize HapMap project

Association mapping to dissect complex traits requires the identification of SNPs in phenotypically diverse populations. Given the size of the maize genome, the anticipated gene content, and the rapid decay of intergenic linkage disequilibrium in maize, Gore et al. (2009b) suggested that several

Table 1 Collections of Next Generation Sequence data for maize (as of August 2012) available and searchable at the at the DNA Data Bank of Japan (http://ddbj.sakura.ne.jp/).

SRA ID	Sequencing Type	Sample	Tissue (if applicable)	Data Size	Pubmed ID or Source
SRA009127	Genomic	UniformMu population		106.5 M	15060129, 16167895, 17490480
SRA009261	Genomic	B73, Mo17		307.4 M	20230488
SRA009340	Genomic	UniformMu stocks		183.9 M	19805815
SRA009397	Genomic	B73		24.7 M	19956743
SRA009756	Genomic	27 inbred line		40.8 G	19965431
SRA010130	Genomics	Zheng58, Mo17, 5003, 478, 178, and Chang7-2		82.2 G	20972441
SRA026308	Genomic	IBM population		8.8 G	21573248
SRA049890	Genomic			95.8 G	JGI
SRA050719	Genomic			85.1 G	JGI
SRA050041	Genomic			54.7 G	JGI
SRA050025	Genomic			54.0 G	JGI
SRA051245	Genomic	103 inbred lines		964.5 G	22660545
SRP012211	Genomic			190.8 G	JGI
SRA053592, SRA053622	Genomic	B73, Mo17		11.4 G	21152036
SRA027365	Genomic	B73, *Zea luxurians*		7.3 G	21296765
SRA008616	Genomic	B73, Mo17		722.5 M	Buckler lab
SRA024031	Genomic	Mo17		241.7 G	JGI
SRA049859	Genomic	278 temperate maize inbred lines		1.2 T	22660547
SRA050081	Genomic			35 G	JGI

Table 1 contd....

Table 1 contd.

SRA ID	Sequencing Type	Sample	Tissue (if applicable)	Data Size	Pubmed ID or Source
SRA050111	Genomic			39.3 G	JGI
SRA054925	Genomic	Mo17		24.8 G	JGI
SRA008289	Transcriptome	Palomero	seedling	164.8 M	19580677
SRA009958	Transcriptome	Mo17	leaf	3.4 G	Brutnell lab
SRA010001	Transcriptome	B73	shoot, root	2.5 G	19376930
SRA010166	Transcriptome	B73xMo17, Mo17xB73	seedling	1.4 G	19965432
SRA010259	Transcriptome	mop1 mutatn and non-mutants	shoot apical meristems	889.4 M	19936292
SRA010261	Transcriptome	B73	root, seedling, tassel, ear, and pollen	1.5 G	19936050
SRA010794	Transcriptome	B73	leaves, ears and tassels	439.2 M	18815367, 19936048
SRA023789	Transcriptome		Ear, Shoot, Meristem	75.6 G	National Center for Genome Resources
SRA029378	Transcriptome	mutant and B73	Ovaries	1.0 G	21325139
SRA037519	Transcriptome	B73	embryo sacs, comparator ovules, mature pollen, seedlings	50.4 G	Maize Gametophyte Project
SRA043751	Transcriptome	W22	anthers	666.2 M	Walbot lab
SRA047032	Transcriptome	Hz32, B73 and Mo17	root	676.6 M	22768123
SRA048055	Transcriptome	B73, Mo17 and B73xMo17, Mo17xB73	endosperm, embryo	132.6 G	22198147
SRA049019	Transcriptome	B73, Mo17 and B73xMo17, Mo17xB73	root	21.5 G	Schnable lab
SRA049062	Transcriptome Transcriptome	B73xMo17, Mo17xB73	seedling, endosperm	20.0 G	22114195
SRA050451	Transcriptome	a panel of samples	shoot apical meristems	49.3 G	Schnable lab

SRA050790	Transcriptome	a panel of samples	ear, root, shoot, tassel	512.3 G	Schnable lab
SRA050928	Transcriptome	teosintes	ear, leaf, tassel	64.2 G	Schnable lab
SRA050934	Transcriptome	Fv2	root, shoot	16.6 G	Schnable lab
SRA050935	Transcriptome	Miami White	leaf	14.5 G	Schnable lab
SRA050957	Transcriptome	B73xEDMX-2233 palomero	leaf	15.3 G	Schnable lab
SRP011929	Transcriptome	W22	ear, tassel	7.8 G	Schnable lab
SRP012162	Transcriptome	rmr2-1 (mutant), Rmr2/rmr2-1 (non-mutant)	developing cobs	1.6 G	22562610
SRP012317	Transcriptome	B73, Mo17, B73xMo17 and Mo17xB73	shoot apex	4.1 G	22689990
SRP012476	Transcriptome	B73	leaf, seed	774.6 M	22853295
SRA053579	Transcriptome	B73*	whole seedling	77.1 G	JGI
SRA012297	Transcriptome	B73	leaf	4.3 G	Cornell University
SRA035621	Transcriptome	B73	Ear, embryo, endosperm, ovule, pollen, seed, silk, tassel	11.6 G	Buell lab
SRA030712	Transcriptome	21 inbred lines	seedling	17.8 G	22438891
SRA049512	Transcriptome			1.7 G	JGI
SRA030599	Transcriptome	B73xMo17 and Mo17xB73	endosperm and kernel	7.9 G	Lai lab
SRA010002	Epigenetics	B73	shoot, root	3.8 G	19376930
SRA037048	Epigenetics	B73	root	2.3 G	22058126
SRA049074	Epigenetics	B73xMo17, Mo17xB73	endosperm	86.6 G	22114195

*includes first PacBio data in maize.
**transcriptome includes RNA-Seq, small RNA sequencing, microRNA sequencing and some array data.

million SNPs within genic regions may be required to properly support association mapping efforts. Significant progress has been made toward building a comprehensive accounting of sequence variation in maize since the first demonstration of NGS mediated SNP discovery in maize (Barbazuk et al. 2007b). Gore et al. (2009b) reported a restriction enzyme based reduced representation SNP discovery procedure targeting the transposable element sparse regions of the maize genome that successfully identified over 120,000 SNPs between the 454 sequences from B73 and Mo17 inbred lines of maize. This result differed from the Barbazuk et al. (2007b) study in that the sequencing was performed on a partial genomic fraction, rather than transcripts, thus providing a more comprehensive collection by avoiding expression bias and including introns and nontranscribed, but repeat poor, regions of the genome. Building on the success of this study, Gore et al. (2009a) sampled sequence and identified SNPs from 27 inbred lines to create the maize HapMap I resource. The 27 inbred lines included the founders of the maize nested association mapping population (NAM), which consists of 25 families of 200 RILs per family; each family was produced by crossing the reference inbred B73 to one of 25 diverse lines chosen to maximize the genetic diversity of the RIL families (McMullen et al. 2009). The population was constructed to enable high power and high resolution through joint linkage-association analysis and, if combined with a comprehensive set of marker SNPs obtained from NAM founders, provides a powerful genetic resource for quantitative trait analysis in maize. Over 32 billion bases (32 Gbp) of Illumina sequence were generated using three genomic sequence libraries that were anchored to complementary restriction enzymes. Estimates suggest that the sequence data sampled over 38% of the maize genome, containing over 32% of the total genic portion, and over 3.3 million SNP markers were identified (Gore et al. 2009a).

The combination of the first generation maize haplotype map (Gore et al. 2009a) and the nested association mapping population (McMullen et al. 2009) has enabled the identification of high resolution QTLs, genes and alleles underlying important maize traits such as flowering time (Buckler et al. 2009; Hung et al. 2012), starch, protein and oil content in kernels (Cook et al. 2012), disease resistance to northern (Poland et al. 2011), and southern leaf blight and leaf architecture (Kump et al. 2011). Recently, Chia et al. (2012) reported on the second generation maize HapMap: HapMap2, which expands from HapMapI by including sequence sampling from 103 inbred lines of maize (Chia et al. 2012). The lines chosen represent broadly the *Z. mays* lineage and include several improved maize lines, the NAM parents, and wild maize relatives. Almost 1 trillion bases pairs were represented in ~ 13 billion DNA sequence reads, representing over 4× coverage of the maize and teosinte genomes. In total, over 55 million SNPs were identified, 21% of which were found to be associated with genic regions, including

825,000 synonymous, 571,000 nonsynonymous and ~ 10,000 nonsense mutations (Chia et al. 2012). The HapMap2 data set promises to provide a major resource for maize association mapping, genomic selection and the mining of genomic regions that have been selected during domestication and improvement, and it is expected that it will play a valuable role in future maize improvement.

Copy number variation and sequence capture

An interesting outcome of the HapMap2 project is observation that the maize genome exhibits a high degree of structural variation. When comparing between lines, it was found that approximately 70% exhibited evidence of copy number variation in at least one line (Chia et al. 2012), confirming previous studies by Schnable et al. (2009) and Springer et al. (2009) suggesting that the maize genome is in flux. High resolution whole genome profiling with microarray based comparative genome hybridization (aCGH) is a powerful method for investigating copy number variation (CNV) and structural variation (SV) within the human genome (for review see, Lockwood et al. 2008; Shinawi and Cheung 2008). In a typical aCGH experiment, total genomic DNA is isolated from two individuals, or test and reference cell populations, differentially labeled with florescence, and cohybridized in equal amounts to an array containing DNA targets derived from a representative genome. Differences in signal intensities between the two samples across groups of probes spatially linked within the reference genome are indicative of structural variations such as duplications and deletions, etc.

In maize, the CGH array was composed of a set of probe sequences that tile across the genomes deriving from the inbred B73 reference genome sequence (Springer et al. 2009). Competitively hybridizing two labeled samples of genomic DNA from the B73 reference and the Mo17 inbred lines to a single B73 target array and examining the ratio of signal intensities revealed unbalanced copy number differences between the B73 and Mo17 genomes. In addition to identifying regions of the genome that have little diversity, presumably a result of selection during domestication, this study revealed thousands of examples of structural variation between the B73 and Mo17 genomes that included over 400 instances of copy number variation (CNV) and thousands of sequences that exhibit present-absent variation (PAV). Springer et al. (2009) noted that sequences accounting for over 20 Mb of DNA are present within B73, but absent in Mo17. Intriguingly, a minimum of 180 gene loci is observed to be present within B73, and entirely absent from the Mo17 genome. The majority of these absent loci are low- or single-copy DNA sequences that occur in B73, are expressed, and over 1/3 of these genes missing from Mo17 do not contain similar sequences located elsewhere in the B73 genome. This observation suggests that a significant fraction of the missing genes likely do not have a functional complement elsewhere in the maize genome. Given this, it may be concluded that these inbreds can tolerate a high level of genome content variation and still develop as "normal" corn plants. Nevertheless the levels of structural variation may contribute to phenotype, and may be implicated in heterosis.

While the aCGH investigation of maize structural variation reported by Springer et al. (2009) was restricted to a comparison between the genomes of the B73 and Mo17, Swanson-Wagner et al. (2010) broadened the study to examine CNV within 19 diverse maize lines, identifying over 3,800 genes that exhibit copy number variation in at least one line relative to B73. Lai et al. (2010) examined CNV using NGS rather than aCGH by performing deep resequencing across Mo17 and an additional five commercial inbred lines. Over 1.26 billion 75 bp paired-end Illumina sequences (83.7 Gbp of DNA) were acquired and aligned to the B73 reference genome, achieving an average depth of coverage of 5.4X for each inbred. In addition to a large number of SNPs and indels (insertion and deletion) identified between the lines, over 296 genes present in B73 were found to be absent in at least one of the inbreds resequenced (Lai et al. 2010). This observation confirmed and extends the work of (Springer et al. 2009), and underscores the possibility that gene content complementation may play a significant role in heterosis in maize.

Another array-based technique that shares some similarity to aCGH is microarray-based sequence capture (Hodges et al. 2007; Okou et al. 2007). Microarray based sequence capture enables targeted resequencing of specific intervals of genome. Essentially, probes are designed to a genomic region of interest based on available reference sequence; these are arrayed on a microarray and used to "capture" equivalent regions within closely related species or from individuals within a species. Captured sequence is "released" from the microarray and sequenced. Captured sequence can either be genomic, or transcriptome (cDNA), depending on what was used as the input samples. The advantage of sequence capture is that it results in a selective enrichment of specific regions, which allows one to assess genotypic changes in a defined set of specific targets relative to the whole genome. For example, to target and cover the exonic regions of a genome (exome) to a given degree of redundancy requires significantly less sequencing reads if they are the products of a capture, than would be required to cover these regions by random sequencing of the whole genome. This makes sequence capture valuable in situations where deep coverage on a few select regions are required across a large number of samples.

Microarray-based sequence capture has been effective for re-sequencing exons, large genomic loci, and candidate gene sets in mammalian genomes (Albert et al. 2007; Hodges et al. 2007; Okou et al. 2007; Kim et al. 2010), and solution-based protocols have also been developed (Fisher et al. 2011). Recently, Fu et al. (2010) have demonstrated microarray based sequence capture in maize for the purpose of saturating a defined interval for SNPs. Typically large numbers of SNPs are not discovered within a set of specific genes or a defined genomic region. However, the enrichment of target sequences achieved via sequence capture makes it possible to identify large numbers of SNPs more efficiently than is possible by re-sequencing whole genomes or even the entire gene space to obtain large numbers of "random" SNPs. Fu et al. (2010) designed probes to a 2.2 Mb interval within the B73 genome and used these to capture sequences from inbred Mo17 maize, which were sequenced with 454 technology. On target sequence recovery levels ranged from 22–36% and provided approximately 3,000-fold enrichment. Over 1,600 SNPs were identified after aligning the captured Mo17 sequences to the maize B73 genome. This demonstrated the feasibility of using capture to enrich a specific region for SNPs, which would aid in identifying candidate genes underlying QTLs of interest.

High-throughput sequencing and gene expression studies

Prior to the emergence of next generation sequence technologies, high-throughput transcriptome analysis to examine the transcriptional state of different cells, tissues, or genotypes in maize, relied on microarray analysis and this, in turn, depended on the availability of EST sequence collections (Stupar et al. 2007; Zhang et al. 2007; Strable et al. 2008; Stupar et al. 2008) and/or serial analysis of gene expression (SAGE) (Gowda et al. 2004). The emergence of next generation sequencing platforms has seen a shift from microarray based gene expression studies to deep sequencing of short EST sequence tags easily and inexpensively generated on platforms such as the Illumina GAIIx and HiSeq. This process, called RNA-Seq, provides several advantages over microarray analysis including a wider dynamic range, less noise, and higher throughput (Kvam et al. 2012) and can provide highly resolved structural information such as transcript initiation, termination and splice junction sites (Wang et al. 2009). In addition, because expression is determined by counting sequence tags generated by sequencing cDNA *de novo* rather than hybridizing to a micro array carrying a pre-defined probe set, RNA-Seq permits novel gene and isoform detection.

The first application of NGS in maize identified novel transcripts associated with the maize shoot apical meristem (SAM) (Emrich et al. 2007). While this study was primarily a gene discovery project, RNA-Seq is now routinely used in maize transcriptome characterization and expression studies. An early study of note utilized RNA-Seq to define the maize leaf transcriptome along a developmental gradient capturing both source and sink tissues (Li et al. 2010). While previous studies had

examined the molecular basis for the partitioning of photosynthetic processes between the bundle sheath and mesophyll in mature leaves, Li et
al. 2010 performed an extensive RNA-Seq survey of four developmental
zones of the maize leaf to examine the transcriptional network that is
associated with the development of C_4 photosynthesis. Over 25,000 genes
were found to be expressed among the developmental zones studied with
transcripts associated with basic cellular functions found predominantly
in the basal region of the leaf. Transcripts associated with establishing the
photosynthetic machinery and transcripts associated with photosynthesis
exhibit elevated abundance in the region of transition from sink to source
tissue and the distal region of the leaf, respectively (Li et al. 2010) (Hansey
et al. 2012) performed extensive RNA-Seq on seedlings from 21 diverse
maize lines that included representatives of North American and exotic
lines. Comparative analysis of this data was conducted to identify novel
maize transcripts within the diverse set of lines, as well as define a core
set of transcripts common to all lines. Over 48% of the annotated genes in
maize were found to be expressed in all 21 inbred lines, while 11,011 genes
were expressed in 1-20 lines leading Hansey et al. (2012) to surmise that this
smaller set represents the "seedling variable transcriptome" and that this
variable set likely contributes to the phenotypic diversity observed in maize.
Eveland et al. (2010) tested a tag-based digital profiling protocol that is a
modification of RNA-Seq. This system was used to examine gene expression
relating to cell fate in axillary meristem, which control ear development
and therefore have relevance to crop improvement strategies.

NGS and the investigation of epigenetics in maize

Epigenetics is the study of changes or alterations in the genetic material,
usually chromosomes, that are not accompanied by changes in the DNA
sequence. Some of the first observations of epigenetics in plants were
observed in maize, and maize remains an excellent model system for the
study of epigenetics. Moreover, epigenetic sources of variation may account,
in some part, for the agronomic success of maize hybrids (Hollick 2009). There
are several epigenetic phenomena in maize that are under investigation,
these include transposable element inactivation, paramutation and
imprinting (for review see Hollick 2009). While characterizing and studying
the mechanisms of epigenetic phenomena is important, the availability of
genome sequence references and the emergence of high throughput NGS
technologies have ushered in a new frontier, namely epigenomics, or the
study and characterization of epigenetic modification across the genome as
a whole. Several techniques that center on NGS to examine the epigenome
have recently emerged and are being used to assess global methylation
states of DNA, histone modifications, DNA-protein interactions and the

transcriptome (for review see Ku et al. 2011; Meaburn and Schulz 2012). While maize research is historically rich with investigations into specific epigenetic mechanisms and phenomena, the study of the maize epigenome is now well underway. Two recent and related studies illustrate the use of RNA-Seq transcriptome analysis and allele specific SNPs to investigate imprinting on a global scale. Waters et al. (2011) performed a global assay for imprinted genes in maize endosperm tissue. Their strategy was to obtain deep RNA-Seq of endosperm and embryo tissues from Mo17 and B73 to assess relative transcript levels and identify SNPs that enable discrimination of Mo17 and B73 alleles. Ultimately the aim is to examine the expression within B73 × Mo17 reciprocal hybrids in order to identify parent of origin effects indicative of imprinting. The study identified 100 putatively imprinted genes, 54 maternally and 46 paternally expressed, in addition to detecting several instances of imprinting conserved among maize, rice and *Arabidopsis* genera. This collection of maize imprinted genes will undoubtedly aid in understanding the role imprinting plays in plant seed development. A research group at the China Agricultural University in Beijing independently conducted a similar study. Zhang et al. (2011) identified a diverse collection of 179 putatively imprinted genes implicated in maize endosperm development. These included 68 maternally and 111 paternally expressed as well as 38 putatively imprinted long non-coding RNAs by employing a similar strategy of RNA-Seq on endosperm tissue from B73, Mo17 and the reciprocal hybrids. It is possible that imprinting is being used to compensate for unbalanced gene dosage in the triploid endosperm. At any rate, further investigation of these imprinted targets will further our understanding of the role of genetic imprinting in endosperm development, which will likely impact agricultural interests.

Concluding Remarks

The release of a reference genome sequence for maize in 2009 has undoubtedly facilitated a better understanding of maize genome structure and content, as well as maize biology. However, with the availability of new sequence platforms and sequence based analysis methods, the collection of maize DNA sequence continues. Recent advances in DNA sequencing technology are providing the platforms to better understand phenomena such as maize genome evolution, gene expression, gene structure and the epigenetic interactions that augment expression. In addition, continued use of NGS technologies in maize are identifying the genome mechanics behind its domestication, allowing a better understanding of its breeding history, and enabling efforts that will lead to improvements in yield and quality.

References

Albert TJ, Molla MN, Muzny DM et al. (2007) Direct selection of human genomic loci by microarray hybridization. Nat Meth 4: 903–905.

Audo I, Bujakowska KM, Leveillard T et al. (2012) Development and application of a next-generation-sequencing (NGS) approach to detect known and novel gene defects underlying retinal diseases. Orphanet J Rare Dis 7: 8.

Bao S, Jiang R, Kwan W et al. (2011) Evaluation of next-generation sequencing software in mapping and assembly. J Hum Genet 56: 406–414.

Barbazuk WB, Emrich S, Schnable PS (2007a) SNP Mining from Maize 454 EST Sequences. CSH Protoc 2007: pdb prot4786.

Barbazuk WB, Emrich SJ, Chen HD et al. (2007b) SNP discovery via 454 transcriptome sequencing. Plant J Cell Mol Biol 51: 910–918.

Bennetzen JL, Schrick K, Springer PS et al. (1994) Active maize genes are unmodified and flanked by diverse classes of modified, highly repetitive DNA. Genome 37: 565–576.

Bentley DR (2006) Whole-genome re-sequencing. Curr Opin Genet Dev 16: 545–552.

Britten RJ, Graham DE and Neufeld BR (1974) Analysis of repeating DNA sequences by reassociation. Meth Enzymol 29: 363–418.

Bruggmann R, Bharti AK, Gundlach H et al. (2006) Uneven chromosome contraction and expansion in the maize genome. Genome Res 16: 1241–1251.

Buckler ES, Holland JB, Bradbury PJ et al. (2009) The genetic architecture of maize flowering time. Science 325: 714–718.

Chan AP, Pertea G, Cheung FL et al. (2006) The TIGR Maize Database. Nucl Acids Res 34: D771–776.

Chardon F, Virlon B and Moreau L (2004) Genetic architecture of flowering time in maize as inferred from quantitative trait loci meta-analysis and synteny conservation with the rice genome. Genetics 168: 2169–2185.

Chen M, SanMiguel P, de Oliveira AC et al. (1997) Microcolinearity in sh2-homologous regions of the maize, rice, and sorghum genomes. Proc Natl Acad Sci USA 94: 3431–3435.

Chia JM, Song C, Bradbury PJ et al. (2012) Maize HapMap2 identifies extant variation from a genome in flux. Nat Genet 44: 803–807.

Coe JEH and Schaeffer ML (2005) Genetic, Physical, Maps, and Database Resources for Maize. Maydica 50: 285–303.

Cook JP, McMullen MD, Holland JB et al. (2012) Genetic architecture of maize kernel composition in the nested association mapping and inbred association panels. Plant Physiol 158: 824–834.

Cooke R, Piègu B, Panaud O et al. (2007) From rice to other cereals: Comparative genomics. In: Upadhyaya NM (ed) Rice Functional Genomics. Springer, New York, USA, pp 429–479.

Cullum R, Alder O and Hoodless PA (2011) The next generation: using new sequencing technologies to analyse gene regulation. Respirology 16: 210–222.

Dong Q, Schlueter SD and Brendel V (2004) PlantGDB, plant genome database and analysis tools. Nucl Acids Res 32: D354–359.

Eid J, Fehr A, Gray J et al. (2009) Real-time DNA sequencing from single polymerase molecules. Science 323: 133–138.

Emberton J, Ma J, Yuan Y et al. (2005) Gene enrichment in maize with hypomethylated partial restriction (HMPR) libraries. Genome Res 15: 1441–1446.

Emrich SJ, Barbazuk WB, Li L et al. (2007) Gene discovery and annotation using LCM–454 transcriptome sequencing. Genome Res 17: 69–73.

Eveland AL, Satoh-Nagasawa N, Goldshmidt A et al. (2010) Digital gene expression signatures for maize development. Plant Physiol 154: 1024–1039.

Fisher S, Barry A, Abreu J et al. (2011) A scalable, fully automated process for construction of sequence-ready human exome targeted capture libraries. Genome biol 12: R1.

Flavell RB, Bennett MD, Smith JB et al. (1974) Genome size and the proportion of repeated nucleotide sequence DNA in plants. Biochem Genet 12: 257–269.

Fu H and Dooner HK (2002) Intraspecific violation of genetic colinearity and its implications in maize. Proc Natl Acad Sci USA 99: 9573–9578.

Fu Y, Emrich SJ, Guo L et al. (2005) Quality assessment of maize assembled genomic islands (MAGIs) and large-scale experimental verification of predicted genes. Proc Natl Acad Sci USA 102: 12282–12287.

Fu Y, Springer NM, Gerhardt DJ et al. (2010) Repeat subtraction-mediated sequence capture from a complex genome. Plant J 62: 898–909.

Fu Y, Wen TJ, Ronin YI et al. (2006) Genetic dissection of intermated recombinant inbred lines using a new genetic map of maize. Genetics 174: 1671–1683.

Gale MD and Devos KM (1998) Comparative genetics in the grasses. Proc Natl Acad Sci USA 95: 1971–1974.

Gaut BS (2001) Patterns of chromosomal duplication in maize and their implications for comparative maps of the grasses. Genome Res 11: 55–66.

Gaut BS and Doebley JF (1997) DNA sequence evidence for the segmental allotetraploid origin of maize. Proc Natl Acad Sci USA 94: 6809–6814.

Gaut BS, Le Thierry d'Ennequin M, Peak A et al. (2000) Maize as a model for the evolution of plant nuclear genomes. Proc Natl Acad Sci USA 97: 7008–7015.

Goff SA, Ricke D, Lan TH et al. (2002) A draft sequence of the rice genome (*Oryza sativa* L. ssp. japonica). Science 296: 92–100.

Goldberg SM, Johnson J, Busam D et al. (2006) A Sanger/pyrosequencing hybrid approach for the generation of high-quality draft assemblies of marine microbial genomes. Proc Natl Acad Sci USA 103: 11240–11245.

Gore MA, Chia JM, Elshire RJ et al. (2009a) A first-generation haplotype map of maize. Science 326: 1115–1117.

Gore MA, Wright MH, Ersoz ES et al. (2009b) Large-scale discovery of gene-enriched SNPs. Plant Genome 2: 121–133.

Gowda M, Jantasuriyarat C, Dean RA et al. (2004) Robust-LongSAGE (RL-SAGE): a substantially improved LongSAGE method for gene discovery and transcriptome analysis. Plant Physiol 134: 890–897.

Haberer G, Young S, Bharti AK et al. (2005) Structure and architecture of the maize genome. Plant Physiol 139: 1612–1624.

Hansey CN, Vaillancourt B, Sekhon RS et al. (2012) Maize (*Zea mays* L.) genome diversity as revealed by RNA-sequencing. PloS one 7: e33071.

Harper LC, Schaeffer ML, Thistle J et al. 2011. The MaizeGDB Genome Browser tutorial: one example of database outreach to biologists via video. Database: bar016.

Hazen SP, Hawley RM, Davis GL et al. (2003) Quantitative trait loci and comparative genomics of cereal cell wall composition. Plant Physiol 132: 263–271.

Hiller NL, Janto B, Hogg JS et al. (2007) Comparative genomic analyses of seventeen Streptococcus pneumoniae strains: insights into the pneumococcal supragenome. J Bacteriol 189: 8186–8195.

Hodges E, Xuan Z, Balija V et al. 2007. Genome-wide in situ exon capture for selective resequencing. Nat Genet 39: 1522–1527.

Hofreuter D, Tsai J, Watson RO et al. (2006) Unique features of a highly pathogenic Campylobacter jejuni strain. Infect Immun 74: 4694–4707.

Hollick JB and Springer N (2009) Epigenetic phenomena and epigenomics in Maize. In: Ferguson-Smith AC, Greally JM and Martienssen RA (eds) Epigenomics. Springer, Netherlands, pp 119–147.

Holt RA and Jones SJ (2008) The new paradigm of flow cell sequencing. Genome Res 18: 839–846.

Hung HY, Shannon LM, Tian F et al. (2012) ZmCCT and the genetic basis of day-length adaptation underlying the postdomestication spread of maize. Proc Natl Acad Sci USA 109: E1913–1921.

Ilic K, SanMiguel PJ and Bennetzen JL (2003) A complex history of rearrangement in an orthologous region of the maize, sorghum, and rice genomes. Proc Natl Acad Sci USA 100: 12265–12270.

Initiative TIB (2010) Genome sequencing and analysis of the model grass Brachypodium distachyon. Nature 463: 763–768.

Jain M (2012) Next-generation sequencing technologies for gene expression profiling in plants. Brief Funct Genom 11: 63–70.

Jiang Q, Turner T, Sosa MX et al. (2012) Rapid and efficient human mutation detection using a bench-top next-generation DNA sequencer. Hum Mutat 33: 281–289.

Johnson DS, Mortazavi A, Myers RM et al. (2007) Genome-wide mapping of *in vivo* protein-DNA interactions. Science 316: 1497–1502.

Kim DW, Nam SH, Kim RN et al. (2010) Whole human exome capture for high-throughput sequencing. Genome 53: 568–574.

Kirihara JA, Petri JB and Messing J (1988) Isolation and sequence of a gene encoding a methionine-rich 10-kDa zein protein from maize. Gene 71: 359–370.

Kridl JC, Vieira J, Rubenstein I et al. (1984) Nucleotide sequence analysis of a zein genomic clone with a short open reading frame. Gene 28: 113–118.

Ku CS, Naidoo N, Wu M et al. (2011) Studying the epigenome using next generation sequencing. J Med Genet 48: 721–730.

Kump KL, Bradbury PJ, Wisser RJ et al. (2011) Genome-wide association study of quantitative resistance to southern leaf blight in the maize nested as sociation mapping population. Nature Genet 43: 163–168.

Kvam VM, Liu P and Si Y (2012) A comparison of statistical methods for detecting differentially expressed genes from RNA-seq data. Am J Bot 99: 248–256.

Lai J, Li R, Xu X et al. (2010) Genome-wide patterns of genetic variation among elite maize inbred lines. Nature genetics 42: 1027–1030.

Langham RJ, Walsh J, Dunn M et al. (2004) Genomic duplication, fractionation and the origin of regulatory novelty. Genetics 166: 935–945.

Lawrence CJ, Harper LC, Schaeffer ML et al. (2008) MaizeGDB: The maize model organism database for basic, translational, and applied research. Int J Plant Genom: 496957.

Lee M, Sharopova N, Beavis WD et al. A (2002) Expanding the genetic map of maize with the intermated B73 x Mo17 (IBM) population. Plant Mol Biol 48: 453–461.

Li P, Ponnala L, Gandotra N et al. (2010) The developmental dynamics of the maize leaf transcriptome. Nature gGenet 42: 1060–1067.

Lippman Z, Gendrel AV, Black M et al. (2004) Role of transposable elements in heterochromatin and epigenetic control. Nature 430: 471–476.

Liu S, Chen HD, Makarevitch I et al. (2010) High-throughput genetic mapping of mutants via quantitative single nucleotide polymorphism typing. Genetics 184: 19–26.

Lockwood WW, Chari R, Coe BP et al. (2008) DNA amplification is a ubiquitous mechanism of oncogene activation in lung and other cancers. Oncogene 27: 4615–4624.

Loman NJ, Misra RV, Dallman TJ et al. (2012) Performance comparison of benchtop high-throughput sequencing platforms. Nat biotechnol 30: 434–439.

Mardis ER (2008) Next-generation DNA sequencing methods. Annu rev genom hum genet 9: 387–402.

Martin JA and Wang Z (2011) Next-generation transcriptome assembly. Nat rev Genet 12: 671–682.

McMullen MD, Kresovich S, Villeda HS et al. (2009) Genetic properties of the maize nested association mapping population. Science 325: 737–740.

Meaburn E and Schulz R (2012) Next generation sequencing in epigenetics: insights and challenges. Sem cell dev biol 23: 192–199.

Messing J, Bharti AK, Karlowski WM et al. (2004) Sequence composition and genome organization of maize. Proc Natl Acad Sci USA 101: 14349–14354.

Messing J, Carlson J, Hagen G et al. (1984) Cloning and sequencing of the ribosomal RNA genes in maize: the 17S region DNA 3: 31–40.

Metzker ML (2010) Sequencing technologies—the next generation. Nat rev Genet 11: 31–46.

Meyers, BC, Tingey SV and Morgante M (2001) Abundance, distribution, and transcriptional activity of repetitive elements in the maize genome. Genome Res 11: 1660–1676.

Miller JR, Koren S, Sutton G (2010) Assembly algorithms for next-generation sequencing data. Genomics 95: 315–327.

Moore G, Devos KM, Wang Z et al. (1995) Cereal genome evolution. Grasses, line up and form a circle. Curr biol: *CB* 5: 737–739.

Morishige DT, Childs KL, Moore LD et al. (2002) Targeted analysis of orthologous phytochrome A regions of the sorghum, maize, and rice genomes using comparative gene-island sequencing. Plant Physiol 130: 1614–1625.

Nielsen R, Paul JS, Albrechtsen A et al. (2011) Genotype and SNP calling from next-generation sequencing data. Nat Rev Genet 12: 443–451.

Okou DT, Steinberg KM, Middle C et al. (2007) Microarray-based genomic selection for high-throughput resequencing. Nat Meth 4: 907–909.

Palmer LE, Rabinowicz PD, O'Shaughnessy AL et al. (2003) Maize genome sequencing by methylation filtration. Science 302: 2115–2117.

Paterson AH, Bowers JE, Bruggmann R et al. (2009) The Sorghum bicolor genome and the diversification of grasses. Nature 457: 551–556.

Pearson BM, Gaskin DJ, Segers RP et al. (2007) The complete genome sequence of Campylobacter jejuni strain 81116 (NCTC11828). J Bacteriol 189: 8402–8403.

Perez-Enciso M and Ferretti L (2010) Massive parallel sequencing in animal genetics: wherefroms and wheretos. Anim Genet 41: 561–569.

Poland JA, Bradbury PJ, Buckler ES et al. (2011) Genome-wide nested association mapping of quantitative resistance to northern leaf blight in maize. Proc Natl acad Sci USA 108: 6893–6898.

Pop M (2009) Genome assembly reborn: recent computational challenges. Brief Bioinformat 10: 354–366.

Quackenbush J, Cho J, Lee D et al. (2001) The TIGR Gene Indices: analysis of gene transcript sequences in highly sampled eukaryotic species. Nucl Acids Res 29: 159–164.

Quackenbush J, Liang F, Holt I et al. (2000) The TIGR gene indices: reconstruction and representation of expressed gene sequences. Nucl Acids Res 28: 141–145.

Rabinowicz PD, Schutz K, Dedhia N et al. (1999) Differential methylation of genes and retrotransposons facilitates shotgun sequencing of the maize genome. Nat Genet 23: 305–308.

Ramakrishna W, Emberton J, SanMiguel P et al. (2002) Comparative sequence analysis of the sorghum Rph region and the maize Rp1 resistance gene complex. Plant Physiol 130: 1728–1738.

Rothberg JM, Hinz W, Rearick TM et al. (2011) An integrated semiconductor device enabling non-optical genome sequencing. Nature 475: 348–352.

SanMiguel P, Gaut BS, Tikhonov A et al. (1998) The paleontology of intergene retrotransposons of maize. Nat Genet 20: 43–45.

Schaeffer ML, Harper LC, Gardiner JM et al. (2011) MaizeGDB: curation and outreach go hand-in-hand. Database: bar022.

Schnable PS Ware D Fulton RS et al. (2009) The B73 maize genome: complexity, diversity, and dynamics. Science 326: 1112–1115.

Shinawi M and Cheung SW (2008) The array CGH and its clinical applications. Drug Discov Today 13: 760–770.

Smith MG, Gianoulis TA and Pukatzki S (2007) New insights into Acinetobacter baumannii pathogenesis revealed by high-density pyrosequencing and transposon mutagenesis. Genes Dev 21: 601–614.

Soderlund C, Descour A, Kudrna D et al. (2009) Sequencing, mapping, and analysis of 27,455 maize full-length cDNAs. PLoS Genetics 5: e1000740.

Song R and Messing J (2002) Contiguous genomic DNA sequence comprising the 19-kD zein gene family from maize. Plant Physiol 130: 1626–1635.

Springer NM, Xu X and Barbazuk WB (2004) Utility of different gene enrichment approaches toward identifying and sequencing the maize gene space. Plant Physiol 136: 3023–3033.

Springer NM, Ying K, Fu Y et al. (2009) Maize inbreds exhibit high levels of copy number variation (CNV) and presence/absence variation (PAV) in genome content. PLoS Genetics 5: e1000734.

Strable J, Borsuk L, Nettleton D et al. (2008) Microarray analysis of vegetative phase change in maize. Plant J 56: 1045–1057.

Stupar RM, Gardiner JM, Oldre AG et al. (2008) Gene expression analyses in maize inbreds and hybrids with varying levels of heterosis. BMC Plant Biol 8: 33.

Stupar RM. Hermanson PJ and Springer NM (2007) Nonadditive expression and parent-of-origin effects identified by microarray and allele-specific expression profiling of maize endosperm. Plant Physiol 145: 411–425.

Swanson-Wagner RA, Eichten SR, Kumari S et al. (2010) Pervasive gene content variation and copy number variation in maize and its undomesticated progenitor. Genome Res 20: 1689–1699.

Tikhonov AP, SanMiguel PJ, Nakajima Y et al. (1999) Colinearity and its exceptions in orthologous adh regions of maize and sorghum. Proc Natl acad Sci USA 96: 7409–7414.

Van Tassell CP, Smith TP, Matukumalli LK et al. (2008) SNP discovery and allele frequency estimation by deep sequencing of reduced representation libraries. Nat Meth 5: 247–252.

Vielle-Calzada JP, Martinez de la Vega O, Hernandez-Guzman G et al. (2009) The Palomero genome suggests metal effects on domestication. Science 326: 1078.

Wang Z, Gerstein M and Snyder M (2009) RNA-Seq: a revolutionary tool for transcriptomics. Nat Rev Genet 10: 57–63.

Ware DH, Jaiswal P, Ni J et al. (2002) Gramene, a tool for grass genomics. Plant Physiol 130: 1606–1613.

Waters AJ, Makarevitch I and Eichten SR (2011) Parent-of-origin effects on gene expression and DNA methylation in the maize endosperm. Plant Cell 23: 4221–4233.

Webb CA, Richter TE, Collins NC et al. (2002) Genetic and molecular characterization of the maize rp3 rust resistance locus. Genetics 162: 381–394.

Wei F, Coe E, Nelson W et al. (2007) Physical and genetic structure of the maize genome reflects its complex evolutionary history. PLoS Genetics 3: e123.

Wei F, Zhang J, Zhou S et al. (2009) The physical and genetic framework of the maize B73 genome. PLoS Genetics 5: e1000715.

Whitelaw CA, Barbazuk WB, Pertea G et al. (2003) Enrichment of gene-coding sequences in maize by genome filtration. Science 302: 2118–2120.

Wurtzel O, Dori-Bachash M, Pietrokovski S et al. (2010) Mutation detection with next-generation resequencing through a mediator genome. PloS One 5: e15628.

Yang L and Bennetzen JL (2009) Distribution, diversity, evolution, and survival of *Helitrons* in the maize genome. Proc Natl Acad Sci USA 106: 19922–19927.

Yu J, Hu S, Wang J et al. (2002) A draft sequence of the rice genome (*Oryza sativa* L. ssp. *indica*). Science 296: 79–92.

Yuan Y, SanMiguel PJ and Bennetzen JL (2003) High-Cot sequence analysis of the maize genome. Plant J 34: 249–255.

Zhang M, Zhao H, Xie S et al. (2011) Extensive, clustered parental imprinting of protein-coding and noncoding RNAs in developing maize endosperm. Proc Natl Acad Sci USA 108: 20042–20047.

Zhang X, Madi S, Borsuk L et al. (2007) Laser microdissection of narrow sheath mutant maize uncovers novel gene expression in the shoot apical meristem. PLoS Genetics 3: e101.

Zhou S, Wei F, Nguyen J et al. (2009) A single molecule scaffold for the maize genome. PLoS Genetics 5: e1000711.

CHAPTER 9

Concerns of and Compliance to Using Corn as a Bioenergy Crop

Siwa Msangi[1], and *Rodomiro Ortiz[2]*

ABSTRACT

Some people may argue that global demand for renewable energy such as maize-derived biofuels, especially in the industrialized countries, could lead to high maize grain prices worldwide. The affordability of maize seeds by the poor may be affected in the short term by high prices, thereby increasing hunger and malnutrition. Alternatively an increase of maize prices can also create new options for investing in farming that may translate into the long term of promoting economic growth, which will help many maize farmers in the developing world to move out of poverty. A spike in maize grain prices could however reduce its demand as raw material for ethanol because of low profit margins. Life cycle analysis (LCA) suggests that maize ethanol can decrease greenhouse gas emissions *vis à vis* petrol in some factory settings. Furthermore, maize stover, which is a major source of biomass residue and among the largest in some highly productive agro-ecosystems, could become an important biomass resource for the generation of renewable liquid fuels (e.g., cellulosic ethanol) without causing soil erosion or the depletion of soil carbon. Nonetheless, maize stover may not be competitive as cellulosic ethanol source because there may be extra high costs for collecting and transporting it over large distances to supply processing

[1]International Food Policy Research Institute (IFPRI), 2033 K Street NW, Washington DC 20006, USA.
Email: s.msangi@cgiar.org
[2]Swedish University of Agricultural Sciences (SLU), Dept. Plant Breeding and Biotechnology, Sundsvagen 14, Box 101, Alnarp, SE 23053, Sweden.
Email: rodomiro.ortiz@slu.se
*Corresponding author

factories. Complicating these scenarios further is the need to balance the bioenergy demands with national food security. These policies will vary country-by-country and will be shaped by infrastructure in addition to technology development. In this regard, crop yields must increase significantly but sustainably using existing farmland to meet this challenge. Data-driven maize breeding, which combines both conv ntional and genomics-based approaches, may assist further to increase acreage under cultivation by modifying these plants to grow on challenged soils not currently under agriculture.

Keywords: maize, ethanol, fuel policy, emissions, land use change

Introduction

Rising prices for fossil fuels (especially crude oil), energy security, climate change and rural development have been main drivers for advocating alternative sources of bioenergy. Policy plays a key role in bioenergy markets worldwide because biofuels cannot be yet competitive with fossil fuels. Governments use subsidy, tax breaks or other incentives to promote investments with the aim of reducing, through blending, fossil fuel use. Such government incentives need however to consider both environmental and social conditions to ensure the long-term viability of this strategy.

In the last decade, maize ethanol got significant attention from policy makers, the industry and researchers. Opposite views about the pros and cons of renewable sources of energy are often noted in popular press or in scientific journals. As noted elsewhere, any biofuel source, including maize ethanol, should not result in rising food prices, produce high energy yields accompanied with low energy inputs (water, land, fertilizers), not affect local populations or the environment, and be steadily produced with a high profit margin. In this regard, the Nuffield Council on Bioethics (2011) concluded that producing biofuels should "not be at the expense of people's essential rights, be environmentally sustainable, develop in accordance with trade principles that are fair, and recognize the rights of people to just reward, distribute costs and benefits in an equitable way." They further state "if biofuels can play a crucial role in mitigating dangerous climate change then, depending on additional key considerations, there is a duty to develop such biofuels." The additional key considerations are absolute cost, availability of alternative energy technologies, alternative uses for biofuels feedstocks (leading to new product development), existing degree of uncertainty in their development, their irreversibility, and the degree of participation in decision-making especially given the overarching notion of proportionate governance.

The Food and Agriculture Organization (FAO) of the United Nations indicate that biofuels will only offset a small percentage of fossil energy use over this decade but will impact significantly on agriculture and food security (FAO 2008). This FAO report also points out that biofuels derived from maize are *per se* one of the many drivers affecting food prices because "weather-related production shortfalls", especially among the major exporters, "low global cereal stocks, increasing fuel costs, the changing structure of demand associated with income growth, population growth and urbanization, operations on financial markets, short-term policy actions, exchange rate fluctuations and other factors also play a role." FAO also acknowledges that biofuel impact on reducing greenhouse gas (GHG) emissions differ worldwide because these depend on feedstock, location, farming system, crop husbandry and processing technology.

The widely shifting prices of both maize (as source of ethanol) and crude oil clearly affect the viability of this biofuel system. In this regard, Tyner and Taheripour (2007) calculated that at a crude oil price of US\$ 60 per barrel, ethanol processors could pay up to US\$ 79.52 per ton of for maize seed to remain profitable, while at crude oil prices of US\$ 100 per barrel, they could pay up to US\$ 162.98 per ton of maize and remain profitable in the USA. Elobeid et al. (2007) estimates at US\$ 4.05, the breakeven maize price in the USA could lead to the production of 31.5 million gallons of ethanol annually. Although the net ethanol cost is sensitive to many factors, the two most important in the USA are maize cost and boiler fuel price (Cassman et al. 2006). Furthermore, the ethanol price tracks that of crude oil in the USA, which roughly accounts for 40% of global maize grain output, and contributes about 60% to global maize grain trade. This situation explains the key role of US maize in international markets.

Governments should aim to safely integrate sustainable food and energy production to ensure security for both as well as simultaneously reducing poverty (Bogdanski 2011). Such integrated food-energy systems will help farmers by reducing the costs for fossil fuel and could be an effective weapon to mitigate climate change. Such an inclusive approach may reduce the need for converting land committed to food production into fields for harvesting biofuel sources. For example, maize stover could be collected, albeit at increased cost as biomass biofuel source. Conservative removal rates of maize crop residues could offer sustainable biomass resources if such an approach takes into account concerns regarding land degradation, wind erosion and soil moisture, as well as nutrient replacement costs (Graham et al. 2007). Dhugga (2007) noted further that modern plant breeding that uses genomic and transgenic tools offer means for increasing stover cellulose content, and for reducing or altering hemicellulose for enhancing ethanol production, respectively.

Potential of Food-Feed vs. Fuel Competition

There are contradictory arguments, sometimes using anecdotal information, on how much the expansion of the maize-derived ethanol industry led to high grain prices and thereby impacting negatively on consumers (Runge and Senauer 2007; Informa Economics 2011). In this regard, von Braun (2007) estimated that by 2020 international maize prices could increase by 26% when considering the available biofuel investment plans. Given this scenario, it is clear that one single factor cannot account solely for food price hikes over time as there are several interrelated factors contributing to such increases. For example, Babcock and Fabiosa (2011) have argued that maize-derived ethanol production would have increased even without any supplemental government assistance. Specifically, maize ethanol expansion is due to both subsidies and market forces. These account for 36% of the average increase in maize prices from 2006 to 2009 in the USA with an accompanying subsidy of 8.5%. A recent report by Informa Economics (2011) for the Renewable Fuels Foundation states further that "corn prices have a relatively weak correlation with food prices, as the farm share is a relatively small portion of the overall retail food dollar and for many products corn is only a portion of the farm value."

FAO (2010) points out that the exact effects of food prices on food security depend on whether households are net food producers or consumers. In this connection, high food prices hurt net food consumers but benefit farmers who are net food producers because their incomes increase. In this regard, Cassman and Liska (2007) suggest that an acceleration in food production capacity, while at the same time protecting natural resources and environmental quality, will be necessary to avoid an excessive rise in food prices and increasing undernourished people. This point is especially true for those in the developing world who face food insecurity regularly, because of either national food shortages or demand that rely in grain imports that never arrive.

The recent drought in the US maize-belt has also raised concerns about whether the US blending mandates, embodied in the Renewable Fuel Standard (RFS) policy, should be relaxed in order to avoid adverse price impacts for livestock producers who depend on this cereal for feed (Babcock 2011). Even prior to the drought, the US livestock producers had been lobbying Congress to change the national ethanol policy by removing the excise tax credit, which has since been allowed to expire as of the end of 2011. Industry lobbyists assert that the flexibility in the implementation in the US ethanol policy would help minimize market shocks by allowing producers to trade allowances among themselves or across time (Cooper 2012). This flexibility is achieved with renewable identification numbers (RINs) that keeps track of blended ethanol volumes and can be accumulated or traded to accommodate market fluctuations and periods of scarcity (McPhail et al. 2011). The fact that maize ethanol production also gives rise to byproducts that are useable as feed, namely dried distillers grains and solubles (DDGS), supports the argument that the impacts of ethanol production on the availability of feed are smaller than what is usually cited in the empirical literature (Taheripour et al. 2008; Beckman et al. 2011).

Adverse Effects on Biodiversity

It has been estimated that about 11 million hectares of land will be necessary to produce 15 billion gallons of maize ethanol (Searchinger et al. 2008). Moreover, De Fraiture et al. (2008) indicated that additional 30 million hectares of farming land may be needed to meet food and biofuel demand in 2030. Such land use may increase greenhouse gas emissions and destroy natural habitats (e.g., rainforests), thereby jeopardizing biodiversity (Mooney and Hobbs 2000). It has been also noted that any additional land planted for maize ethanol in the USA will be at the expenses of decreasing land for barley, cotton, sorghum, soybean and wheat as well as for the Conservation Research Program (Cassman et al. 2006). Hence, a steady increase of maize grain yield is mandatory for expanding ethanol production sustainably, while avoiding disturbing other cropping systems and natural habitats.

Diversifying farming systems can protect farmers economically and the natural resource base on which they depend. Sound crop rotations may further reduce external inputs by promoting nutrient cycling efficiency and effective use of natural resources that includes both agrobiodiversity and water. This strategic approach should maintain long-term land productivity, while increasing yields sustainably in the agro-ecosystems for both food and energy (Zegada-Lizarazu and Monti 2011). Tuomisto et al. (2012) further indicate that integrated farming systems have a great potential to enhance energy and GHG balances and biodiversity *vis à vis* both organic and conventional farming systems.

Relatively few quantitative, model-based studies have been able to directly draw links between the growth of biofuels production and loss of ecosystem quality, natural habitat and species biodiversity. Most of the recent literature has looked more broadly at the impacts of biofuel expansion on land cover conversion and land use change, without identifying the specific environmental or ecosystem impacts. This situation is mostly due to the difficulty of measuring and quantifying losses in biodiversity within a model-based framework, as well as the current policy focus on GHG emissions as the primary indicator of relevance within the context of climate change mitigation-focused discussions. The European Commission, in particular, is focused closely on these issues as it seeks to better-define the sustainability criteria of its Renewable Energy Directive and the Fuel Quality Directive that is embedded within it.

There has been a good deal of scientific inquiry as to the magnitude of these land use change-driven impacts and the sensitivity of those predicted impacts to model assumptions, which has been documented in some comparative studies (Edwards et al. 2010; Witzke et al. 2010). The impacts of biofuels on biodiversity would then have to be deduced through the impacts of biofuels on land use change, which then would serve as a key driver to predict impacts on biodiversity. An example of this approach is the GLOBIO3 model, which quantifies biodiversity loss globally, in response to a number of model-driven changes in land use (Alkemade et al. 2009). In principle, the marginal contribution of maize ethanol production to biodiversity loss could be derived by examining the partial effects implied between scenario changes within this kind of a framework. However, no study has undertaken this exercise as of yet.

Concerns Related to Climate Change and Global Warming

The use of biofuels has been advocated elsewhere as an alternative energy that could mitigate climate change because of their promise for reducing GHG emissions. Von Blottnitz and Currant (2006) provided a summary of the available literature at the time that presents arguments for and against using biofuels to address climate change. The review concluded that making ethanol from sugar crops, e.g., sugarcane or maize, in tropical countries should approach with extreme caution any expansion of current agricultural land usage for biofuels. Fischer et al. (2009) estimated net GHG savings resulting from expansion of biofuels can only be expected after 30 to 50 years because in the short term (2020–2030), net GHG balances are dominated by carbon debts due to direct and indirect land use changes. Nonetheless, the US Environmental Protection Agency (EPA 2009) indicates that biofuel displacement of petroleum can "pay back" in the long term the early land conversion impacts.

Life-cycle analysis (LCA) can assist on assessing the efficiency of biofuel feedstocks as well as their impact on GHG emissions (Bessou et al. 2011). Nonetheless, Davis et al. (2009) indicate that inputs and outputs of agro-ecosystem differ and therefore will affect LCA estimates that may disagree with respect to GHG emissions and energy balances. For example, Hill et al. (2006) estimated the energy requirements for maize cropping, produce transport and fuel processing that were 46%, 12% and 48% lower, respectively of than in that of Pimentel and Patzek (2005). Furthermore, as noted by Tuomisto et al. (2012), LCA interpretations could be flawed if alternative land use options are not taken into account. Some studies also deduced that maize-based ethanol did not reduce GHG emissions relative to petrol (Patzek et al. 2005). This conclusion is not universally shared as Brewer (2007) determined that maize-based ethanol can reduce GHG emissions at least by 20%. It should be noted here however that maize remains as the least effective biofuel feedstock vis-à-vis sugarcane, sugarbeet sugar, and various cellulose sources (Fulton 2004).

When considering the choice of renewable energy sources that might best reduce GHG emissions and the overall carbon footprint of the fuel pool, maize ethanol also tends to compare less favorably to other alternatives. A recent study that evaluated the replacement or supplementation of the current US biofuels policy, e.g., the Renewable Fuel Standard (RFS), with a more carbon-focused one, concluded that there would be an overall reduction in GHG emissions and land use change with a 'low-carbon fuel standard' (LCFS), such as the one implemented in California (Farrell et al. 2007; Sperling and Yeh 2009; Yeh et al. 2012). This kind of LCFS policy has also been evaluated by other US subregional regulatory authorities, and is closer to the spirit of the European Fuel Quality Directive and the standards for environmental sustainability that are laid out therein (MGA 2011; NESCAUM 2011). According to the analysis of the BEPAM model, the reduction of GHG emissions in the implementation of the LCFS on a national scale in the USA is achieved by reducing the use of maize ethanol, increasing use of sugar-based ethanol from Brazil, or the longer-term development of cellulosic-based ethanol from *Miscanthus* or switchgrass (Khanna et al. 2012; Yeh et al. 2012). The attribution of GHG emissions to maize ethanol, within this analysis, is done with the same indirect land use change, e.g., 'iLUC' factors, that are applied in the environmental impact evaluation of the current national US policy by EPA and is subjected to the some sensitivity analysis. Within the literature, the determination of iLUC factors has remained a key point of scientific uncertainty that examines the environmental impacts of biofuels, and is one of the factors that both the California-based LCFS policy as well as the existing RFS enforced by EPA are committed to periodically revising and re-evaluating as better measurements emerge (Farrell et al. 2007; Kim and Dale 2011; Nassar et al. 2011).

Adverse Effects on Nature Habitat, Ecology and the Environment

For any biofuel to be considered as an alternative to fossil fuels it must provide a net energy gain, show environmental benefits, be economically competitive, and be produced in large amounts without affecting food supply (Hill et al. 2006). In this regard, Hill et al. (2006) show that maize-derived ethanol yields 25% more energy than the energy invested in its production. However, the California Air Resources Board using a controversial approach for calculating the environmental impact of alternative transportation fuels indicated that corn ethanol is worse than petrol (Charles 2009). Nevertheless, Farrel et al. (2006) demonstrate that some assessments showing a negative net energy for biofuels incorrectly ignored co-products and used some obsolete data. For example, about 1/3 of maize grain, which includes protein, oil, bran, and cellulosic seed coat material, ends as animal feed (i.e., a byproduct), after its processing in ethanol plants. Hence, LCA needs to include as a factor this accounting for this nutritious animal feed.

Kauffman et al. (2011) proposed measuring GHG emissions considering a unit of land instead of energy inputs and recovery provided by a biofuel pathway of only one feedstock. Following their approach, 1 ha of maize gives two biofuel feedstocks, grain for ethanol and stover for both biochar and bio-oil, which can be further upgraded and used as fuel for internal combustion engines. As a result, such a biofuel pathway leads to a 52.1% reduction in GHG emissions, which is sufficient to qualify maize as an advanced biofuel.

Early research by Wang et al. (1999) shows a greater energy and GHG emission reduction benefits by transitioning from maize's grain-derived ethanol to cellulosic ethanol. Further research by Wu et al. (2006) indicates that for each Btu of maize stover-derived ethanol produced and used, 0.09 Btu of fossil fuel is required and that this cellulosic ethanol pathway avoids 86 to 89% of GHG emissions in the USA. Nonetheless, as noted by Wilhem et al. (2007), the amount of maize stover (5.25–12.50 Mg ha^{-1}) to keep soil organic carbon, and thus agro-ecosystem productivity, remains as an important constraint for using this biomass source as an environmentally sustainable cellulosic feedstock. Increasing biomass yield in agro-ecosystems may therefore provide sustainably a supply of cellulosic feedstock that does not affect the soil.

There is still much research progress to be made for evaluating the impacts of maize-based or other biofuels within a broader environmental and landscape quality context. The most important driver of change that affects land cover, landscape quality, habitat and the species biodiversity within natural and managed landscapes is that of land use change. It should be possible, to the extent to which it is possible, to quantify the effects of land use change on landscape and habitat quality through monitoring shifting plant species specifically dedicated to biofuel production. Studies, such as the Global Environmental Outlook (GEO) of the United National Environment Programme (UNEP 2007), try and measure impacts to environmental quality and biodiversity within the context of a rather broad set of drivers, which makes the specific attribution of biodiversity loss due to biofuel expansion rather difficult. The Millennium Ecosystem Assessment

(MEA 2005) also tried to pay particular attention to environmental quality and ecosystem functionality. MEA attributed the key drivers of change to an increasing demand for food, feed, fiber and biomass for fuel. Like the GEO analysis, the MEA did not attempt to identify the partial effects of any particular demand-side driver such as the increased feedstock demand for crop-based biofuels, although this could have been possible, had it been an explicit part of the research design. In subsequent studies, a number of quantitative elements that were used within the MEA have been extended to do more specific work on the environmental impacts of biofuels using the IMAGE model. No definitive conclusions regarding the effects of biofuels on habitats and their underlying ecologies has emerged (Prins et al. 2011).

Conclusions

In summary, a number of issues have been discussed in which the use of maize for ethanol raises concerns among the environmental and policy communities. Some of these concerns might be allayed by more advanced feedstock production and conversion technologies that would make the energy and carbon balances of maize more favorable, from a life-cycle perspective. Other concerns might be mitigated, in the future, as the science and methodology develop measuring the indirect impacts of biofuel production for specific feedstocks, including maize, are improved with better data and modeling approaches.

What will likely remain an issue well into the future is that of overall land use change from agricultural expansion, and its impact on natural land cover, habitat and the interdependent ecologies and ecosystems that are embedded within the landscape. Maize—being both a major food and feed crop, aside from its use as a biofuel feedstock—is one of many agricultural commodities that drive those changes, as the global demand for food, feed, fiber and fuel products increases and requires higher levels of production and trade. The avoidance of land use change through enhancing per hectare productivity is a key mitigating factor that applies to maize, as well as many other crops. The avoidance of land use change through stricter environmental regulation and safeguards—as well as better monitoring and enforcement of them—is an issue that extends beyond maize cultivation and production technology, and involves a host of other policies and institutional interventions.

A key policy innovation that might affect the use of maize as a biofuel feedstock might be that of more low-carbon-focused types of renewable energy policies, which would tend to favor other feedstocks such as sugar-based ones or those coming from cellulosic feedstocks. The proposed national low-carbon fuel standard currently being discussed within the

USA is one example of this innovation—although even under the case where this policy completely replaces the existing RFS (rather than just supplementing it), there would still be a sizable US production of ethanol from maize. One reason for this is the sheer volume of transportation fuel demand within the USA and the fact that no other type of ethanol could completely replace maize as a blending agent with gasoline. A study by the United States Department of Agriculture (Shapouri et al. 2006) showed that sugar-based ethanol was not economically viable within the USA, and current efforts at producing ethanol from cellulosic sources have continue to incur unit costs that are too high to be competitive with maize ethanol as a blending substitute (Campiche et al. 2010).

The continued use of maize ethanol as a biofuel feedstock is therefore almost assured within the USA, for the foreseeable future. The mitigation of environmental impacts coming from its conversion to biofuels can come from higher levels of efficiency in both its production as well as its conversion into ethanol. The by-products that come from maize ethanol production (i.e., DDGS) will continue to provide feed for intensively-fed livestock, while the residues from maize harvest—namely the stalk—might eventually prove economic for conversion into cellulosic ethanol, as well. Clearly there is need for better science to achieve the efficiency gains needed to make this happen, but there is also a need for more forward-looking policies to provide greater incentives for cellulosic biofuel technologies through emphasizing carbon reduction as a performance target. If the "body politic" of the USA can "stomach" it, perhaps a combination of tax-related penalties on fossil fuels and direct credits or subsidies to low-carbon technologies might bring the incentives in line, in order for cellulosic-based fuels to be competitive at current costs, or for faster progress towards this outcome to occur within the renewable fuel sector.

References

Alkemade R, van Oorschot M, Miles L et al. (2009) GLOBIO3: A framework to investigate options for reducing global terrestrial biodiversity loss. Ecosystems 12: 374–390.

Babcock BA (2011) The impact of US biofuel policiess on agricultural price levels and volatility. ICTSD Issue Paper 35. International Center for Trade and Sustainable Development, Geneva, Switzerland.

Babcock BA and Fabiosa JF (2011) The impact of ethanol and ethanol subsidies on corn prices: revisiting history. CARD Policy Brief 11-PB 5. Iowa State University, Ames, Iowa, USA, 10 p.

Beckman J, Keeney R and Tyner W (2011) Feed demands and co-product substitution in the biofuel era. Agribusiness 27(1): 1–18.

Bessou C, Ferchaud F, Gabrielle B et al. (2011) Biofuels, greenhouse gases and climate change. A review. Agron Sustain Dev 31: 1–79.

Bogdanski A, Dubois O, Jamieson C et al. (2011) Making integrated food-energy systems work for people and climate. An overview. Environment and Natural Resources Management Working Paper 45. Food and Agriculture Organization of the United Nations, Rome, Italy. 116 p.

Brewer TL (2007) Biofuels for climate change mitigation: international trade issues for the G8+5 countries. Paper for the G8+5 Climate Change Dialogue. Washington DC, USA 12 p.

Campiche JL, Bryant HL and Richardson JW (2010) Long-run effects of falling cellulosic ethanol production costs on the US agricultural economy. Env Res Lett 5: 014018 doi: 10.1088/1748-9326/5/1/014018.

Cassman K, Eidman V and Simpson E (2006) Convergence of agriculture and energy: implications for research and policy. CAST Commentary QTA2006-3.

Cassman KG and Lyska AJ (2007) Food and fuel for all: realistic or foolish? Biofuels Bioprod Biorefin 1: 18–23.

Charles D (2009) Corn-based ethanol flunks key test. Science 324: 587.

Cooper G (2012) The 2012 corn crop: Implications for ethanol and the RFS. Webinar presentation given on 14 August 2012. Renewable Fuels Association, Washington DC, USA, Available at: http: //www.adkinsenergy.com/pdfs/Renewable%20Fuels%20Association%20 drought_RFS%20webinar.pdf (accessed on 8 Sept 2012).

Davis SC, Anderson-Texeira KJ and deLucia EH (2009) Life-cycle analysis and the ecology of biofuels. Trends Plant Sci 14: 40–46.

de Fraiture C, Giordano M and Liao Y (2008) Biofuels and implications for agricultural water use: Blue impacts of green energy. Science 10: 67–81.

Dhugga KS (2007) Maize biomass yield and composition for biofuels. Crop Sci 47: 2211–2227.

Edwards R, Mulligan D and Marelli L (2010) Indirect land use change from increased biofuels demand: Comparison of models and results for marginal biofuels production from different feedstocks. JRC Scientific and Technical Reports. Joint Research Center of the European Commission, Ispra, Italy.

Elobeid A, Tokgoz S, Hayes DJ et al. (2007) The long-run impact of corn-based ethanol on the grain, oilseed, and livestock sectors with implications for biotech crops. AgBioForum10(1): 11–18.

EPA (2009) EPA lifecycle analysis of greenhouse gas emissions from renewable fuels. EPA-420-F-09-024. Office of Transportation and Air Quality, Environmental Protection Agency, Washington DC, USA, 5 p.

FAO (2008) The state of food and agriculture. Biofuels: prospects, risks and opportunities. Food and Agriculture Organization of the United Nations, Rome, Italy, 128 p.

FAO (2010) Bioenergy and food security. The BESF analytical framework. Environment and Natural Resources Management Working Paper 16. Food and Agriculture Organization of the United Nations, Rome, Italy, 116 p.

Farrell AE, Plevin RJ, Turner BT et al. (2006) Ethanol can contribute to energy and environmental goals. Science 311: 506–508.

Farrell AE, Sperling D, Brandt AR et al. (2007) A low-carbon fuel standard for California Part II: Policy Analysis. California Air Resources Board. Available at: http: //www.arb.ca.gov/fuels/lcfs/lcfs_uc_p2.pdf (accessed 8 Sept 2012).

Fischer G, Hizsnyik E, Prieler S et al. (2009) Biofuels and food security Implications of an accelerated biofuels production. Summary of the OFID study prepared by IIASA. OFID Pamphlet Series 38. OPEC Fund for International Development, Vienna, Austria, 41 p.

Fulton L, Howes T and Hardy J (2004) Biofuels for transport: an international perspective. International Energy Agency, Paris, France, 210 p.

Graham RL, Nelson R, Sheehan J et al. (2007) Current and potential U.S. corn stover supplies. Agron J 99: 1–11.

Hill J, Nelson E, Tilman D et al. (2006) Environmental, economic, and energetic costs and benefits of biodiesel and ethanol biofuels. Proc Natl Acad Sci USA 103: 11206–11210.

Informa Economics (2011) Analysis of corn, commodity and food prices. The Renewable Fuels Foundation, Washington DC, USA, 49 p.

Kauffman N, Hayes D and Brown R (2011) A life cycle assessment of advanced biofuel production from a hectare of corn. Fuel 90: 3306–3314.

Khanna M, Chen X, Huang H et al. (2011) Land use and greenhouse gas mitigation effects of biofuel policies. University of Illinois Law Review 2: 549–588.

Kim S and Dale BE (2011) Indirect land use change for biofuels: Testing predictions and improving analytical methodologies. Biomass Bioenergy 91(3): 1–6.

McPhail L, Westcott P and Lutman H (2011) The renewable identification number system and US biofuel mandates. Report BIO-03 from the Economic Research Service. United States Department of Agriculture, Washington DC, USA.

MGA (2010) Energy security and climate stewardship platform for the Midwest: Low carbon fuel policy advisory group recommendations. Report of the Policy Advisory Group for the Midwestern Governors Association (MGA). Accessible at: http: //www. midwesterngovernors.org/Publications/LCFPagDoc.pdf.

Millennium Ecosystem Assessment [MEA] (2005) Ecosystems and human well-being: Scenarios. Findings of the scenarios working group. Island Press, Washington DC, USA.

Mooney HA and Hobbs RJ (eds) (2000) Invasive species in a changing world. Island Press, Washington DC. USA, 384 p.

Nassar AM, Harfuch L, Bachion LC et al. (2011) Biofuels and land-use changes: searching for the top model. Interface Focus doi: 10.1098/rsfs.2010.0043.

NESCAUM (2011) Economic analysis of a program to promote clean transportation fuels in the Northeast/Mid-Atlantic Region. Final Report produced by Northeast States for Coordinated Air Use Management.

Nuffield Council on Bioethics (2011) Biofuels: ethical issues. Nuffield Press, Abingdon, Oxfordshire, UK, 187 p.

Patzek TW, Anti S-M, Campos R et al. (2005) Ethanol from corn: clean renewable fuel for the future, or drain on our resources and pockets? Env Dev Sustain 7: 319–336.

Pimentel D and Patzek TW (2005) Ethanol production using corn, switchgrass, and wood; biodiesel production using soybean and sunflower. Nat Resour Res 14: 65–76.

Prins AG, Eickhout B, Banse M et al. (2011) Global impact of European agricultural and biofuel policies. Ecology and Society 16(1): 49 http: //www.ecologyandsociety.org/vol16/iss1/art49/ (accessed on 8 Sept 2012).

Runge CF and Senauer B (2007) How biofuels could starve the poor. Foreign Affairs 3: 41–53.

Searchinger T, Heimlich R, Houghton RA et al. (2008) Use of U.S. croplands for biofuels increases greenhouse gases through emissions from land use change. Science 319(5867): 1238–1240.

Shapouri H, Salassi M and Fairbanks JN (2006) The economic feasibility of ethanol production from sugar in the United States. United States Department of Agriculture, Washington DC, USA, Accessible at: http: //www.usda.gov/oce/reports/energy/EthanolSugarFeasibilityReport3.pdf.

Sperling D and Yeh S (2009) Low carbon fuel standards. Issues Sci Tech Winter 57–66.

Taheripour F, Hertel TW, Tyner WE et al. (2008) Biofuels and their by-products: Global economic and environmental implications. Working paper, Department of Agricultural Economics, Purdue University, West Lafayette, Indiana, USA, Accessible at: https: //www.gtap.agecon.purdue.edu/resources/download/3974.pdf (accessed 8 Sept 2012).

Tuomisto HL, Hodge ID, Riordan P et al. (2012) Comparing energy balances, greenhouse gas balances and biodiversity impacts of contrasting farming systems with alternative land uses. Agri Syst 108: 42–49.

Tyner WE and Taheripour F (2007) Biofuels, energy security, and global warming policy interactions. Paper presented at the National Agricultural Biotechnology Council Conference, 22–24 May 2007, South Dakota State University, Brookings, South Dakota, USA.

UNEP (2007) Global environmental Outlook: environment for development (GEO-4). United Nations Environment Programme. Accessible at: http: //news.bbc.co.uk/2/shared/bsp/hi/pdfs/15_10_2007_un.pdf (accessed 8 Sept 2012).

von Blottnitz H and Curran MA (2007) A review of assessments conducted on bio-ethanol as a transportation fuel from a net energy, greenhouse gas, and environmental life-cycle perspective. J Cleaner Prod 7: 607–619.

von Braun J (2007) The world food situation: new driving forces and required actions. IFPRI's Bi-Annual Overview of the World Food Situation presented to the CGIAR Annual General Meeting, Beijing, 3 December 2007. International Food Policy Research Institute, Washington DC, USA, 25 p.

Wang M, Saricks C and Santini D (1999) Effects of fuel ethanol use on fuel-cycle energy and greenhouse gas emissions. ANL/ESD-38. Center for Transportation Research, Energy Systems Division, Argonne National Laboratory, Argonne, Illinois, USA, 31 p.

Wilhelm WW, Johnson JMF, Karlen DL et al. (2007) Corn stover to sustain soil organic carbon further constrains biomass supply. Agron J 99: 1665–1667.

Witzke HP, Fabiosa JF, Gay SH et al. (2010) A decomposition approach to assess ILUC results from global modeling efforts. No 91430, Proceedings Issues, 2010: Climate Change in World Agriculture: Mitigation, Adaptation, Trade and Food Security, June 2010, Stuttgart-Hohenheim, Germany, International Agricultural Trade Research Consortium. Accessible at: http: //EconPapers.repec.org/RePEc: ags: iatr10: 91430. (accessed 8 Sept 2012).

Wu M, Wang M and Huo H (2006) Fuel-cycle assessment of selected bioethanol production pathways in the United States. ANL/ESD/06-7. Univ. Chicago Argonne, LLC, Energy Systems Division, Argonne National Laboratory, Argonne, Illinois, USA, 52 p.

Yeh S, Sperling D, Batka M et al. (2012) National Low Carbon Fuel Standard: Technical Analysis Report. Accessible at: http: //ssrn.com/abstract=2102817.

Zegada-Lizarazu W and Monti A (2011) Energy crops in rotation. A review. Biomass Bioenergy 15: 12–25.

CHAPTER 10

Legal and Regulatory Issues in the US

Jay P. Kesan[1],* and Timothy A. Slating[2]

ABSTRACT

Legal and regulatory frameworks in the US play an increasingly
important role in the successful deployment of corn-based biofuels and
biotechnology innovations. Through legal and regulatory regimes, the
US federal government mandates the commercialization of biofuels,
controls the lawful commercialization of biofuels, and provides a
means to protect intellectual property associated with biofuels and
their feedstocks. In this Chapter, we take a thorough look at the main
legal and regulatory frameworks affecting the commercialization of
corn-based biofuels and biotechnology innovation in the US. First,
we address the role of the US Clean Air Act in not only governing the
lawful commercialization of biofuels, but also in controlling their lawful
blending limits. Second, we detail the effects of the federal Renewable
Fuel Standard, which mandates the commercialization of different
types of biofuels in the US. Finally, we address the options for the
legal protection of intellectual property rights associated with corn-
based biotechnology innovations. Throughout the Chapter, we focus
on three main categories of corn-based biofuels: (1) ethanol derived
from corn starch; (2) butanol derived from corn starch; and (3) biofuels

[1]College of Law, University of Illinois at Urbana-Champaign. 504 E. Pennsylvania Ave.,
Champaign, IL 61820.
[2]Energy Biosciences Institute, University of Illinois at Urbana-Champaign. 1206 W. Gregory
Dr., Urbana, IL 61801.
Email: slating2@illinois.edu
*Corresponding author: kesan@illinois.edu

(whether ethanol or butanol) derived from the cellulosic components of the corn plant.

Keywords: Biofuels, Biofuel Blending, Biofuel Mandates, Biofuel Incentives, Renewable Fuel Standard, Intellectual Property, Plant Variety Protection

Introduction

Legal and regulatory frameworks in the US play an increasingly important role in the successful deployment of corn-based biofuels and biotechnology innovations. Through legal and regulatory regimes, the US federal government mandates the commercialization of biofuels, controls the lawful commercialization of biofuels, and provides a means to protect intellectual property associated with biofuels and their feedstocks. As such, no comprehensive book on the bioenergy-related issues associated with maize would be complete without a chapter on relevant legal and regulatory issues.

In this Chapter, we take a thorough look at the main legal and regulatory frameworks affecting the commercialization of corn-based biofuels and biotechnology innovation in the US. First, we address the role of the US Clean Air Act in not only governing the lawful commercialization of biofuels, but also in controlling their lawful blending limits. Second, we discuss the federal Renewable Fuel Standard, which mandates the commercialization of certain types of biofuels in the US. Finally, we will address the options for the legal protection of intellectual property rights associated with maize-based biotechnology innovations. Throughout the Chapter, we will focus on three main categories of corn-based biofuels: (1) ethanol derived from corn starch; (2) butanol derived from corn starch; and (3) biofuels (whether ethanol or butanol) derived from the cellulosic components of the corn plant.

The Lawful Commercialization of Biofuels

The federal Clean Air Act (CAA) (US Congress 1970) and its implementing regulations (US EPA 2010a, 2010b), which are promulgated and administered by the US Environmental Protection Agency (EPA), govern the lawful commercialization of biofuels in the US. At a base level, the CAA requires that all fuels and fuel additives be registered with the EPA before they can be lawfully commercialized (US Congress 1970; US EPA 2010a). As it is currently the norm for corn-based biofuels to be blended with gasoline as fuel additives, this means that the biofuels themselves must first be registered as fuel additives and then the finished fuels (i.e., the gasoline/biofuel blends) must also be separately registered with the EPA. As a requirement for both of these types of registration, the manufacturer seeking registration must specify a recommended blending concentration for the corn-based biofuel that complies with the CAA's so-called "substantially similar" regulatory framework (US Congress 1970; US EPA 2008). Since the lawful blending limit for a biofuel effectively caps the available commercial market for that biofuel, this regulatory framework has a significant effect on the successful deployment of corn-based biofuels in the US.

Lawful Blending Limits Under the CAA

The CAA broadly seeks to improve air quality throughout the US by regulating emissions from both stationary sources (e.g., a coal-fired power plant) and mobile sources (e.g., a passenger vehicle) (US Congress 1970). In regards to mobile sources, the CAA and its regulatory frameworks address the emission of air pollutants through measures such as setting emissions limitations for different classes of vehicles, setting miles-per-gallon targets for different classes of vehicles, and specifying the lawful chemical content of transportation fuels. But in order to understand the way in which the CAA regulates the lawful blending limits for corn-based biofuels, we must begin with a brief discussion of the emissions certification process for vehicles and engines.

Before any vehicle or engine can be lawfully sold in the US, it must first be certified to be in conformity with the CAA's relevant emissions standards (US Congress 1970). To achieve certification, vehicle and engine manufacturers must subject their products to the very detailed emissions testing protocols specified by the EPA. In carrying out this detailed emissions testing, vehicles and engines must be run on a very specific fuel with a precise chemical make-up established by the EPA. This specific fuel has come to be known as "certification fuel."

All of the foregoing affects the lawful blending limits for corn-based biofuels through operation of the CAA's "substantially similar" prohibition (US Congress 1970). Specifically, the CAA prohibits any fuel manufacturer from commercializing any finished fuel that is not "substantially similar" to certification fuel. Likewise, fuel additive manufacturers are prohibited from commercializing their products for use in a concentration that results in a finished fuel that is not "substantially similar" to certification fuel. The catch is that the CAA does not specify what qualifies as being substantially similar to certification fuel and, instead, leaves the task to the discretion of the EPA (US Congress 1970; US EPA 2008).

The Substantially Similar Rule

In order to notify fuel and fuel additive manufacturers as to what qualifies as being "substantially similar" to certification fuel, the EPA issues detailed fuel guidelines via what is referred to as its Substantially Similar Rule (US EPA 2008). The most recent Rule, which was issued in 2008, sets out the specific parameters that any given fuel must comply with for it to be considered substantially similar (i.e., for it to be sold lawfully in the US). While any fuel must meet all of the Rule's detailed requirements in order to be lawfully sold, the most important provision affecting corn-based biofuels states that "fuels containing . . . alcohols . . . must contain no more than 2.7% oxygen by weight." Since traditional gasoline contains no oxygen and all current forms of corn-based biofuels are oxygen-containing alcohols (e.g., ethanol and butanol), this 2.7% oxygen by weight limitation effectively caps the amount of corn-based biofuels that lawful finished fuels in the US can contain. Nonetheless, since the Rule opts to merely specify an oxygen weight limitation, it is impossible to translate this limitation into a volumetric limit since the densities of finished fuels remain in a state of flux that is influenced by pressure and temperature. Thus, despite the fact that it is the norm in the transportation fuels industry to speak of volumetric blending limits for biofuels, the EPA's Substantially Similar Rule only allows us to specify the volumetric blending limits for alcohol-based biofuels in terms of a permissible range that satisfies the Rule's oxygen by weight limitation at various densities. In regards to ethanol (whether derived from corn starch or the cellulosic portions of the corn plant), the Substantially Similar Rule allows finished fuels to contain roughly 6.5%–7.5% by volume. Since butanol possesses a lower oxygen weight than ethanol, the Rule allows lawful finished fuels to contain roughly 11.5%–12.5% by volume biobutanol (Bu11.5–Bu12.5), whether derived from corn starch or cellulosic materials.

Fuel Waivers

While the CAA prohibits the commercialization of fuels and fuel additives that are not substantially similar to certification fuel, it also provides a mechanism whereby fuel and fuel additive manufacturers can seek a waiver of the prohibition in order to lawfully commercialize fuel blends that do not comply with the EPA's Substantially Similar Rule (US Congress 1970; US EPA 2008). Specifically, the EPA is authorized to grant a fuel waiver if the fuel or fuel additive manufacturer can demonstrate that the waiver fuel (i.e., the fuel blend that is the subject of the fuel waiver application) "will not cause or contribute to a failure of any emissions control device or system (over the useful life of the motor vehicle, motor vehicle engine, nonroad engine or nonroad vehicle in which such device or system is used) to achieve compliance by the vehicle or engine with the emissions standards with respect to which it has been certified pursuant to" the CAA. In order to carry this burden, the waiver applicant is required to submit data that demonstrates the waiver fuel's effects on exhaust emissions (i.e., tailpipe emissions), evaporative emissions (i.e., those coming off of a fuel tank in the absence of combustion), materials compatibility (which can substantially increase exhaust and evaporative emissions), and the drivability and operability of vehicles and engines (based on the notion that if a waiver fuel has adverse effects, drivers and operators might attempt to modify their vehicles and engines into a configuration that differs from the one certified by the EPA). Once the EPA grants a fuel waiver for a given fuel blend, the fuel requirements set out in the official waiver supersede the Substantially Similar Rule and all manufacturers are permitted to legally commercialize the fuel blend.

To date, the EPA has considered and either granted or conditionally granted two distinct fuel waivers for ethanol/gasoline fuel blends (US EPA 1979, 2011a). The first, granted in 1978 by operation of law based on a now removed provision of the CAA that automatically granted fuel waivers if the EPA failed to issue a decision within 180 days of the waiver application being submitted, allows for the lawful commercialization of a 10% ethanol/90% gasoline blend (E10) (US EPA 1979). Since the granting of this waiver, the use of ethanol fuel blends has steadily increased and E10 now comprises well over 90% of the vehicle fuel sold in the US. In early 2011, the EPA conditionally granted another ethanol fuel waiver that allows for the lawful commercialization of a 15% ethanol/85% gasoline blend (E15) for use in light-duty motor vehicles (i.e., passenger cars, light-duty trucks, and medium-duty passenger vehicles) produced in model years 2001 and newer (US EPA 2011a). Finally, an 85% ethanol/15% gasoline blend (E85) is also approved for use only in flex-fuel vehicles (since E85 is used in the emissions certification process for flex-fuel vehicles, it is obviously considered "substantially similar" to the certification fuel for these vehicles and can lawfully be commercialized in the absence of a specific fuel waiver). In summary, ethanol (whether derived from corn starch or the cellulosic portions of the corn plant) can currently be lawfully commercialized in the US in blends up to E10 for use in all gasoline powered vehicles and engines (US EPA 1979), blends up to E15 for use only in model year 2001 and newer light-duty vehicles (US EPA 2011a), and blends up to E85 for use only in flex-fuel vehicles.

In regards to corn-based biobutanol, the EPA has yet to consider any fuel waivers that specifically focus on this type of biofuel. Nonetheless, in the 1980s, the EPA conditionally granted two different fuel waivers for methanol-based fuel blends that both contain requirements that arguably allow for the lawful commercialization of 16%–17% by volume biobutanol blends (Bu16–Bu17) (US EPA 1985, 1988). Specifically, these waivers provide that their allowable fuel blends can contain a "maximum" amount of methanol and a "minimum" amount of cosolvent (such as "butanol") so long as the finished fuel contains no more than 3.7% oxygen by weight. As such, even though these waivers were originally directed at methanol-based fuel blends, it is arguable that these two fuel waivers allow for the lawful commercialization of a blend that contains no methanol and as much butanol as can be achieved within the confines of the finished fuel containing no more than 3.7% oxygen by weight. Again, it is impossible to calculate a precise volumetric blending limit based only on this oxygen weight limitation, but it can be said that Bu16-Bu17 blends would arguably be in compliance with either of these fuel waivers. Nonetheless, when the EPA granted these waivers, it never considered the emissions effects of 16%–17% biobutanol blends and it is highly uncertain whether it would allow a fuel manufacturer to register a Bu16-Bu17 blend pursuant to one of these waivers. As such, any fuel manufacturer seeking to commercialize corn-based biobutanol blends can either register and commercialize a Bu11.5-Bu12.5 blend pursuant to the Substantially Similar Rule (US EPA 2008), attempt to register and commercialize a Bu16-Bu17 blend pursuant to one of the aforementioned fuel waivers (US EPA 1985, 1988), or attempt to seek a new fuel waiver that allows for the lawful commercialization of a different biobutanol/gasoline fuel blend.

The Renewable Fuel Standard

The federal Renewable Fuel Standard (RFS) (US EPA 2011b), which was enacted as a provision within the Energy Independence and Security Act of 2007 (EISA) (US Congress 2007), is arguably the most important regulatory framework affecting the deployment of corn-based biofuels. It not only supports the market for these biofuels, but it essentially creates their market by mandating that gasoline refiners, blenders, and importers must commercialize minimum volumes of biofuels each year through 2022 (Fig. 1). Specifically, the RFS establishes unique mandates for four nested categories of biofuels (Table 1) that are each defined by the feedstocks they are derived from and their ability to reduce lifecycle greenhouse gas (GHG) emissions from a baseline, which is defined as the lifecycle GHG emissions from gasoline or diesel produced in 2005 (whichever the biofuel is replacing).

The broadest overarching category of biofuels mandated by the RFS is simply referred to as "renewable fuel." In order to qualify as a renewable fuel, a given biofuel must be "produced from renewable biomass" and

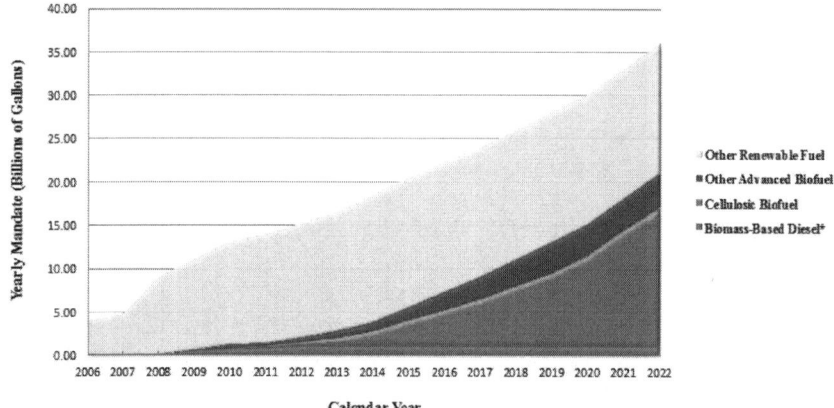

Figure 1 The RFS's yearly volumetric mandates.

Color image of this figure appears in the color plate section at the end of the book.

have "lifecycle [GHG] emissions that are at least 20 percent less than baseline lifecycle [GHG] emissions." While the RFS's complex definition for "renewable biomass" might create unique issues for other types of biofuels, for the purposes of corn-based biofuels, the definition clearly encompasses any material from corn plants grown on land that was either cleared or cultivated as of December 19, 2007 (this latter requirement was implemented so that the RFS would not incentivize land-use changes). Finally, the RFS defines lifecycle GHG emissions as "the aggregate quantity of [GHG] emissions (including direct emissions and significant indirect emissions such as significant emissions from land use changes), as determined by the Administrator [of the EPA], related to the full fuel lifecycle, including all stages of fuel and feedstock production and distribution, from feedstock generation or extraction through the distribution and delivery and use of the finished fuel to the ultimate consumer, where the mass values for all greenhouse gases are adjusted to account for their relative global warming potential."

The second category of biofuels mandated by the RFS is referred to as "advanced biofuel" and is a sub-category of renewable fuel (i.e., any fuel qualifying as an advanced biofuel also qualifies as a renewable fuel, but not *vice versa*). Specifically, an advanced biofuel is defined as "renewable fuel, other than ethanol derived from corn starch, that has lifecycle [GHG] emissions that are at least 50 percent less than baseline lifecycle [GHG] emissions." Within the category of advanced biofuel, two unique sub-categories exist. The first sub-category is referred to as "cellulosic biofuel" and is defined as "renewable fuel derived from any cellulose, hemi-

Table 1 The RFS's biofuel categories

Category	GHG-Reduc0on-Threshold	Permissible-Feedstock	Example
Renewable Fuel	20%*	Renewable Biomass	Ethanol or butanol derived from corn starch
Advanced Biofuel	50%	Renewable Biomass, with the exception of maize starch if the fuel is ethanol	Butanol derived from corn starch using advanced process technologies
Biomass-Based Diesel	50%	Renewable Biomass	Biodiesel derived from soy bean oil or canola oil
Cellulosic Biofuel	60%	Cellulose, Hemi-Cellulose, or Lignin	Ethanol or butanol derived from cellulosic components of corn

* with the exception of ethanol produced at facilities that commenced construction on or before December 19, 2007

cellulose, or lignin that has lifecycle [GHG] emissions that are at least 60 percent less than the baseline lifecycle [GHG] emissions." The final subcategory is called "biomass-based diesel" and is defined as "either biodiesel or non-ester renewable diesel" that is a "renewable fuel that has lifecycle [GHG] emissions that are at least 50 percent less than baseline lifecycle [GHG] emissions."

As previously mentioned, the RFS mandates minimum volumes for each category of biofuels that must be commercialized each year, through 2022, by gasoline refiners, blenders, and importers (Fig. 1). Specifically, for each calendar year, the RFS specifies: (1) how much total renewable fuel must be commercialized; (2) how much of this total renewable fuel must be comprised of advanced biofuel; (3) how much of this advanced biofuel must be cellulosic biofuel; and (4) how much of the advanced biofuel must be biomass based diesel. To use 2022 as an example, 36 billion gallons (bg) of renewable fuel must be commercialized and at least 21 bg of this must be advanced biofuel. If only the minimum 21 bg of advanced biofuel is commercialized, a 15 bg renewable fuel gap will exist that can be satisfied by any biofuel that merely meets the definition of a renewable fuel. Of the 21 bg of advanced biofuel mandated for 2022, at least 16 bg must be cellulosic biofuel and an as yet to be determined amount must be biomass-based diesel (the RFS only specifies biomass-based diesel volumes through 2012 and thereafter the EPA sets the yearly mandate, which must be at least 1 bg per year). If only the minimum amount of cellulosic biofuel is commercialized and less than 5 bg of biomass-based diesel is mandated and used in 2022, then an advanced biofuel gap will exist that can only be satisfied by the commercialization of biofuels that meet the definition for advanced biofuel.

Treatment of Corn-based Ethanol

The EPA has currently established four unique "fuel pathways" for ethanol derived from corn starch that result in biofuels capable of satisfying the RFS's definition for renewable fuel (US EPA 2011b). The first pathway requires a dry mill fermentation process that uses biogas, biomass, or natural gas for process energy and implements at least two types of "advanced technologies" (e.g., corn oil fractionation, corn oil extraction, membrane separation, raw starch hydrolysis, or the use of combined heat and power in production). The second pathway is essentially the same as the first, but it only requires one advanced technology and that the producer dry "no more than 65% of the distillers grains and solubles that it markets annually." Likewise, the third pathway also requires a dry mill fermentation process that uses biogas, biomass, or natural gas for process energy, but this pathway requires no advanced technologies so long as the producer dries "no more than 50% of the distillers grains and solubles it markets annually." In contrast, the only requirements for the fourth pathway are that a wet mill fermentation process is used with either biogas or biomass providing the process energy. While any ethanol derived from maize starch that satisfies the requirements of one of these fuel pathways has been deemed to meet the requisite 20% GHG reduction threshold necessary to be classified as a renewable fuel (US EPA 2010c), it is important to note that any ethanol produced at a facility that commenced construction before December 19, 2007 (i.e., the enactment date of EISA) has been grandfathered in to the RFS and does not need to satisfy the GHG reduction threshold.

Finally, it is also important to note that ethanol derived from corn starch will never be able to qualify for any of the RFS's biofuel categories except renewable fuel. It is explicitly excluded from the definition for advanced biofuel, it violates the feedstock requirements for cellulosic biofuel, and it is not a form of renewable diesel. As such, ethanol derived from corn starch will only be able to be used to satisfy the portion of the RFS's mandate that equals the difference between the total renewable fuel mandate and the advanced biofuel mandate (Fig. 1). So long as only the minimum mandated amount of advanced biofuel is commercialized, the amount of ethanol derived from corn starch that can be used to satisfy the RFS's renewable fuel mandate peaks in 2015 at 15 bg and remains at 15 bg per year through 2022. As the US corn ethanol industry is approaching 15 bg of yearly production capacity, this 15 bg limitation will no doubt have a significant impact on the growth of the industry in the coming years.

Treatment of Corn-based Biobutanol

While biobutanol derived from corn starch is not predicted to reach the commercial market until sometime in 2013 (Slating and Kesan 2011), the EPA has already established a fuel pathway for purposes of the RFS that requires a dry mill fermentation process that utilizes either biogas, biomass, or natural gas for process energy (US EPA 2011b). Although the EPA used data from a significantly outdated and nearly abandoned production process when it modeled this pathway, it did find that the process results in a biofuel that satisfies the 20% GHG reduction threshold necessary to qualify as a renewable fuel (US EPA 2010c). Nonetheless, it is unlikely that the RFS will incentivize the increased production of biobutanol derived from corn starch so long as it remains classified as only a renewable fuel. This is due to the fact that current biobutanol commercialization strategies involve retrofitting existing ethanol facilities to output biobutanol (http://www.gevo.com/) and, as previously mentioned, the existing corn ethanol industry is on track to saturate the portion of the RFS's mandates that can be satisfied by fuels categorized only as renewable fuel.

Despite biobutanol derived from corn starch currently being only classified as a renewable fuel, the fact that any form of biobutanol is assigned an "equivalence value" of 1.3 for purposes of satisfying the RFS's mandates might nonetheless incentivize its increased commercialization. When the EPA developed the official rules for the RFS, it opted to assign equivalence values to different types of biofuels that reflect their energy content compared to ethanol. The idea was to equalize the gasoline reduction potential of all biofuels to a common denominator. Because one gallon of butanol possesses roughly 30% more energy than one gallon of ethanol, it has been assigned an equivalence value of 1.3. The practical effect of this is that if a given gasoline refiner was required by the RFS to commercialize 1,300 gallons of renewable fuel in a given year, it could either purchase and blend 1,300 gallons of ethanol or, instead, 1,000 gallons of biobutanol. As such, it is the costs saved from having to purchase, blend, and transport fewer gallons that could potentially stimulate demand for biobutanol derived from corn starch.

Before moving on, we must note that as of the writing of this Chapter, one producer of corn-based biobutanol is currently in the process of seeking out the establishment of an advanced biofuel pathway for its production process (Voegele 2011). Since the RFS's definition for advanced biofuel only excludes "ethanol derived from corn starch," corn-based biobutanol could qualify so long as it meets the requisite 50% GHG reduction threshold. While it is unknown when the EPA will issue its determination regarding this new pathway application, the chances of a new production process meeting the 50% GHG reduction threshold is quite promising if you consider the fact that the EPA found an extremely outdated corn-based biobutanol production process to result in a 31% GHG reduction (US EPA 2010c). Moreover, this 31% GHG reduction was actually the midpoint of a 20%–40% range that resulted from uncertainties in the indirect land-use change values used by the EPA and recent studies are beginning to question whether increased biofuel production in the US results in any indirect land-use change (Kim and Dale 2011). If an advanced biofuel pathway were established for biobutanol derived from corn starch, the RFS could be a significant driver for increased production since its advanced biofuel mandate is far from saturated.

Treatment of Corn-based Cellulosic Biofuels

In regards to biofuels derived from the cellulosic components of the corn plant (e.g., its stalks, leaves, and cobs), the EPA has only established one relevant pathway for purposes of the RFS (US EPA 2011b). This pathway requires the production of ethanol through any production process and since it allows for crop residues to be utilized as a feedstock, the cellulosic portions of corn clearly qualify. Based on its modeling, the EPA has concluded that biofuels produced in accordance with this pathway satisfy the 60% GHG reduction necessary to be classified as cellulosic biofuels for purposes of complying with the RFS's mandates (US EPA 2010c). As there is currently so little cellulosic biofuel production, the RFS's mandates for cellulosic biofuels are nowhere near saturated and, in fact, the EPA has had to exercise its statutory authority to reduce the cellulosic biofuel mandates in 2011 (US EPA 2010d) and 2012 (US EPA 2011c). In order to capitalize on this captive market for cellulosic biofuels, at least one major US ethanol producer is currently in the process of constructing a facility to convert corn residues into cellulosic ethanol (http://www.poet.com/). As far as cellulosic biobutanol is concerned, production has currently only been demonstrated at the lab scale (Higashide et al. 2011) and if a manufacturer attempts to commercialize the technology at some point, a new cellulosic biofuel pathway would have to be established for purposes of the RFS.

Intellectual Property Regimes

In addition to incentivizing and governing the commercialization of corn-based biofuels in the US, legal and regulatory frameworks also provide a means of protecting intellectual property (IP) associated with corn-based biotechnology innovations. Several IP regimes exist to protect agricultural biotechnology and each may be used alone or in combination with others. In general, the ease of obtaining a particular form of IP protection is proportional to the strength of protection afforded by that mechanism. In other words, the more robust the protection afforded, the more stringent the requirements for obtaining that particular form of protection. This section provides an overview of the three most important forms of IP protection available for agricultural biotechnology innovations associated with the corn plant. Specifically, we will address utility patents, plant variety protection, and trade secret protection.

4.1 Utility Patents

The utility patent provides the most extensive coverage for inventions and, in the context of corn-based biotechnology, may be obtained to protect everything from genetically modified seeds and genetically modified plants, to transformation methods. Under the US statute governing utility patents, "[w]hoever invents or discovers any new and useful process, machine, manufacture, or composition of matter, or any new and useful improvement thereof, may obtain a patent therefor, subject to the conditions and requirements of this title" (US Congress 2010a). The corn plant is eligible subject matter for utility patent protection under the category of "compositions of matter."

Scope of protection

In the US, utility patents grant a broad right to exclude others from making, using, offering to sell, selling, or importing into the US the patented invention (US Congress 2010b). Unauthorized exploitation of the patented invention by others within the patent term constitutes patent infringement. The broad scope of protection afforded by utility patents provides great flexibility for tailoring protection to various corn-based biotechnology innovations.

Utility patents may cover individual components of a corn plant, including the plant's genome, cells, cell culture, and tissues, as well as methods for making the plant. To provide an example from another agricultural context, the agrichemical corporation Monsanto holds US patents on ROUNDUP READY® soybeans, which are genetically modified to withstand ROUNDUP®, the company's broad-spectrum herbicide. The company creates ROUNDUP READY® soybeans by inserting a gene sequence that allows the plant to survive the herbicide. Monsanto's utility patents allow the company to claim protection for not only methods of producing the ROUNDUP READY® soybeans, but also for the DNA molecule that encodes the herbicide-resistant trait, for the herbicide-resistant plant cell, for the seed of the herbicide-resistant plant, and for the final ROUNDUP READY® soybean plant itself.

A utility patent may also cover multiple varieties of corn at once, as well as an entire species or genus if the applicant meets the disclosure requirements discussed below. Moreover, the scope of the protection is broader than the specific corn variety developed. Under the patent law's doctrine of equivalents, trivial variations to an invention that may not fall within the literal terms of the claims of the patent may nevertheless infringe as an equivalent of the claimed invention.

Requirements for obtaining a utility patent

An applicant must meet the highest threshold for acquiring IP protection in order to obtain the protection of a utility patent. To qualify as a subject of a patent, an invention must be "new," "useful," and "nonobvious" (US Congress 2010c, 2010d). First, an invention must be "new," or in other words, not already known to the public (US Congress 2010c). An invention fails to meet this requirement if it was in public use, was described in a printed publication, or was covered in a preexisting patent. In the United States, there is a one-year grace period on the bar on public use and printed publication. Second, an invention must be "useful" in that it is capable of providing a specific benefit (US Congress 2010a). For example, failure to identify a specific use for a gene sequence renders the gene sequence ineligible for patent protection. Finally, an invention must be "non-obvious" in that the invention is not obvious to one of ordinary skill in the art (US Congress 2010d). The nonobvious requirement takes into account the scope and content of the prior art, and the level of ordinary skill in the pertinent art. For example, a patent may be denied if the invention is a combination of previously known components A, B, and C, and the idea to combine the components A, B, and C was obvious to a person of ordinary skill in the art.

At minimum, a patent application must contain a specification and at least one claim. In the specifications, an applicant must disclose in writing what the applicant believes he has invented and must describe the invention in sufficient detail to enable others of ordinary skill in the pertinent art to practice the invention (US Congress 2010e). For example, in an application claiming DNA as the invention, a description of the DNA is adequate if it includes a definition of the physical properties, formula, chemical name, or structure of the claimed invention; a description that merely states that DNA is involved in the invention falls short of the requirement. In situations where the starting materials required to practice the invention are not readily available to the public, the applicant may also be required to deposit the materials in a depository in order to fulfill the enablement requirement. The written description of the invention must also reveal what the inventor believes to be the best way to practice the invention.

The claims in a patent define the boundaries of a patentee's right to exclude. The patent application must describe what the inventor claims as his invention by including "one or more claims particularly pointing out and distinctly claiming the subject matter which the applicant regards as his invention" (US Congress 2010e). Ideally, claims should be both broad enough to afford the patentee a wide scope of protection, and yet narrow enough to avoid invalidation by any prior art. In general, claim language that contains fewer limitations provides broader patent protection than claim language that includes many limitations. Take for example the following two simplified claims for a bucket: a claim that reads, "a bucket comprising a wooden circular bottom, wooden side walls, and a stainless steel handle," provides narrower protection than a claim that reads, "a bucket comprising a bottom, side walls, and a handle." A competitor's metal bucket with a square bottom would fall outside the claim language of the first example, but would infringe the second example by falling within its claim language.

Additionally, claims may be classified as either independent or dependent claims. Independent claims generally are the broadest claims and do not refer to any other claim in the patent. Dependent claims on the other hand incorporate other claims by reference and add additional limitations. Consequently, dependent claims provide a narrower scope of protection than independent claims. Consider an example where an inventor claims: (1) a bucket comprising a bottom and side walls; (2) the bucket of claim 1 further comprising a handle; (3) the bucket of claim 2 wherein the bottom and side walls are wooden; and (4) the bucket of claim 2 wherein the bottom is circular. In this example, claim 1 is the independent claim and claims 2 through 4 are the dependent claims that rely on claims that have come before. Ultimately, a patent covers only that which the applicant describes and claims in his patent application.

Procedure for obtaining a patent

In the US, the Patent and Trademark Office (PTO), an arm of the US Department of Commerce, administers utility patents. The PTO receives and examines applications and has power to grant patents if it is convinced that the invention is "new," "useful," and "non-obvious," and meets other conditions and requirements as set forth by statute (US Congress 2010c, 2010d). The first step in acquiring a patent is to file a patent application with the PTO. Thereafter, a series of communications between the applicant and the PTO follows. Six months to two years after the filing date of the patent application, the PTO will send communications to the applicant known as an "Office Action." This communication notifies the applicant of whether the claims have been allowed and in the event that claims have been rejected, reasons are provided. The applicant then has a chance to respond to the PTO within a time specified in the Office Action, typically three months. The applicant may amend the application to overcome the rejections. Two to six months after the PTO receives the applicant's response to the Office Action, it may send another Office Action to the applicant or it may send a Notice of Allowance, which indicates that the PTO has allowed all of the claims in the application. A patent will issue after the applicant pays an issue fee. Once granted, utility patent protection lasts for a term of 20 years, measured from the date the patentee filed the application and is not subject to exemptions from enforcement. During the term of the issued patent, the patent holder must pay periodic maintenance fees to the PTO.

In the current global economy, an inventor may wish to procure patent protection for his invention in more than one country. A patent confers rights only in the jurisdiction in which the patent application was filed and outside of the country where the patent is issued, others are free to use the invention without incurring patent infringement liability. A patent that issues in the US, for example, confers no automatic patent protection for the invention in France. To protect an invention internationally, an inventor must secure a patent in each country in which he desires protection.

Many nations have adopted international agreements that make the process of obtaining multiple patents easier. One of these agreements is the Patent Cooperation Treaty (PCT). The PCT, administered by the World Intellectual Property Organization, is an international agreement that streamlines the process of securing patents for an invention in multiple countries. A patent applicant may seek simultaneous patent protection in multiple countries by filing a single application and designating the countries where protection is desired. While the PCT does not alter the substantive requirements of patentability in each country, it does eliminate the duplicative effort wasted in filing separate patent applications for the same invention.

An inventor who wishes to take advantage of the PCT first files an application in his or her home patent office, which is designated as the "Receiving Office." The home office conducts an initial prior art search and gives the applicant the opportunity to request an international preliminary examination. The preliminary examination, while not binding, provides an indication of the patentability of the invention, which may assist the applicant in deciding whether to commit to an expensive filing abroad. In the next step, called the "national stage," an applicant has 30 months to convert the PCT application into parallel patent applications in the countries in which he desires patent protection. From there, the patent application process proceeds according to the procedures established by each designated country.

Rights of the inventor

A patent grants its owner the right to exclude others from making, using, offering for sale, and selling the patented invention without the patent owner's permission. Patents are personal property and therefore may be licensed or assigned to others, including to companies. An assignment transfers the rights of the patent from the current owner to a new owner. In contrast, a license grants a revocable permission to engage in conduct that would otherwise constitute patent infringement without transferring ownership of the patent. Licenses may be either exclusive, and issued strictly to one licensee, or nonexclusive, and issued to several licensees at once.

Plant Variety Protection

While utility patents provide the most robust protection for corn-based innovations, only a minority of countries afford utility patent protection for agricultural biotechnology. A more common regime is referred to as plant variety protection, also known as plant breeder's rights. In general, plant variety protection provides a *sui generis* form of intellectual property protection to breeders of new variety of plants.

International protection: the international convention for the protection of new varieties of plants

Many countries that have introduced a system for protecting new varieties of plants have based their system on the International Convention for the Protection of New Varieties of Plants (UPOV Convention) (International Union for the Protection of New Varieties of Plants 1991). Originally adopted in Paris in 1961 with the objective of providing intellectual property protection for new plant varieties, the UPOV Convention has undergone several revisions, first in 1972, again in 1978, and most recently in 1991. The International Union for the Protection of New Varieties of Plants (UPOV), an intergovernmental organization headquartered in Geneva, Switzerland, administers the UPOV Convention.

The UPOV Convention defines a minimum scope of protection that enables plant breeders to prohibit the unauthorized exploitation of their protected variety. Under the UPOV Convention, the authorization of the breeder of an eligible plant variety is required to produce or reproduce, condition for the purpose of propagation, offer for sale, sell, export, import, and stock the propagating material of the protected variety. Where the plant breeder has not had a reasonable opportunity to exercise his rights as to the propagating material, the same rights are extended to the harvested material of the protected variety. The rights also attach to varieties "essentially derived" from the protected variety, varieties "not clearly distinguishable from the protected variety," and varieties that "require the repeated use of the protected variety." The Convention explains that "essentially derived varieties" are those that "may be obtained for example by the selection of a natural or induced mutant, or of a somaclonal variant, the selection of a variant individual from plants of the initial variety, backcrossing, or transformation by genetic engineering."

To obtain plant variety protection, the UPOV requires examination of an application to ensure that the proposed variety meets the conditions for protection. To qualify for UPOV protection, a plant variety must be "(i) distinct from existing, commonly known varieties, (ii) sufficiently uniform, (iii) stable and (iv) new in the sense that they must not have been commercialized prior to certain dates established by reference to the date of the application for protection." Once granted, the UPOV dictates that a plant breeder's rights shall last for at least 20 years.

The UPOV also defines acts that are exempt from the plant breeder's rights. The plant breeder's permission is not required for acts done privately and for non-commercial purposes, experimental use of the protected variety, and acts done for the purpose of breeding other varieties. In addition to the compulsory exceptions, an optional exception allows farmers to save harvested seeds for replanting.

Member nations to the UPOV agree to adopt all measures necessary to implement the plant breeder's rights as outlined in the Convention and to extend to foreign nationals the same rights it provides to its own citizens. Implementation of the Convention entails the establishment of legal remedies and enforcement mechanisms for breeder's rights and the designation of an authority entrusted with the power to grant such rights to applicants. The UPOV provides the basic framework for plant variety protection. However, since countries are free to tailor their laws to domestic circumstances when implementing the provisions of the UPOV Convention, different countries have adopted slightly different versions of the plant variety protection regime.

Protection in the United States: The Plant Variety Protection Act (PVPA)

The US is a member of the UPOV and it implemented the UPOV Convention in 1981. Plant variety protection certificates, which are issued by the Plant Variety Protection Office of the US Department of Agriculture, supply patent-like protection for new varieties of seed-bearing plants and may be obtained to protect new plant varieties. Governed by the Plant Variety Protection Act (PVPA), rights are granted to "[t]he breeder of any sexually reproduced or tuber propagated plant variety (other than fungi or bacteria) who has so reproduced the variety, . . . subject to the conditions and requirements of this Act" (US Congress 2010f).

The PVPA protects discrete varieties from unauthorized exploitation by others. Following the UPOV Convention, a PVP certificate grants its holder the right to exclude others from selling, offering for sale, reproducing, importing, or exporting the protected variety, and from using the protected variety to produce (as distinguished from to develop) a hybrid or different variety. Per the UPOV Convention, protection under the PVPA extends not only to the protected plant variety, but also to "essentially derived varieties," narrowly defined in the PVPA to include two generations of derivation. The PVPA defines the term as "a variety that—(i) is predominantly derived from another variety (referred to in this paragraph as the 'initial variety') or from a variety that is predominantly derived from the initial variety, while retaining the expression of the essential characteristics that result from the genotype or combination of genotypes of the initial variety; (ii) is clearly distinguishable from the initial variety; and (iii) except for differences that result from the act of derivation, conforms to the initial variety in the expression of the essential characteristics that result from the genotype or combination of genotypes of the initial variety." Inclusion of "essentially derived varieties" within the metes and bounds of the breeder's rights guards against acts that boarder on blatant copying. "Essentially derived varieties" delineate a zone of protection around the protected variety that captures plants produced by inducing minor changes to a protected variety. As an example, a hybrid variety of maize produced from a protected variety may exhibit cosmetic differences that make the hybrid distinct from its parent, but as an "essentially derived variety," the hybrid nevertheless falls within the scope of PVPA protection for the parent.

As required by the guidelines of the UPOV Convention, the PVPA includes several exceptions that shield certain acts from infringement liability. Private noncommercial use of a protected variety does not constitute infringement, nor does the saving of seed for replanting "of a crop for use on the farm" and sale of such seeds "for other than reproductive purposes." Additionally, the PVPA explicitly provides a research exemption by stating that the "use and reproduction of a protected variety for plant breeding or other bona fide research shall not constitute an infringement of the protection provided under this Act." Furthermore, though not an exemption from infringement liability, the PVPA is subject to a requirement that allows the US Secretary of Agriculture to declare a compulsory license allowing use of the protected variety for two years in exchange for a royalty, if such action is deemed necessary in the public interest to maintain a sufficient food supply. The many exceptions to the PVPA allow others under certain circumstances to exploit a protected plant variety without the owner's authorization and therefore diminish the strength of plant variety protection.

As a trade-off for a narrower scope of protection, the PVPA demands a lower threshold for obtaining protection. Unlike the utility patent, the PVPA does not call for rigorous disclosure of the claimed invention, nor does it impose a non-obvious requirement. Applicants for a plant variety protection certificate must show that the variety qualifies for protection, provide a description of the variety, and deposit seed in a repository.

To qualify for protection under the PVPA, a plant variety must be new, distinct, uniform, and stable. The statute defines each of these terms. First, a variety is "new" if "the variety has not been sold or otherwise disposed of to other persons" more than 1 year before the date the applicant filed the application for the PVPA. Second, a variety is "distinct" if "the variety is clearly distinguishable from any other variety the existence of which is publicly known or a matter of common knowledge at the time of the filing of the application," and in determining distinctiveness, "[t]he distinctness of one variety from another may be based on one or more identifiable morphological, physiological, or other characteristics (including any characteristics evidenced by processing or product characteristics, such as milling and baking characteristics in the case of wheat) with respect to which a difference in genealogy may contribute evidence." Third, a variety is "uniform" when "any variations are describable, predictable, and commercially acceptable." Finally, a variety is "stable" if "the variety, when reproduced, will remain unchanged with regard to the essential and distinctive characteristics of the variety with a reasonable degree of reliability commensurate with that of varieties of the same category in which the same breeding method is employed."

Once a plant protection certificate issues, the term of protection lasts for 20 years from the date of issue of the certificate. Unlike utility patents and plant patents, which must issue under an individual inventor's name, a plant variety protection certificate may issue in the name of a corporation. Additionally, as a requirement for maintaining PVPA protection, the certificate holder must periodically replenish the repository of seeds of the protected plant variety, but the PVPA does not require payment of maintenance fees for the certificate. When compared to a utility patent, the scope of protection under the plant variety protection regime is limited. However, one advantage of the PVPA is the immediacy of protection. As soon as a plant variety protection application is filed and the fee is paid, provisional protection attaches to the plant variety. By marking the seed with protection notices "Unauthorized Propagation Prohibited" or "Unauthorized Seed Multiplication Prohibited," the seed owner acquires protection prior to issuance of the plant variety protection certificate.

Trade Secret Protection

Along with utility patents and plant variety protection, trade secret protection represents another essential tool for protecting innovations related to the corn plant. Most significantly, trade secret protection is available for inventions that do not otherwise qualify for plant variety protection or protection under a patent. In general, the purpose of trade secret protection is to uphold commercial morality by preventing the unauthorized use and disclosure of secret information, but leaves other parties free to independently develop the same matter. The subject matter protected by a trade secret coincides with the subject matter protected under patent regimes. Typically, protection attaches to information that is used in business, that gives a competitive advantage, and that has been kept confidential.

Unlike patents, trade secret protection arises instantly and requires no formal application or review process. Once trade secret protection arises, it grants recourse against one who wrongfully acquires the secret information. However, to recover damages for trade secret misappropriation, the trade secret owner must show that the information was protected by reasonable measures to ensure the secrecy of the trade secret. The requirement of maintaining the confidentiality of the information is critical; trade secret protection evaporates if the underlying information is no longer a secret. As such, the cost of maintaining a trade secret is essentially the cost of maintaining secrecy measures, which typically involve continuous and costly expenditures on measures to prevent the unauthorized use or disclosure of the information.

Unlike other intellectual property regimes, trade secret protection provides protection for an indefinite period rather than for a fixed term. So long as the underlying information continues to be a secret, the information remains protected as a trade secret. Some trade secrets, most notably the secret formula for the beverage Coca-Cola, have been maintained as trade secrets for very long indeed. However, trade secret protection can end at any time, since once the underlying information is no longer a secret, the trade secret protection disappears. Loss of trade secret protection may result from disclosure, successful reverse engineering, or independent development by others. Unlike patent protection, trade secret protection provides no recourse against one who reverse engineers or independently discovers the same matter. The uncertainty of protection is the risk borne by one who chooses trade secret protection.

In the context of innovations involving the corn plant, trade secret protection is a mixed bag. For seed companies, protection of corn varieties under trade secrets alone may prove difficult. Maintaining the secrecy of information is challenging since corn is grown in open fields and seed is sold on the open market with no assurances of confidentiality. Hybrid seed varieties are the easiest to maintain as a trade secret. Since the exact characteristics of the parental lines of a hybrid cross are difficult for others to ascertain, the owner of the hybrid plant variety may maintain the parental lines as a trade secret and sell only the seed resulting from the cross of the parental lines. Trade secret protection might also be employed to protect "know-how," or the methods and techniques of the plant breeder. Additionally, trade secret protection may be used to protect an invention during the patent examination period, until a patent issues.

Most importantly, trade secret protection is instrumental for protecting innovations that do not otherwise qualify for protection under patent and patent-like protection regimes. Trade secret protection extends to the same subject matter covered by patents, and requires only secrecy. Consequently, trade secret protection is vital for protecting matter where patent and patent-like protection is unavailable.

Conclusion

Legal and regulatory regimes play an integral role in the successful deployment of corn-based biofuels and biotechnology innovations in the US. First of all, corn-based biofuels cannot be lawfully commercialized in the US without complying with the detailed requirements of the CAA's regulatory frameworks, some of which prescribe their lawful blending limits. Additionally, the successful commercialization of corn based biofuels in the US is significantly bolstered by the RFS's mandates. Finally, through various types of IP regimes, legal and regulatory frameworks provide a means of protecting IP associated with corn-based biotechnology innovations.

But in concluding, we would be remiss to not caution that these current legal and regulatory regimes are not carved in stone and remain in a potential state of flux. The lawful blending limits for corn-based biofuels under the CAA could be legislatively altered or affected by future fuel waivers. Likewise, new biofuel pathways could be approved for the RFS. Moreover, the current RFS regime could be legislatively revised or even repealed. As a prime example of how the legal and regulatory landscape can be altered, when we first drafted this Chapter, it included an entire section on US tax incentives that related to corn-based biofuels. At that time, the US government provided a so-called blenders' credit to fuel refiners who blended alcohols with gasoline prior to commercialization (US Congress 2010g). This tax credit equaled $0.45 per gallon for every gallon of ethanol blended into finished fuels and $0.60 per gallon for other alcohols. Additionally, special tax credits were available for ethanol producers whose production capacity did not exceed 60 million gallons per year (US Congress 2010h) and also for owners of retail fueling stations that installed E85 fuel pumps (US Congress 2010i). Due to efforts to reduce the US government's budget and raise federal revenues, all of these tax incentives were allowed to expire at the close of 2011 and none are currently predicted to be reenacted or reinstated. As of this writing, the only remaining major biofuel-related tax incentive in the US involves a special tax credit for producers of cellulosic biofuels (US Congress 2010j). While this credit could work to incentivize the production of corn-based cellulosic biofuels, it is set to expire on December 31, 2012 (likely long before this Chapter appears in print) and it remains unclear whether or not it will be extended. With all of this in mind, we have tried to provide a snapshot of the current legal and regulatory landscape. While a thorough treatment of every legal and regulatory regime impacting the successful deployment of corn-based biofuels and biotechnology innovations could fill an entire book of its own, we hope that this Chapter provides an informative introductory primer on the most significant current regimes.

Acknowledgments

This work was funded by the Energy Biosciences Institute at the University of Illinois at Urbana-Champaign. The authors would like to thank Yulia A. Chembulatova for her impeccable research assistance.

References

Higashide W, Li Y, Yang Y et al. (2011) Metabolic engineering of Clostridium cellulolyticum for isobutanol production from cellulose. Appl Environ Microbiol 77: 2727–2733.

International Union for the Protection of New Varieties of Plants (1991) International Convention for the Protection of New Varieties of Plants, 1991: http://www.upov.int/en/publications/conventions/1991/w_up911_.htm (accessed 16 Nov 2011).

Kim S and Dale BE (2011) Indirect land use change for biofuels: testing predictions and improving analytical methodologies. Biomass Bioenergy 35: 3235–3240.

Slating TA and Kesan JP (2011) Making Regulatory Innovation Keep Pace with Technological Innovation. The Wisconsin Law Review 2011: 1109–1179.

US Congress (1970) Clean Air Act of 1970. United States Code, Title 42, Section 7545 et seq.

US Congress (2007) Energy Independence and Security Act of 2007. Public Law 110–140.

US Congress (2008) Food, Conservation and Energy Act of 2008. Public Law Number 110–267.

US Congress (2010a) Inventions Patentable. United States Code, Title 35, Section 101.

US Congress (2010b) Infringement of Patent. United States Code, Title 35, Section 271.

US Congress (2010c) Conditions for Patentability; Novelty and Loss of Right to Patent. United States Code, Title 35, Section 102.

US Congress (2010d) Conditions of Patentability; Non-obvious Subject Matter. United States Code, Title 35, Section 103.

US Congress (2010e) Specification. United States Code, Title 35, Section 112.

US Congress (2010f) Protectability of Plant Varieties. United States Code, Title 7, Section 2401 et seq.

US Congress (2010g) Alcohol Fuel Mixture Credit. United States Code, Title 26, Section 6426.

US Congress (2010h) Small Ethanol Producer Credit. United States Code, Title 26, Section 40(b)(4).

US Congress (2010i) Alternative Fuel Vehicle Refueling Property Credit. United States Code, Title 26, Section 30C.

US Congress (2010j) Cellulosic Biofuel Producer Credit. United States Code, Title 26, Section 40(b)(6).

US EPA (1979) Fuels and Fuel Additives: Gasohol; Marketability. Federal Register 44: 20777–20778.

US EPA (1985) Fuels and Fuel Additives; Waiver Decision; E.I. Du Pont de Nemours & Co., Inc. Federal Register 50: 2615–2616.

US EPA (1988) Fuels and Fuel Additives; Waiver Application. Federal Register 53: 3636–3638.

US EPA (2008) Regulation of Fuels and Fuel Additives: Revised Definition of Substantially Similar Rule for Alaska. Federal Register, 73: 22277–22281.

US EPA (2010a) Registration of Fuels and Fuel Additives. US Code of Federal Regulations, Title 40, Section 79 et seq.

US EPA (2010b) Regulation of Fuels and Fuel Additives. US Code of Federal Regulations, Title 40, Section 80 et seq.

US EPA (2010c) Renewable Fuel Standard Program (RFS2) Regulatory Impact Analysis: http://www.epa.gov/otaq/renewablefuels/420r10006.pdf (accessed 8 Nov 2011).

US EPA (2010d) Regulation of Fuels and Fuel Additives: 2011 Renewable Fuel Standards. Federal Register 75: 76790–76830.

US EPA (2011a) Partial Grant of Clean Air Act Waiver Application Submitted by Growth Energy to Increase the Allowable Ethanol Content of Gasoline to 15 Percent; Decision of the Administrator. Federal Register 76: 4662–4683.

US EPA (2011b) Renewable Fuel Standard. US Code of Federal Regulations, Title 40, Chapter 1, Subchapter C, Subpart M.

US EPA (2011c) Regulation of Fuels and Fuel Additives: 2012 Renewable Fuel Standards. Federal Register 76: 1320–1358.

Voegele E (2011) An advanced future for Gevo. Biorefin Magaz :http://www.biorefiningmagazine.com/articles/5789/an-advanced-future-for-gevo (accessed 8 Nov 2011).

CHAPTER 11

Social Issues: Biofuel Use of Corn and other Foods are Causing Malnutrition in the World

David Pimentel* and Michael N. Burgess

ABSTRACT

Malnutrition in the world today is causing more illness and deaths than any other factor affecting humans. According to the World Health Organization and the Food and Agriculture Organization of the United Nations, an estimated 4.8 billion people, more than 66% of the population, are at risk today, by far the largest number of malnourished in history. Iron and iodine deficiencies are affecting perhaps as many as 4 billion (38%) people. In addition to the deaths from iron, iodine and vitamin A deficiencies, these deficits cause mental retardation in young children and decrease the capacity of adults to work and participate in family life. Driving the caloric insufficiency is the conversion of nearly 400 million tons of food worldwide into biofuels that results concomitantly in multiple tragedies affecting species diversity, the carbon footprint, to say nothing of the questionable practices affecting land management resources.

Keywords: malnutrition, biomass production, solar energy capture, input energy costs, output energy recovered

College of Agriculture and Life Sciences, Cornell University, Ithaca, New York 14853, USA.
*Corresponding author: dp18@cornell.edu

Introduction

Global shortages of fossil energy especially of oil and natural gas have encouraged governmental and corporate interest in biomass as an energy source (Pimentel and Pimentel 2008). As interest in biofuels as a renewable energy source has developed globally, this focus has emphasized biofuels made mainly from corn grain, soybeans, canola, and sugarcane. Such changes have rarely led to responsible alterations in land management practices. For example, using crop residues as an energy source has a devastating impact on agriculture and food production because removing crop residues increases soil erosion and with it the removal of vital nutrients from the soil (Pimentel and Pimentel 2008; Lal 2009).

A wide number of conflicts continue to develop over the use of land, water, energy and other environmental resources for food versus biofuel production. Each is dependent on these same resources for production. In the US, about 19% of the fossil energy is utilized in the food system: about 7% for agricultural production, 7% for processing and packaging of foods, and about 5% for distribution and preparation of foods (Pimentel et al. 2009). In developing countries about 2 kcal is required to cook 1 kcal of food; and wood or crop residues provide most of this energy (Pimentel and Pimentel 2008). The objective of this chapter is to examine the interrelationship between malnutrition and the use of food crops for biofuel and the elements that effect it.

World Malnutrition

Malnutrition, which includes inadequate intake of calories, protein, iron, and numerous essential vitamins, is a major disease related to the environment (Myers and Kent 2001; Pimentel et al. 2009). The World Bank World Development Report estimated that deficiencies of vitamin A, iron, and iodine waste as much as 5% of global gross domestic product (GDP), while providing the needed nutrients would cost only 0.3% of the global GDP (World Bank 1995). Malnutrition exists in regions where the overall food supply is inadequate, where people lack adequate economic resources to purchase needed nutrients, have inadequate access to dietary information and where political unrest and instability interrupt food supplies. In addition, the rapid growth in the use of food for biofuels in the US, Europe, and other nations has increased world malnutrition and hunger (Pimentel 2009).

In 1950, about 500 million people, approximately 20% of the world population, were malnourished (Grigg 1993). Today, more than 4.8 billion people, 66% of the world population, suffer from undernourishment, the largest number ever in history. Each year, an estimated 6 million children under the age of five die from a lack of proper nutrition (FAO 2002). Even in the US, over 14.5% of all households experienced food insecurity during 2010 and in 5.4% of these households "the food intake of some household members was reduced and normal eating patterns were disrupted due to limited resources" (Coleman-Jensen et al. 2011). Notably too, malnutrition at an early age can lead to physical and mental retardation in addition to developmental problems in adulthood. According to the WHO (2009), every year over 200 million children fail to reach their learning potential due to malnutrition, chronic and severe enough to retard mental and physical growth. These results in inadequate stimulation or learning opportunities, due to iodine deficiency or iron deficiency anemia. Developmental anomalies exacerbate a poverty trap where people are stuck at a low-level of productivity at a major cost to society and the environment (WHO 2009). The data is summarized in Table 1.

Vitamin A malnutrition diminishes and impairs the immune response to infectious diseases in children. Vitamin A deficiency contributes in a major way to 2.5 million childhood deaths per year (Centers for Disease Control and Prevention 2010). Vitamin A supplements have been shown to decrease mortality by 30% in vitamin A deficient children of ages 6 months to 5 years (Stephensen 2001). Vitamin A and iron shortages can also cause mental disabilities in children (Uzendu 2004; WHO 2009a). In addition, more than 9.8 million pregnant women and 5.2 million preschool children suffer from night blindness due to a lack of vitamin A (WHO 2009a).

Similarly, iron intake per person has been declining during the past 10 years, especially in the developing countries (WHO 2000). Globally, from 4 to 5 billion people are iron-deficient and 2 billion people suffer from anemia (WHO 2000; GAIN 2012). In addition, about 1.6 billion people live in iodine-deficient environments and almost 2 billion suffer from iodine deficiency disease (UN 2004; GAIN 2012). Iron and iodine deficiency also cause mental disabilities in children (UN 2004; UNICEF 2008). Both iron deficiency and iodine deficiency in pregnant women and children impair brain development in children; while the number of children impaired by iron deficiency is suspected to be high, the actual number is unknown (Table 1). Iodine deficiency in pregnant women causes 18 million children per year to be born with impaired mental abilities (Table 1). While maize does not serve either as a major source of vitamin A or iron, increasing acreage devoted maize biofuel, and indeed to cereal production in general,

Table 1 The number of people who are malnourished in terms of iron and iodine deficiencies worldwide (WHO 2000; GAIN 2012).

Nutrients	Deficiencies Worldwide (millions)	Mentally Retarded Children (millions)
Iron	2,000	Widespread but Unknown
Iodine	2,000	18 million/year
Total	4,000	Unknown

creates a competition whereby individuals who rely on plant resources for these supplements become compromised. A responsible land policy use must be put into place to regulate what is cultivated. Packaged nutritional supplements are not readily available in underdeveloped countries as they are in the developed world.

Energy Resources

Approximately 50% of the solar energy captured worldwide by photosynthesis is used by humans for food, fiber, forest products, and pastures grazed by domesticated animals. This amount of energy is still inadequate to meet all human food and other biomass needs (Pimentel et al. 2009). To make up for the shortfall, about 473 quads (1 quad = 1 x 10^{15} BTU) of energy, mostly oil, natural gas, and coal (US EIA 2006; BP 2011) are needed. Of this total fossil energy used, the US, with only 4.5% of the world population, consumes 22% (USCB 2008).

Each year the US population uses three times more fossil energy than the total solar energy captured by all the US crops, forests, rangelands and pastures (Tables 2 and 3). Indeed, industry, transportation, home heating and cooling, and the food production account for most of the fossil energy consumed in the US (USCB 2008). Per capita use of fossil energy in the US amounts to 9,500 liters of oil equivalents (USCB 2008). This said, worldwide, oil supplies are estimated to last for about 40 years (IEA/OECD 2008), natural gas hopefully about 60 years (IEA/OECD 2008) and

Table 2 Total amount of biomass and solar energy captured each year in the United States. An estimated 1.8 x 10^7 BTU of sunlight reaching the US per year suggests that the green plants (crops, grasses, and forests) in the US are collecting 0.1% of the solar energy reaching these plants (Jolli and Giljum 2005; US Forest Service 2007; NASS 2011; Stöckle and Nelson 2011).

Total Energy Collected

	Million ha	tons/ha	x 10^6 tons	x 10^{15} BTU
Crops	160	5.5	901	14.4
Pasture	300	1.1	333	9.6
Forests	264	2.0	527	8.4
Total	724		1,761	32.4

Table 3 Total amount of biomass and solar captured each year in the world. An estimated 26 x 10^{18} BTU of sunlight reaching the earth is captured as biomass by green plants (crops, grasses, and forests) each year in the world suggesting that green plants are collecting about 0.1% of the solar energy reaching these plants (Pimentel et al. 2009).

Total Energy Collected

	Billion ha	Tons/ha	x 10^9 tons	x 10^{18} BTU
Crops	1.5	5.0	750	12.0
Grasses	3.5	1.1	385	6.2
Forests	3.0	1.6	629	7.8
Total	8.0		1,764	26.0

coal optimistically about 100 years (IEA/OECD 2008). At present the US is importing about 70% of its oil at an annual cost of about $700 billion (USCB 2008). Unfortunately there are no known full substitute replacements for the current fossil energy sources of oil, natural gas, and coal at our current level of energy consumption. The best estimate employing wood biomass, photovoltaics, solar thermal, geothermal, and conservation is that we would be fortunate to obtain one half of our current fuel use (Pimentel 2008).

Biomass Resources

The amount of biomass produced in the US totals 1.8 million metric tons per year (Table 2). Our estimate of biomass produced in the world totals 1.8×10^9 metric tons per year (Table 3)-1,000 times greater than the biomass produced in the US. Most of the biomass is forest biomass, totaling 629 billion metric tons (Table 3). The Intergovernmental Panel on Climate Change (IPCC) reported that world biomass energy flow for 2004 was 20×10^{15} kcal (Nabuurs 2007). Most of the forest biomass was used for fuel and some for building materials and in addition wood and crop residues have long been used as preferred energy sources for cooking (Nabuurs 2007).

Global forest area removed each year totals 15 million ha (Sundquist 2007). Global forest biomass harvested each year is just over 1,430 billion kg, of which 60% is industrial roundwood and 40% fuelwood (FAOSTAT 2011). About 90% of the fuelwood is utilized in the developing countries (Parikka 2004). A significant portion, 26% of all forest wood, is converted into charcoal (Arnold and Jongma 2007). In the production of charcoal, all the nutrients in the wood are lost during its production. Approximately 2,000 trees are required to produce 5,000 kg of charcoal (Pimentel and Burgess, submitted). Production of charcoal causes between 30% and 50% of the wood energy to be lost during the charcoal-making process and produces enormous quantities of smoke (Demirbas 2001). Charcoal is both light weight and dirty, but easy to handle, and it is clean burning and produces little smoke when used for cooking (Arnold and Jongma 2007). While charcoal production is still necessary for food preparation in the third world countries, it is inefficient, thus calling for more equitable, coordinated, land use policies. This is especially so in Asia, Africa, and portions of the Southern Hemisphere where grain production practices are still evolving in rapidly expanding populations where availability of land and potable water are at risk.

As mentioned, about 40% of the forest wood is used for fuel. In developing countries, about 2 kcal of wood are utilized in cooking 1 kcal of food (Fujino et al. 1999). Thus, more biomass and therefore more land and water are needed to produce the biofuel for cooking than what is needed to produce the food. Biomass can also be used to produce electricity. Assuming an optimal yield, such as in the northeast of the US, 3 dry tons per hectare per year can be harvested sustainably (Ferguson 2001). This would provide a gross energy yield of 13.5 million kcal/ha. Harvesting this wood biomass requires an energy expenditure of approximately 30,000 kcal per hectare, plus the embodied energy for cutting and collecting wood for transport to an electric power plant. Thus, the energy input per output ratio for such a system is calculated to be 1:25. Woody biomass has the capacity to supply the US with about 5 quads of its total gross energy supply by the year 2050, provided that the amount of forest land remains constant (Pimentel 2008). A city of 100,000 people using biomass from a sustainable forest, producing 3 t/ha/yr for electricity requires about 200,000 ha of forest area, based on an average electrical demand of slightly more than 1 billion kWh of electrical energy (Pimentel 2008). Per capita consumption of woody biomass for heat in the US amounts to 625 kg/yr (Kitani 1999). The diverse biomass resources (wood, crop residues, and dung) used in developing nations averages about 630 kg per capita (Kitani 1999).

The impact on air quality from burning biomass is less harmful than those associated with coal, but more harmful than that associated with natural gas (Pimentel 2008). Biomass combustion releases more than 200 different chemicals, including 14 carcinogens into the atmosphere (Burning Issues 2006). As a result of cooking with biomass, especially in the developing countries, about 4 billion people suffer from various respiratory diseases due to continuous exposure to smoke (Smith 2006). In the US, wood smoke is reported to kill about 30,000 people each year (Burning Issues 2003). Various pollution controls can be installed in wood fired stoves. Of course, low income people, even those living in small homes, may not be able to afford to install such pollution controls.

An estimated 1.8 billion metric tons of biomass is produced per year on US land area (Table 2). This translates into nearly 30 quads of energy, which means that the solar energy captured by all the green plants in the US per year equates to less than 30% of fossil energy consumed (Pimentel et al. 2009). Clearly there is insufficient US biomass produced per year to make the US fossil fuel independent.

Biofuel Systems and Net Returns

Corn Ethanol

Ethanol produced from corn grain is the largest producer of biofuels in the world. However, even if we totally ignore corn ethanol's negative energy balance and high economic cost, we still find that it is absolutely impossible to use corn ethanol as a replacement for US oil consumption (Pimentel and Patzek 2008). If all of the 340 billion kilograms of corn produced in the US each year were converted into ethanol, then only 130 billion liters of ethanol could be produced (USDA 2008). This would provide only 5% of total oil consumption in the US. Of course, in this situation there would be no corn available for livestock production and other human needs (Pimentel and Patzek 2008).

Corn production utilizes more insecticides and more herbicides than any other crop grown in the US (Unnevehr et al. 2003). In addition, maize cultivation causes more soil erosion than any other crop in the US, about 15 t/ha/yr. The degree of reduced biodiversity due to corn production is unknown. However, if one considers that 26,000 species of microbes and invertebrates are present in a hectare of rich topsoil with abundant biomass, corn production's impact has to be significant with respect to species change as a result of erosion and fertilizer and pesticide applications (Pimentel et al. 1995, 2006).

Cellulosic Ethanol

Tilman et al. (2006) assumed that about 1,032 liters of ethanol could be produced through the conversion of 4 t/ha/yr of harvested grasses like switchgrass. However, Pimentel and Patzek (2008) report a negative 68% return in ethanol produced relative to the total fossil energy inputs needed to raise and process the switchgrass into ethanol. Furthermore, converting all 235 million hectares of US grassland into ethanol at this optimistic rate would provide the US with only 12% of annual consumption of oil (Tilman et al. 2006; USCB 2008).

To achieve converting 235 million hectares of grassland into ethanol would mean that US farmers have to displace 97 million head of cattle, 6 million sheep, and 7.3 million horses that are currently grazing on US grassland and rangeland (USDA 2008; AVMA 2011). These statistics do not even begin address that there is a serious overgrazing problem on US grasslands (Brown 2002; USDA 2008). Thus, the projections of Tilman et al. (2006) are unduly optimistic.

In addition, several millions of wild animals including white-tailed deer, mule deer, elk, mice, woodchucks, and numerous other mammals depend on these grasses for food. It has been reported that 17 grasshoppers per sq. meter in a 16.2 hectares can almost consume a metric ton of hay daily (Ratcliffe et al. 2004). For example, as few as 10 grasshoppers per square meter are reported to consume 60% of the forage whereas cattle consume only about 20% (Calpas and Johnson 2009). In New York State pastures, pasture insects consumed twice as much forage as cows in the pasture per day (Wolcott 1937). The cows ate about 180 kg during a study period whereas the insects, for the same period, consumed 364 kg per hectare. Not only is it clear that conversion of cellulosic mass today remains expensive and insufficient, but it can irreversibly damage the ecosystem resulting in species shifts whose impact cannot be estimated on a global scale.

The consequences for using cellulosic biomass for ethanol production are four fold: 1) More than twice as much cellulosic biomass has to be grown, harvested, and processed as cellulosic biomass compared with corn grain. 2) The cellulose biomass conversion technology for ethanol production is not sufficiently developed for the cost-effective production of ethanol. 3) The conversion process currently has several and inherently limiting factors that may be overcome with future research. 4) The lignin byproduct can be burned as a fuel if it were possible to separate it from the water in its diluted form but at present producing burnable lignin is energetically and economically prohibitive. There is not a single commercial ethanol plant currently using cellulosic biomass functioning in the world although a recent New York Times article giving no real operational data says that there is one plant operating now (Patzek and Pimentel 2008; Howarth and Bringezu 2009; Patzek 2010; Wald 2013).

Soybean Biodiesel

Throughout the 2000s, the US subsidizes biodiesel at about $11 billion per year for the production of about 900 million liters of biodiesel, which is 75 times greater than the subsidies per liter of diesel fuel (Koplow and Steenblik 2008). The $1/gallon biodiesel tax credit is slated to be discontinued at the end of 2013 but is more than made up for by the Renewable Fuel Standard mandates which will keep the price of soybeans for biodiesel and food high (D. Koplow, per. comm., Sept 8, 2013). The environmental impacts of producing soybean biodiesel fuel are second only to that of corn ethanol production (Pimentel and Patzek 2008). If all 71 billion tons of soybeans were converted into biodiesel, it would provide the US with only 2.6% of the total US oil consumption (Pimentel 2008).

Rapeseed and Canola

The European Biodiesel Board estimates a total biodiesel production of nearly 9 million metric tons for the year 2009 (European Biodiesel Board 2011). Well suited to the colder climates, rapeseed is the dominant crop in European biodiesel production. Often confused with canola, rapeseed is an inedible crop of the Brassica family yielding oil seeds high in erucic acid.

The choice of rapeseed/canola as candidate bioenergy plants is compromised by the need for fertilizer and pesticide during cultivation. The input energy required to manufacture these chemicals must be subtracted from the overall energy produced (Frondel and Peters 2007). Although soybeans contain less oil than canola, about 18% soy oil versus 30% oil for canola seeds, soybeans can be grown with nearly zero nitrogen inputs (Pimentel et al. 2009). The biomass yield of rapeseed/canola per hectare is also lower than that of soybeans, approximating 1,600 kg/ha for rapeseed/canola as compared with 2,900 kg/ha for soybeans (Pimentel et al. 2009). The production of 1,600 kg/ha of rapeseed requires the input of about 4.4 million kcal per hectare (Pimentel et al. 2009). About 3,333 kg of rapeseed/canola is required to produce 1,000 kg of biodiesel (Pimentel et al. 2009). The total energy input to produce the 1,000 liters of rapeseed/canola oil biodiesel is 13 million kcal. This calculates to a net energy loss of 58% (Pimentel et al. 2009). The cost per kilogram (or per liter) of biodiesel is also high at $1.52 or more than six dollars/gallon and does not address either transportation or distribution costs leaving one to speculate the price at the pump. Given this data it may be concluded that rapeseed and canola are energy intensive and economically inefficient biodiesel crops.

Oil Palm

A major effort worldwide exists to plant and harvest oil palms for biodiesel in several tropical developing countries, such as Indonesia, Malaysia, Thailand, Columbia, and some West African countries (Thoenes 2007). In the last 20 years, the production of vegetable oils has more than doubled. Palm oil makes up 33% of the biological oils produced worldwide (USDA Foreign Agricultural Service 2011a, 2011b). In 2011–2012, worldwide palm oil production is projected at 50.23 million metric tons, mostly from Indonesia and Malaysia (USDA Foreign Agricultural Service 2011a, 2011b).

In considering palm oil as an energy source, attention must be given to the fact that the tree requires years to mature to become fully productive. The oil palm tree, once established will produce about 4,000 kg of oil per hectare per year (Carter et al. 2007). Approximately 7.4 million kcal are required to produce 26,000 kg of oil palm bunches (Pimentel et al. 2009). This 26,000 kg of oil palm bunches supplies sufficient palm nuts to produce 4,000 kg palm oil. Thus, a total of 6.9 million kcal are required to process 6,500 kg of palm nuts to produce 1 ton of palm oil (Pimentel et al. 2009). This is clearly a better energy return than corn ethanol or soybean biodiesel. This calculation, however, does not take into account that an estimated 200 ml, 2,080 kcal of methanol must be added to the 1,000 kg of palm oil, for transesterfication. This results in a negative 8% net energy output for palm oil and raises a legitimate question as for the sustainability of palm oil given current technology (Pimentel et al. 2009).

There are several negative environmental and social issues associated with the oil palm plantations. First, the removal of tropical rainforests to plant the oil palm results in an increase in CO_2 (Thoenes 2007). Second, the removal of tropical rain forests and the planting of oil palms reduce the biodiversity of the ecosystem (Pimentel et al. 2006). Finally, using oil palm for biofuel reduces the availability of palm oil for human use and increases the cost of the oil for all other uses (Thoenes 2007).

Algae for Oil

Some cultures of microalgae may consist of 30% to 50% oil (Dimitrov 2007). Thus, there is increasing interest in using single-cell algae as a source for US biofuels. A study by Dimitrov (2007) reported that a barrel of oil produced using algae costs an estimated $800 per barrel (1 barrel = 42 US gallons; 1 gallon of algae biodiesel costs about $19). This said, algae cultured in closed systems do not raise the ecological concerns that cultivating crops do because the requirements for pesticides, fertilizers and insecticides are eliminated.

In contrast, a second investigation calculated the cost per gallon of biodiesel using electricity in a bioreactor. With electricity costing 7¢ per kWh the cost rises to $32.81 per gallon, $8.67 per liter, or about $1,378 per barrel (Kanellos 2009). An algae biofuel startup company Solix can produce biofuel from algae for about $32.81 per gallon and the cofounder of Solix reports that the costs are so high because of the energy required to circulate gases and other materials inside the photo bioreactors (Rapier 2009). A third study reported that a kilogram of microalgae costs about $3.45 wet ($115 dry) (Lane 2009). These data suggest cost problems in producing algae for biodiesel once again indicating that this industrial technology is in its infancy.

Conclusion

The rapidly growing world population and rising consumption of fossil fuels is increasing the demand for both food and biofuels (Pimentel et al. 2009). This is exacerbating both food and fuel shortages. Producing biofuels today sacrifices nearly 400 million tons of food per year.

Using food to produce ethanol and biodiesel raises major nutritional and ethical concerns. Given that the FAO has stated that every person has a right to adequate food, how can consuming food by using it to fuel motor vehicles be justified when over 70% of the world human population suffers from some form of malnutrition and millions die from the direct and indirect effects of malnutrition every year (Table 2; FAO 2010)? Thus the need for grains and other basic food is critical. Producing crops for biofuels squanders land, water, and fossil energy resources vital for the production of food for people. Using food and feed crops for biofuels has increased the price of some foods, like bread, 100% to 200% (World Bank 2008). Overall, according to the US Congressional Budget Office, grocery bills have increased in the US from $6 to $9 billion per year. Both the Secretary General of the UN and Director General of the UN Food and Agricultural Organization report that using food grains to produce biofuels is increasing starvation worldwide and Jean Ziegler, the UN Special Rappoteuron, the "Right to Food", states

"It is a crime against humanity to convert agricultural productive soil into soil which produces food stuff that will be burned into biofuel." (Deen 2007; Tenenbaum 2008). Clearly, the use of food and feed crops is exacerbating worldwide starvation by turning food into biofuel.

Recent policy decisions have mandated increased production of biofuels in the US and worldwide. For instance, in the Energy Independence and Security Act of 2007, President Bush set "a mandatory renewable fuel standard (RFS) requiring fuel producers to use at least 36 billion gallons of biofuel in 2022". Such a move would require harvesting more than 80% of all biomass produced each year in the US. With literally taking a "lawnmower" to the total vegetation produced in the US, biodiversity and food supplies in this country alone would be decimated.

Most problems associated with biofuels have been ignored by some scientists and policy makers. The production of biofuels that are supposed to diminish dependence on imported fossil fuels are actually increasing our dependence on fossil fuels. All conversion of biomass into ethanol and biodiesel result in a negative energy return based on careful up-to-date analysis of all the fossil energy inputs. The negative energy returns in producing biofuels using food and feed crops range from 8% to 100%. Clearly then an increased use of biofuels will further damage the global environment and especially the world food system causing serious, if not catastrophic malnutrition.

Acknowledgments

We wish to express our sincere gratitude to the Cornell Association of Professors Emeriti for the partial support of our research through the Albert Podell Grant Program.

References

AVMA (2007) US Pet Ownership and Demographics Sourcebook, 2007 edn. American Veterinary Medical Association (AVMA), Member and Field Services, Schaumburg, Illinois, USA.

Arnold JEM and Jongma J (1978) Fuelwood and charcoal in developing countries: An economic survey. Unasylva 29(118): http://www.fao.org/docrep/l2015e/l2015e01.htm.

BP (2011) Statistical Review of World Energy 2011. Primary Energy Consumption. British Petroleum (BP): http://www.bp.com/sectionbodycopy.do?categoryId=7500&contentId=7068481

Brown LR (2002) Plan B Updates: World's Rangelands Deteriorating Under Mounting Pressure. Earth Policy Institute: Providing a Plan to Save Civilization. Washington DC, USA, Feb 5, 2002: http://www.earth-policy.org/plan_b_updates/2002/update6.

Burning Issues (2003) References for Wood Smoke Brochure [from 2003]: http://burningissues.org/car-www/science/wsbrochure-ref-3-03.html.

Burning Issues (2006) Burning Issues: What are the medical effects of wood smoke: http://burningissues.org/health-effects.html.

Calpas J and Johnson D (2003) Grasshopper Management. Government of Alberta, Agriculture and Rural Development. Agdex 622–27: http://www1.agric.gov.ab.ca/$department/deptdocs.nsf/all/agdex6463.

Carter C, Finley W, Fry J et al. (2007) Palm oil markets and future supply. Eur J Lipid Sci Technol 109(4): 307–314.

Centers for Disease Control and Prevention (2010) IMMPaCt—International Micronutrient Malnutrition Prevention and Control Program. Centers for Disease Control and Prevention, Atlanta, GA, USA: http://www.cdc.gov/immpact/micronutrients/index.html

Coleman-Jensen A, Nord M, Andrews M et al. (2011) Household Food Security in the United States in 2010. U.S. Department of Agriculture, Economic Research Report Number 125: http://www.ers.usda.gov/Publications/ERR125/err125.pdf.

Deen T (2007) Food to Biofuels a "Recipe for Disaster". Inter Press Service (IPS): The Story Underneath. Nov 6, 2007: http://ipsnews.net/news.asp?idnews=39942.

Demirbas A (2001) Biomass resource facilities and biomass conversion processing for fuels and chemicals. Energy Convers Manag 42(11): 1357–1378.

Dimitrov K (2007) Green Fuel Technologies: A Case Study for Industrial Photosynthetic Energy Capture, 2007: http://www.nanostring.net/Algae/CaseStudy.pdf.

European Biodiesel Board (2011) Statistics: The EU Biofuel Industry. European Biofuel Board: http://www.ebb-eu.org/stats.php.

FAOSTAT (2011) Food Balance Sheets. Food and Agriculture Organization. United Nations, Rome, Italy: http://faostat.fao.org/site/354/default.aspx.

FAO (2002) Food Insecurity in the World 2002. Food and Agriculture Organization of the United Nations, Rome, Italy: ftp://ftp.fao.org/docrep/fao/005/y7352e/y7352e00.pdf

FAOSTAT (2011) ForesSTAT, Food and Agricultural Organization, United Nations. Rome, Italy.

FAO (2010) The Right to Food. The human right to adequate food.Food and Agriculture Organization of the United Nations, Rome, Italy: http://www.fao.org/righttofood/.

Ferguson ARB (2001) Biomass and Energy. Optim Popul Trust J 4: 14–18.

Frondel M and Peters J (2007) Biodiesel: A new oildorado. Energy Policy 35: 1675–1684.

Fujino J, Yamaji K and Yamamoto H (1999) Biomass-balance table for evaluating bioenergy resources. Appl Energy 63(2): 75–89.

Grigg DB (1993) The World Food Problem. Blackwell, Oxford, UK.

Howarth RW and Bringezu S (eds) (2009) Biofuels: Environmental Consequences and Interactions with Changing Land Use. Proc Scientific Committee on Problems of the Environment (SCOPE). International Biofuels Project Rapid Assessment. 22–25 Sept 2008. Grummersback, Germany.

IEA/OECD (2008) World Energy Outlook 2008 Edition. International Energy Agency/Organization for Economic Cooperation and Development, Paris, France: http://www.iea.org/textbase/nppdf/free/2008/weo2008.pdf.

Jölli D and Giljum S (2005) Unused Biomass Extraction in Agriculture, Forestry and Fishery. SERI Studies No 3. Sustainable Europe Research Institute, Vienna, Austria

Kanellos M (2009) Algae biodiesel: It's $33 a gallon. Greentechmedia http://www.greentechmedia.com/articles/read/algae-biodiesel-its-33-a-gallon-5652/.

Kitani O (1999) Biomass resources. In: Kitani O, Jungbluth T, Peart RM and Ramdami A (eds) Energy and Biomass Engineering. American Society of Agricultural Engineers, St. Joseph, Michigan, USA, pp 6–11.

Koplow D and Steenblik R (2008) Subsidies to ethanol in the United States. In: Pimentel D (ed) Biofuels, Solar and Wind as Renewable Energy Systems: Benefits and Risks. Springer, Dordrecht, Netherlands, pp 79–108.

Lal R (2009) Soil quality impacts of residue removal for bioethanol production. Soil Till Res 102(2): 233–241.

Lane J (2009) Algaeventure Systems announces process to reduce cost of dewatering by 99 percent; potential breakthrough for algae commercialization. BiofuelsDigest: The world's most widely read biofuels daily. March 10, 2009: http://www.biofuelsdigest.com/blog2/2009/03/10/algaeventure-systems-announces-process-to-reduce-cost-of-dewatering-by-99-percent-potential-breakthrough-for-algae-commercialization/

Myers N and Kent J (2001) Perverse Subsidies. Island Press, Washington, DC, USA.

Nabuurs GJ, Masera O, Andrasko K et al. (2007) Forestry. In: Metz B, Davidson OR, Bosch PR, Dave R and Meyer LA (eds) Climate Change 2007: Mitigation. Contribution of Working Group III to the Fourth Assessment Report of the Intergovernmental Panel on Climate Change. Cambridge Univ Press, Cambridge, UK; New York, USA, 862 pp.

National Agricultural Statistics Service (NASS) (2011) Crop Production. National Agricultural Statistics Service, US Department of Agriculture, Economics, Statistics, and Market Information System: http://usda.mannlib.cornell.edu/MannUsda/viewDocumentInfo.do?documentID=1046.

Parikka M (2004) Global biomass fuel resources. Biomass Bioenergy 27: 613–620.

Patzek TW (2010) A probabilistic analysis of the switchgrass-ethanol cycle. Sustainability 2: 3157–3194.

Pimentel D, Harvey C, Resosudarmo P et al. (1995) Environmental and economic costs of soil erosion and conservation benefits. Science 267: 1117–1123.

Pimentel D (2008) Renewable and solar energy technologies: Energy and environmental issues. In: Pimentel D (ed) Biofuels, Solar and Wind as Renewable Energy Systems: Benefits and Risks. Springer, Dordrecht, Netherlands, pp 1–17.

Pimentel D (2009) Biofuel food disasters and cellulosic ethanol problems. Bull Sci Technol Soc 29(3): 205–212.

Pimentel D, Petrova T, Riley M et al. (2006) Conservation of biological diversity in agricultural, forestry, and marine systems. In: Burk AR (ed) Focus on Ecology Research. Nova Science Publishers, New York, USA, pp 151–173 (Also, 2007. In: Schwartz J (ed) Focus on Biodiversity Research. Nova Science Publishers, New York, USA, pp 1–25.)

Pimentel D and Patzek T (2008) Ethanol production: energy and economic issues related to U.S. and Brazilian sugarcane. In: Pimentel D (ed) Biofuels, Solar and Wind as Renewable Energy Systems: Benefits and Risk. Springer, Dordrecht, Netherlands, pp 357–371.

Pimentel D, Marklein A, Toth MA et al. (2009) Food versus biofuels: environmental and economic costs. Hum Ecol 37(1): 1–12.

Pimentel D and Pimentel M (2008) Food, Energy and Society. CRC Press. Boca Raton, FL, USA, 400 pp.

Pimentel D and Burgess M. Biochar and soil organic matter. Unpublished.

Rapier R (2009) More Reality Checks for Algal Biodiesel. Consumer Energy Reports: R Squared Energy Blog by Robert Rapier: http://www.consumerenergyreport.com/2009/02/28/more-reality-checks-for-algal-biodiesel/.

Ratcliffe ST, Gray ME and Steffey KL (2004) Grasshoppers. Integrated Pest Management (IPM), University of Illinois at Urbana-Champaign Extension. C:\Users\mnb2\Documents\Data often used\IPM Field Crops Grasshoppers.mht.

Smith KR (2006) Health impacts of household fuelwood use in developing countries. Unasylva 224: 57, 41–45: ftp://ftp.fao.org/docrep/fao/009/a0789e/a0789e09.pdf.

Stephensen CB (2001) Vitamin A, infection, and immune function. Annu Rev Nutr 21: 167–192.

Stöckle, CO and Nelson R (2011) Cropping Systems Simulation Model User's Manual. Biological Systems Engineering Department, Washington State University Pullman, Washington. September 19, 2007. http://www.sipeaa.it/tools/CropSyst/CropSyst_manual.pdf.

Sundquist B (2007) Chapter 4: Forest Degradation Data, Edition 6, July 2007: http://home.windstream.net/bsundquist1/df4.html.

Tanenbaum D (2008) Food vs. fuel: diversion of crops could cause more hunger. Environ Health Perspect 16(6): A254–A257.

Tilman D, Hill J and Lehman C (2006) Carbon negative biofuels from low-input, high-diversity grassland biomass. Science 314: 1598–1600.

Thoenes P (2007) Biofuels and Commodity Markets—Palm Oil Focus. FAO Commodities and Trade Division. Food and Agriculture Organization of the United Nations, Rome, Italy: http://www.fao.org/es/ESC/common/ecg/122/en/full_paper_English.pdf.

UN (2004) Independent Expert on Effects of Structural Adjustment. Special Rapporteur on Right to Food Present Reports: Commission Continues General Debate on Economic,

Social and Cultural Rights. Press Release HR/CN/1064: http://www.un.org/News/Press/docs/2004/hrcn1064.doc.htm.

UNICEF (2008) UNICEF in action. Nutrition: Micronutrients—Iodine, Iron and Vitamin A: http://www.unicef.org/nutrition/index_iodine.html.

Unnevehr LJ, Lowe FM, Pimentel D et al. (2003) Frontiers in Agricultural Research: Food, Health, Environment, and Communities. National Academies of Science, Washington DC, USA, 268 p.

USCB (2008) Statistical Abstract of the United States 2009. US Government Printing Office, Washington DC, USA.

USDA (2008) Agricultural Statistics 2008. US Department of Agriculture. Washington DC, USA.

USDA Foreign Agricultural Service (2011) Table 6. Major Vegetable Oils: World Supply and Distribution (Country View). United States Department of Agriculture, Foreign Agricultural Service, Circular Series, World Agricultural Production, WAP 09-11: http://www.fas.usda.gov/psdonline/psdReport.aspx.

USDA Foreign Agricultural Service (2011a) Table 16. Copra, Palm Kernel and Palm Oil Production.United States Department of Agriculture, Foreign Agricultural Service, Circular Series, World Agricultural Production, WAP 09-11: http://www.fas.usda.gov/psdonline/psdreport.aspx.

US Forest Service (2007) Forest Inventory and Analysis National Program. National Assessment —Resources Planning Act (RPA). 2007 RPA Resource Tables. Last Modified 2009. U.S. Forest Service: http://www.fia.fs.fed.us/program-features/rpa/.

US EIA (Energy Information Administration) (2006) International Energy Annual 2006.

Uzendu M (2004) Nigeria: 350,000 Kids Delivered With Mental Impairment Yearly—Stella Obasnajo. Daily Champion, Lagos, Nigeria, 29 Oct 2004. Story archived at allAfrica.com: http://allafrica.com/stories/200410290482.html (Accessed June 2, 2011).

Wald ML (2013) Milestone Claimed in Creating Fuel From Waste. New York Times, July 31, 2013. http://www.nytimes.com/2013/08/01/business/energy-environment/company-says-its-the-first-to-make-ethanol-from-waste.html?_r=0.&pagewanted=print.

World Energy Overview. Energy Information Administration: http://www.eia.gov/emeu/iea/contents.html.

WHO (2000) Turning the tide of malnutrition: responding to the Challenge of the 21st century. Geneva: World Health Organization. 2000 (WHO/NHD.007): http://www.who.int/mip2001/files/2232/NHDbrochure.pdf.

WHO (2009a) Global prevalence of vitamin A deficiency in populations at risk 1995–2005: WHO global database on vitamin A deficiency. World Health Organization of the United Nations, Geneva.http://whqlibdoc.who.int/publications/2009/9789241598019_eng.pdf.

WHO (2009) Early child development. World Health Organization of the United Nations, Media Centre, Fact Sheet No. 332, August 2009. http://www.who.int/mediacentre/factsheets/fs332/en/index.html.

WHO (2013) Nutrition: Micronutrient Deficiencies - Iron deficiency anaenia. World Health Organization. http://www.who.int/nutrition/topics/ida/en/.

Wolcott GN (1937) An animal census of two pastures and a meadow in Northern New York. Ecol Monogr 7(1): 1–90.

World Bank (1995) Enriching Lives: Overcoming Vitamin and Mineral Malnutrition in Developing Countries. World Bank. Executive Summary from this work appears on line at: http://web.worldbank.org/WBSITE/EXTERNAL/TOPICS/EXTHEALTHNUTRITIONANDPOPULATION/EXTNUTRITION/0,,contentMDK:20205924~menuPK:282594~pagePK:148956~piPK:216618~theSitePK:282575,00.html.

World Bank (2011a) Vitamin A Deficiency: http://web.worldbank.org/WEBSITE/EXTERNAL/TOPICS/EXTHEALTHNUTRITION.

World Bank (2008) Rising Food Prices Threaten Poverty Reduction. World Bank, News & Broadcast, Press Release No: 2008/264/PREM http://web.worldbank.org/WBSITE/EXTERNAL/NEWS/0,contentMDK:21722688~pagePK:64257043~piPK:437376~theSitePK:4607,00.html.

CHAPTER 12

Maize, Ethanol and US Policies
A Volatile Mixture

Lehman B. Fletcher

ABSTRACT

Maize (corn), one of the oldest grains, is the world's largest volume and most widely grown crop. The United States is the country with largest production and exports of the grain. The U.S. crop is predominantly yellow maize used heavily as animal feed but also for food and industrial products. Since 2001 production of corn-based fuel ethanol (ethyl alcohol denatured by gasoline) has become the single largest use of corn in the US. Interest in ethanol as a gasoline substitute rose in the 1970s with the oil crises. Its use picked up in the 1990s when it became popular as a gasoline oxygenate and replacement for the methyl tertiary butyl ether (MTBE) additive. Its use was encouraged first by oil price increases after 2000 and then by strong government support policies after 2005. Higher corn yields and improved conversion efficiency have improved corn ethanol's net energy value but greenhouse gas (GHG) emissions resulting from indirect land use effects of increasing corn production have challenged the environmental benefits claimed for the fuel. The U.S. government pays subsidies for ethanol blended with gasoline, protects domestic producers with an import tariff and mandates a stair-step scale of annual ethanol use. The number of plants, and production volumes, has soared since 2005 in response to investment incentives created by these policies. Corn ethanol plants face an uncertain future. They now use 40 percent or more of the U.S. corn crop and are criticized for competing with food uses of corn. As

Iowa State University, Ames, Iowa USA.
Email: lbf@iastate.edu

the subsidy is reduced or eliminated, profitability of the existing plants will be compromised and incentives for new plants eroded. As the tariff is lowered or removed, all U.S. ethanol production will face severe competition from ethanol produced more efficiently from sugar cane feedstock in Brazil and similar countries. The corn ethanol industry in the U.S. expanded rapidly and matured with government support, and now seems destined to become a declining industry facing eventual extinction as that support is ratcheted downward.

Keywords: Maize, ethanol, government subsidies, quantity mandates, tariff protection, indirect land use effects

Introduction

Production of ethanol from maize feedstock has grown enormously since the beginning of this century. This growth has been most rapid in the United States, which is at the same time the world's largest producer of maize and of maize ethanol. For that reason this chapter concentrates on the growth of, and prospects for, the conversion of maize to ethanol in the US. It emphasizes the role of government policies in the rapid expansion of the industry and assesses the consequences of withdrawal of that support.

Maize Production and Utilization in Context

Maize, a grain known in the United States and some other countries as corn, was first domesticated in Mesoamerica in prehistoric times. It is thought to have been cultivated initially by the Mayan and Olmec indigenous peoples in central and southern Mexico. Prior to the CE, the crop spread through much of the North, Central and South Americas. The grain was carried back to Europe by early explorers in the late 15th and early 16th centuries. From there it was taken by trade to other continents becoming widely cultivated due to its adaptability to diverse geoclimatic conditions. It is now grown in more countries than any other crop, and is the world's largest volume crop.

Maize has long been used as a food for people and animals and recently has become a renewable energy crop in the form of fuel ethanol, thus providing the motivation for this chapter. Maize was not only the main staple food of Native American people but also was important in their cultural and social lives. Arriving in Mexico the Spanish Conquistadores found that the Aztecs had a well-developed system of land reclamation and irrigation for growing the crop.

Maize, especially in the form of cornmeal produced by grinding the dried grain, is a major staple food in many countries of the world. In the Americas and Sub-Saharan Africa, the mainly white maize is used (Mexico and Central America) or eaten in the form of mush (Eastern and Southern Africa). Introduced into Africa by the Portuguese in the 16th century, maize has become the most important staple food in those African countries.

In the United States corn is consumed directly as bread, grits, popcorn and cornflakes. Sweet corn, a genetic variety high in sugar and harvested while the kernels are immature, is consumed on the cob or cut from the cob to be eaten as a vegetable or used as an ingredient in salads, soups or other cooked dishes.

The vast majority of the corn grown in the United States is yellow "dent" corn, known as field corn. This yellow corn is used for animal feed and also for many food and industrial products. In the second half of the 20th century corn milling to produce high fructose corn syrup (HFCS), other sweeteners, corn starch, beverage ethanol and many other products, grew to become an important competing use with animal feed. More recently, production of fuel ethanol (ethyl alcohol denatured by gasoline) has become the largest industrial use of corn. Currently, corn processed industrially exceeds the amount of corn used for animal feed in the United States, as will be discussed in more detail later in this chapter.

World Maize Production

Maize is widely cultivated throughout the world. A greater weight of maize is produced each year than any other grain crop. World production in the 2010/2011 marketing year totaled 32,125 million bushels (Table 1). The United States, the world's largest producer of maize, accounted for 39 percent of the total. Other major producing countries are listed in Table 1 along with the percentage of world production each produced in the 2010 grain marketing year. Together, countries of the Americas produce half of the world's maize crop.

Maize is also an important commodity in world trade. Japan is consistently the largest importer of maize, most of which comes from the United States. Mexico, South Korea, Taiwan, Egypt, and Colombia are also usually major importers of maize used predominantly as animal feed. China has just recently joined the list of major importing countries. The United States is the leading exporter. Argentina, Brazil and Ukraine are other major exporting countries.

Table 1 World Production of Maize 2010–2011.

Countries	Bushels (Millions)	Percent of total
U.S.A	12,447	38.7
China	6,614	20.6
EU-27	2,173	6.8
Brazil	2,008	6.3
Mexico	965	3.0
Argentina	925	2.9
India	827	2.6
South Africa	492	1.5
Ukraine	472	1.5
Canada	461	1.4
Others	4,741	14.8
Total	32,125	100.0

Source: USDA, FAS, Grain: World Markets and Trade, January 14, 2011

Corn Production in the United States

Corn is the leading agricultural crop in terms of both volume and value of production. It is grown on over 400,000 farms and accounts for about one-quarter of cropland harvested each year. The Midwestern states of Iowa, Illinois, Nebraska and Minnesota usually produce more than half of the nation's corn crop. Indiana, Ohio, South Dakota, Missouri and Kansas are also important corn-producing states (NCGA 2011).

The numbers of acres planted to corn were at their lowest levels in the 1960s and 1970s when government policies attempted to reduce the planted area in order to curtail the accumulation of surpluses. In the closing decades of the 20th century policies were changed to give farmers more freedom in their choices of what crops to plant, and corn plantings increased. Plantings fluctuate annually depending on producers' expectation of the profitability of corn relative to competing crops, especially soybeans in the Corn Belt states. Harvested acres are more variable than planted acres due to the effects of drought, floods, diseases and pests. This historical pattern persisted until a big jump in area and production began in 2007, highly correlated to the growing use of corn as a feedstock for fuel ethanol (Table 2).

Corn production depends on the number of acres harvested, and also on the yield per acre. Publicly supported research beginning in the 1920s was directed to breeding hybrid corn varieties with higher yields. This

Table 2 Corn Production and Value in the United States 1991–2010.

Year	Planted Acres (000)	Harvested Acres (000)	Yield (Bushels Per Acre)	Production (000) Bushels	Market Year Average $/Bushel	Value of Production (000 $)
1991	75,957	68,822	108.6	7,474,765	2.37	17,860,947
1992	79,311	72,077	131.5	9,476,698	2.07	19,723,258
1993	73,239	62,933	100.7	6,337,730	2.50	16,035,515
1994	78,921	72,514	138.6	10,050,520	2.26	22,874,154
1995	71,479	65,210	113.5	7,400,051	3.24	24,202,234
1996	79,229	72,644	127.1	9,232,557	2.71	25,149,013
1997	79,537	72,671	126.7	9,206,832	2.43	22,351,507
1998	80,165	72,589	134.4	9,758,685	1.94	18,922,084
1999	77,386	70,487	133.8	9,430,612	1.82	17,103,991
2000	79,551	72,440	136.9	9,915,051	1.85	18,499,002
2001	75,702	68,768	138.2	9,502,580	1.97	18,878,819
2002	78,894	69,330	129.3	8,966,787	2.32	20,882,448
2003	78,603	70,944	142.2	10,089,222	2.42	24,476,803
2004	80,929	73,631	160.4	11,807,086	2.06	24,381,294
2005	81,759	75,107	147.9	11,112,072	1.90	21,040,707
2006	78,327	70,648	149.1	10,534,868	3.20	33,837,454
2007	93,600	86,542	151.1	13,073,893	4.00	52,090,108
2008	85,985	78,640	153.9	12,101,238	3.90	47,377,576
2009	86,482	79,620	164.9	13,130,632	3.70	48,588,665
2010	88,192	81,446	152.8	12,446,865	5.30	65,968,384

Source: National Corn Growers Association

research, increasingly done by private seed companies, continues to the present. The hybrid varieties began to be adopted in the Corn Belt prior to WW II. Technology developed during the war years made it possible to greatly increase the availability and lower the cost of nitrogen fertilizer. The higher levels of fertilization were essential for realizing the yield potential of the improved seeds and intensive cultivation methods. These varietal and management improvements increased average yields from less than 40 bushels per acre in 1950 to more than 100 bushels per acre by 1990.

Developments since 1990, involving higher plant populations and genetic modifications to create plant resistance to insects, diseases and herbicides, continue to push yields higher and make corn production more profitable. Currently, average yields have risen to 160 bushels per acre or more, with annual variations reflecting production conditions (Table 2). Although most corn in the U S is rainfed, as much as 15 percent of the crop is now irrigated, which also contributes to higher average yields. Continuing genetic modifications and more intensive cultivation are expected to maintain the upward trend in corn yields.

Corn acres harvested, average yields and production during 1991–2010 are summarized graphically in Figs. 1–3. These figures provide a context for later discussion of economic issues that arise around the rapid growth in the use of corn as a feedstock for fuel ethanol production in the last decade. The positive trends in the three output dimensions are verified by the linear trend line shown in each of the figures.

Changing production practices also need to be considered in assessing the impacts of corn-based ethanol production. Most corn farmers have adopted low-tillage or no-tillage methods involving spring seed-bed preparation or not plowing the fields at all. Use of herbicide-resistant varieties permits chemical weed control. These practices require less use of machinery and energy and make corn more profitable. Leaving plant materials on the fields improves organic content of the soil and reduces soil loss due to wind and water erosion.

The trends in corn acres, yields and production can be discerned even more acutely in Iowa, identified earlier as the leading corn-producing state (Table 3). The corn area was relatively stable in the state during 1991–2006, while yields were rising more rapidly than in the country as a whole. Iowa

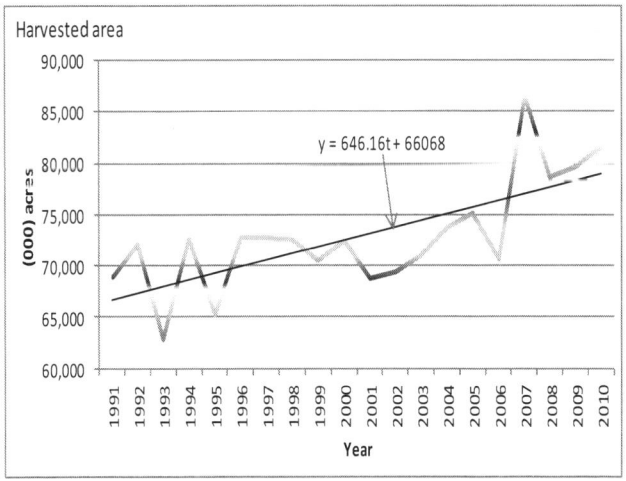

Source: National Corn Growers Association

Figure 1 Corn Area Harvested in the United States 1991–2010.

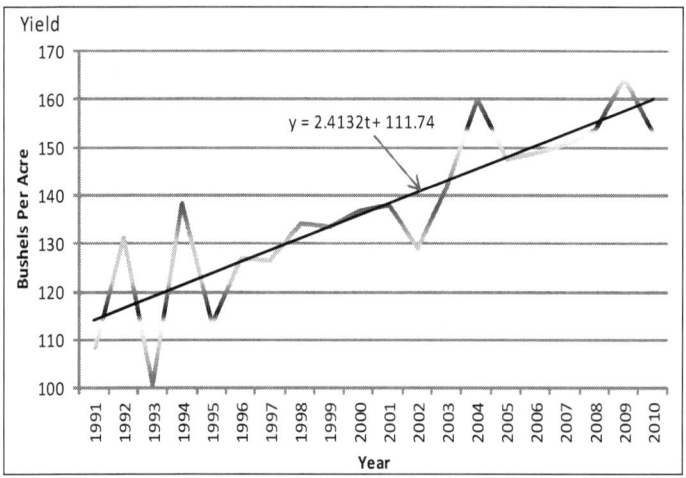

Source: National Corn Growers Association

Figure 2. Corn Yield in the United States 1991–2010.

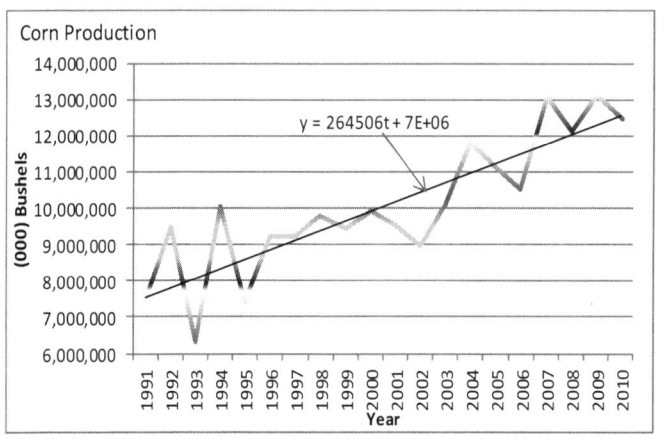

Source: National Corn Growers Association

Figure 3 Corn Production in the United States 1991–2010.

became the epicenter of the US ethanol boom in the early 2000s. For that reason it is an illuminating *avatar* portraying the issues arising from corn-based ethanol production.

Table 3 Corn Production and Value in Iowa, 1991–2010.

Year	Planted Acres (000)	Harvested Acres (000)	Yield Bushels Per Acre	Production (000) Bushels
1991	12,500	12,200	117	1,427,400
1992	13,200	12,950	147	1,903,650
1993	12,000	11,000	80	880,000
1994	12,900	12,600	152	1,915,200
1995	11,900	11,600	123	1,426,800
1996	12,700	12,400	138	1,711,200
1997	12,200	11,900	138	1,642,200
1998	12,500	12,200	145	1,769,000
1999	12,100	11,800	149	1,758,200
2000	12,300	12,000	144	1,728,000
2001	11,700	11,400	146	1,664,400
2002	12,200	11,850	163	1,931,550
2003	12,300	11,900	157	1,868,300
2004	12,700	12,400	181	2,244,400
2005	12,800	12,500	173	2,162,500
2006	12,600	12,350	166	2,050,100
2007	14,200	13,850	171	2,368,350
2008	13,300	12,800	171	2,188,800
2009	13,700	13,400	180	2,438,800
2010	13,400	13,050	165	2,153,250

Source: National Corn Growers Association

US Corn Utilization

Historically feed for animals has been the predominant use of the US yellow corn crop, both domestically and in importing countries. Prior to WW II only a relatively small part of the crop was consumed directly as human food, processed for use in prepared foods, or used for beverage ethanol.

In the second half of the 20th century corn milling expanded to produce starch, corn oil, high fructose corn syrup, and other sweeteners and products, as well as ethyl alcohol. Rapid growth is the use of corn sweeteners was promoted by policies that used import quotas to protect the domestic market and provide higher prices to US sugarcane and sugarbeet farmers. Many food processors and beverage companies turned to high fructuose corn syrup HCFS and other corn sweeteners as replacements for higher priced cane and beet sugars.

Using 2001 as a reference year, a total of 2,284.8 million bushels of corn was used for wet-mill processing, which represented 30 percent of that year's production. Feed use and exports were the other uses for the remaining 70 percent of the crop. After 2001, a change in utilization began, involving an increase in the industrial use category. These increases primarily reflected the growing use of corn for production of fuel ethanol that grew dramatically after 2004, which together with other corn processing equaled the use of corn for feed by 2009 and then exceeded feed use in 2010 (Fig. 4). The upward trajectory of corn processed in dry-mill plants for ethanol after 2001 is charted in Fig. 5, reaching almost 40 percent of the US corn crop in 2010. A detailed breakdown of corn availability and utilization by year during 2006–2010 is provided in Table 4.

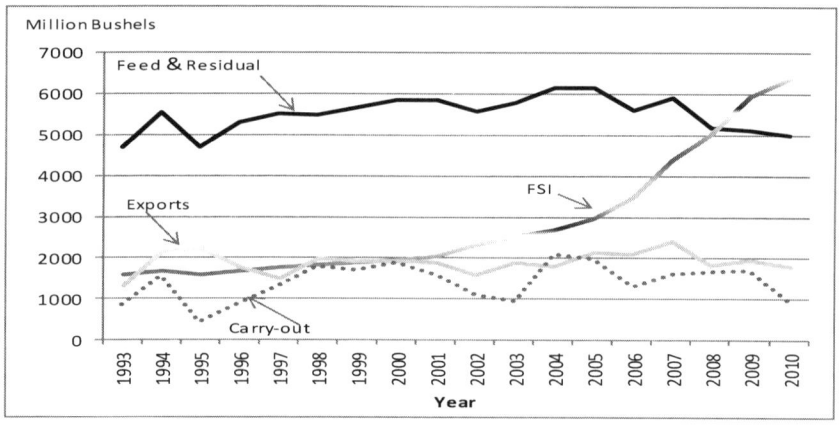

Source: World Agricultural Supply and Demand Estimates, various years, USDA.
FSI: food, seed and industrial
Carry-out: stocks at end of marketing year

Figure 4 Corn Utilization by segment in the United States 1993–2010.

Early Ethanol

Prior to the oil price shocks of the 1970s only a few million gallons of ethanol were produced each year in the US. The initial impetus for ethanol expansion came from the oil crises of that decade, which exposed the vulnerability of the US to disruptions in petroleum imports. Fuel ethanol was put forward

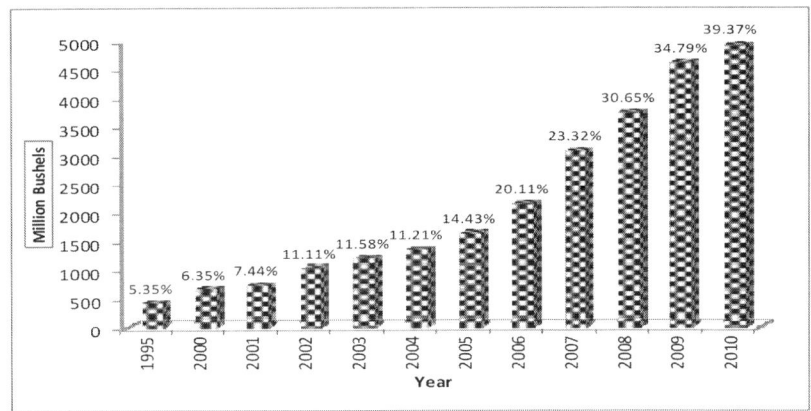

Note: The Value on the bar represents percentage of U.S total corn production
Source: USDA, ERS, Feed Outlook, 1/11

Figure 5 Corn Used for Ethanol Production 2000–2010.

Table 4 U.S. Corn Balance Sheet.

	2003	2004	2005	2006	2007	2008	2009	2010
(Million Acres)								
Planted Area	78.6	80.9	81.8	78.3	93.5	86.4	86.4	88.2
Harvested Area	70.9	73.6	75.1	70.6	86.5	78.6	79.5	81.4
Harvested Yield (Bu./Acre)	142.2	160.3	147.9	149.1	150.7	153.8	164.7	152.8
(Million Bushels)								
Beginning Stocks	1,087	958	2,114	1,967	1,304	1,624	1,673	1,708
Production	10,089	11,806	11,112	10,531	13,038	12,092	13,092	12,447
Imports	14	11	9	12	20	14	8	25
Total Availability	11,190	12,775	13,235	12,510	14,362	13,730	14,773	14,180
Use:								
Feed and Residual	5,798	6,135	6,115	5,540	5,858	5,182	5,141	5,150
Food, Seed and Industrial	2,537	2,707	3,019	3,541	4,442	5,025	5,939	6,450
Ethanol for Fuel	1,168	1,323	1,603	2,119	3,049	3,709	4,568	5,050
Total Domestic Use	**8,335**	**8,843**	**9,134**	**9,081**	**10,300**	**10,208**	**11,079**	**11,600**
Exports	1,897	1,818	2,134	2,125	2,437	1,849	1,987	1,900
Total Utilization	**10,232**	**10,661**	**11,268**	**11,206**	**12,738**	**12,057**	**13,066**	**13,500**
Ending Stocks	958	2,114	1,967	1,304	1,624	1,673	1,708	680
Ending Stocks, % of Use	9.4	20	17	12	13	14	13	5
Average Farm Price, $/Bu.	2.42	2.06	2.00	3.04	4.2	4.06	3.55	5.30

Source: USDA, CBOT and Informa Economics

as a gasoline substitute such that blending it with gasoline for vehicle use would lower petroleum imports and make the US less dependent on foreign oil.

This early use of ethanol was severely criticized on the grounds that it suffered from a negative energy balance. Conceptually, the energy balance evaluation takes into account all the energy used to produce the corn feedstock plus all the energy required to make the ethanol and move it through the distribution system to the point of retail sale, to establish the total amount of energy use attributable to the ethanol. This amount is compared to the energy released by burning the ethanol fuel to arrive at a net energy gain or loss. Some studies from the 1970s contended that fuel ethanol was "energy negative" meaning that more energy was used to produce it than was contained in fuel itself. The general view at the time was that at best the blended product was just about equivalent to gasoline in carbon intensity. In any event interest waned in ethanol use with the retreat in oil prices in the 1980s.

Later in the decade of the 1980s interest in ethanol rebounded around its role as an octane enhancer and was spurred when the additive methyl tertiary butyl ether (MTBE) became recognized as a contaminant of ground water, leading to a ban on its use in California and other states. Ethanol was the favored replacement for MTBE. Its use also received a major boost in the Clean Air Act of 1990 that led to its popularity as an oxygenate replacing lead in gasoline. In these uses ethanol acts as a complement to gasoline, meaning the fuel products are blended together to create higher quality fuels meeting legal standards.

An Ethanol Tsunami

Higher oil prices in the 1990s rekindled interest in ethanol as a gasoline substitute. This interest intensified on the basis of new studies showing a more favorable ethanol energy balance. Higher corn yields, lower energy use per unit of fertilizer produced, and more efficient conversion ratios helped create a positive net energy balance for the fuel. Another important difference was the inclusion of co-products, especially distillers' grains, in the calculations. Ethanol was credited with the energy saved in feedstuffs replaced by use of the coproducts for animal feed. While results of the studies varied depending on the definitions of the ethanol life cycle, their thrust was to raise the net energy balance from slightly positive (1.06) to 1.67 or more. Ethanol advocates used these results to argue that fuel ethanol indeed offered a net energy gain and could economically displace petroleum imports while benefitting the environment.

One of the major responses involved government policies to promote fuel ethanol use. Ethanol subsidies date to the 1970s, varying between 40 and 60 cents per gallon, and have been provided continuously. Legislation introduced use mandates in 2005, which were then amended in 2007. The Energy Independence and Security Act (EISA) of 2007 increased the mandates for biofuels, set the subsidy for ethanol at $.45 per gallon, and fixed a specific tariff of $.54 per gallon on ethanol imports. While the subsidies and the tariff are supposed to disappear at the end of 2011 they could be extended perhaps at a reduced level. In any event it is the use mandate that has been the primary driver of ethanol consumption, which has been produced domestically due to the import tariff.

Production of fuel ethanol has soared in the US since 2001 (Table 5). Corn purchased by ethanol producers has also grown dramatically, both in bushels and as a proportion of the country's corn crop (Fig. 4, Fig. 5).

Table 5 Fuel Ethanol Production in the United States 1980–2010.

Year	Millions Gallons
1980	175
1981	215
1982	350
1983	375
1984	430
1985	610
1986	710
1987	830
1988	845
1989	870
1990	900
1991	950
1992	1,100
1993	1,200
1994	1,350
1995	1,400
1996	1,100
1997	1,300
1998	1,400
1999	1,470
2000	1,630
2001	1,770
2002	2,130
2003	2,800
2004	3,400
2005	3,904
2006	4,855
2007	6,500
2008	9,000
2009	10,600
2010	13,230

Source: Renewable Fuels Association

This proportion reached almost 40 percent in 2010 compared to less than 10 percent just 10 years ago. Ethanol's share of US gasoline rose to 9% in 2010, up from 1.4% at the beginning of the decade. Most of the consumption has been in the form of a fuel blend of 10% ethanol with gasoline (E10). Although a blend with 15 percent ethanol has been approved recently, that fuel is not widely available due to a lack of pump and storage infrastructure at retail gasoline outlets. E10 remains the most widely utilized blended fuel. Some states mandate its use, but in most of the country consumers must choose E10 over other unblended fuels. E10 is cheaper per gallon to reflect the lower energy content of the fuel. Were the E10 fuel used universally in the US it would still replace less than a quarter of imported petroleum.

Using pure ethanol causes engine performance problems but vehicles can be engineered to utilize an 85 percent ethanol blend (E85). This fuel is becoming more widely available, but again a distribution infrastructure for it is not widespread. Flex fuel vehicles that can use this fuel are being produced but are still relatively few in number. Use of this fuel in the future will depend on growth of the flex fuel fleet.

Greenhouse Gas Emissions and Land Use Changes

The evolution of the carbon intensity debate, summarized succinctly as "point ethanol," has now been overtaken by a controversy concerning conversion of grassland and tropical forests to cropland. The context for this controversy is based on the increasing use of corn for fuel ethanol in the US, which has shifted the demand for corn outward, raised corn prices and encouraged higher corn production.

Additional output can come from higher yields per acre and from using more cropland for corn. The positive trend in corn yields is shown in Fig. 2 and the positive trend in corn acreages in Fig. 1. However, more careful analysis of the contributions of the area and yield changes to output growth leads to the important conclusion that changes in production after 2001 are more due to expansion in corn acreages than in corn yields.

Growth rates in both areas and yields were calculated for the 1991–2010 periods and also for the 1991–2001 and 2002–2010 subperiods (Table 6). For yields, the average annual growth rate in the recent period was lower than in the earlier period. In contrast, area growth was quite slow in the earlier period but was sharply higher in the more recent period. These growth rates imply that below average growth in yields accounted for less than half of output growth while higher than average growth rates in area planted contributed fully 55 percent of increased production in 2002–2010.

The additional crop acreage used for corn came from where? Some marginal land not previously farmed was brought into production. Some land held under conservation programs was taken out to use for corn. Both

Table 6 Average Annual Percentage Growth Rates.*

Time Period	Harvested Area (acres)			Corn yield (bushels per acre)		
	1991–2010	1991–2001	2002–2010	1991–2010	1991–2001	2002–2010
Growth rate (r)	.008748	.004249	0.021324	0.01826	0.0208136	0.017654
Annual Percentage growth rate (g)	0.87	0.42	2.13	1.83	2.08	1.77

*Calculated from the semi-log equation $\ln Y_t = a + bt$

of these conversions of unfarmed land involve release of greenhouse gases when put to crop use, which offsets some of the energy gains when the resulting ethanol is burned. The majority of the new corn acres, however, came from shifting land to corn from other crops. In the Corn Belt states more corn acreage was mainly at the expense of soybean acres, thus reducing production of that oilseed crop. This effect was intensified because the energy mandates also included use of biodiesel for which soybean oil is the main ingredient. In the US this fuel is mostly used by commercial vehicles but it is widely used in Europe where diesel engines are much more popular and vegetable oils are the main ingredient in the fuel.

The additional land use for corn has intensified environmental concerns. Some of the marginal land is more subject to wind and water soil erosion, raising the issue of soil loss. More continuous corn cultivation implies less legume crops in the rotation, which requires higher fertilization rates to maintain or increase yields. The result is more chemicals in the water runoff going into streams and rivers that cause water quality problems downstream.

The land use argument was elevated to an international scale after it was recognized that diversion of land from soybeans in the US, in conjunction with the rising world demand for vegetable oils, was resulting in the conversion of grasslands and tropical forests for soybean production in Brazil and other countries. Such land conversion results in large GHG emissions, which require many years of corn ethanol use to offset, thereby challenging the reductions in carbon intensity claimed for the fuel (Matthews and Tan 2009; Thompson 2010).

The advocates of fuel ethanol attack this argument claiming that the increase in corn production has at most a very tenuous connection to land use internationally making the scientific basis for indirect land use consequences doubtful. On the other hand, clearly corn acreage in the US has expanded since 2001 and land in Brazil and similar countries has been converted to soybeans and other crops. While more definitive measurement of the causes and consequences of these land use changes is undoubtedly needed, a clear connection between ethanol use and large GHG emissions due to land conversion has been established (Oladosu et al. 2011).

Growth of the US Ethanol Industry

A gradual buildup in the number of operating ethanol plants began around 2000, and then accelerated after 2005 (Table 7). The yearly ethanol production capacities of the plants grew seven-fold during 1999–2011.

In January 2011, 204 ethanol plants were operating in 29 states (Table 7). The distribution of ethanol production capacity by state is closely related to the geographic distribution of corn production. Three-quarters of the total 2011 capacity was in the six states of Iowa, Nebraska, Illinois, Minnesota, Indiana and South Dakota (Table 8). Iowa has the largest number of plants and the largest production capacity, just as it is the leading corn-producing state.

Fuel ethanol production has followed a pattern similar to changes in the number of plants. Starting from a low base in 1980 production grew quite slowly in the 1980s and 1990s (Table 8). Comparable percentage increases continued through 2005 after which production growth greatly accelerated. By 2010 production had more than doubled to 2.5 times the 2005 level. Moreover, yearly output rose in tandem with the total capacities of the plants: production in both 2005 and 2010 exceeded the recorded capacities of the plants, 107 percent in 2005 and 111% in 2010. These output figures suggest that plant owners generally found it profitable to operate their plants continuously at full capacity.

A Panoply of Government Policies

Ethanol producers buy corn, process it into fuel ethanol and coproducts, and sell the ethanol and coproducts to buyers in markets. Given that the ethanol industry has only recently undergone sustained growth, little empirically implemented research is available on firm behavior and market outcomes in the industry. Government policies have also injected additional complexity into the operation of the markets.

One of the important government policies is the subsidy paid to blenders for each gallon of ethanol utilized. Although blenders receive the subsidies, it cannot be automatically assumed that the benefits go only to the buying side of the ethanol market. The microeconomic theory of competitive markets can be used as a framework for answering the crucial question. "Which side of the market benefits from the subsidies?" (Taheripour and Tyner 2007).

The basic demand and supply model is shown in Fig. 6, which assumes pure competition between buyers and sellers in the domestic market. Imports/exports are not included because tariff protection has virtually isolated the domestic market from cheaper foreign ethanol. The demand (D) for ethanol by blenders depends on many factors assumed constant,

Table 7 Fuel Ethanol: Operating Plants and Production Capacity in the United States 1999–2011.

	Year*												
	1999	2000	2001	2002	2003	2004	2005	2006	2007	2008	2009	2010	2011
Total Ethanol Plants	50	54	56	61	68	72	81	95	110	139	170	189	204
Ethanol Production Capacity mg**	1701.1	1748.7	1921.9	2347.3	2706.8	3100.8	3643.7	4336.4	5493.4	7888.4	10,569.4	11,877.4	13,507.9
Number of States with Ethanol Plants	17	17	18	19	20	19	18	20	21	21	26	26	29

*Number of Plants Operating in January
**Million gallons per year
Source: Renewable Fuels Association

Table 8 Fuel Ethanol: Distribution of Production Capacity by State 2011.

States	Capacity* mg
Iowa	3,595
Nebraska	1,839
Illinois	1,480
Minnesota	1,119
Indiana	906
South Dakota	1,016
Ohio	424
Kansas	437
Wisconsin	498
Texas	250
North Dakota	343
Michigan	265
Missouri	261
California	123
Tennessee	177
New York	164
Oregon	40
Colorado	125
Georgia	100
Pennsylvania	110
Virginia	0
North Carolina	0
Arizona	55
Idaho	54
Mississippi	54
Kentucky	35
New Mexico	30
Wyoming	7
Louisiana	2
Total	13,509

Source: Renewable Fuels Association
*Operating Plants, January; million gallons per year

a) More Elastic Supply

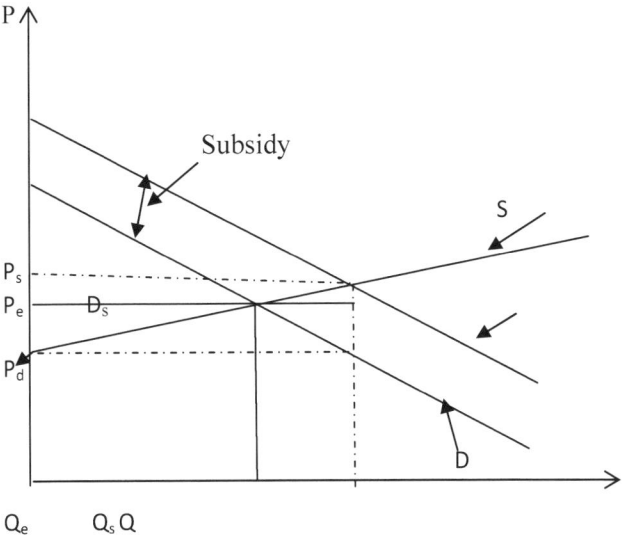

b) Less Elastic Supply

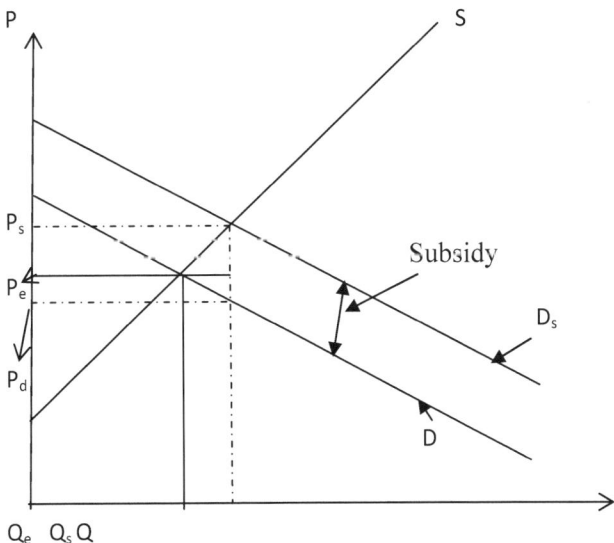

Figure 6 Effect of a Per-unit Subsidy in a Competitive Market.

especially the price of oil. This demand curve can be shifted outward or inward if the price of oil changes: the higher the price of oil, the greater the demand for ethanol. The supply (S) of ethanol shows the amounts ethanol producers are willing to supply at different prices. Many factors affecting this supply curve are also assumed constant, the number of plants perhaps the most prominent of them. P_e and Q_e are the market-clearing price and quantity equilibrium outcomes.

The subsidy paid per gallon to blenders can be represented by a parallel outward shift in the demand curve to D_s. The subsidy causes increases in both the price and quantity of ethanol, to P_s and Q_s respectively. The total subsidy paid to blenders is equal to the area $(P_s - P_d) \times Q_s$, where P_d is the price at which Q_s would be taken by buyers with no subsidy.

Examination of Fig. 6 indicates that the subsidy will be divided between buyers and sellers depending on the responsiveness of sellers to price changes, measured by the elasticity of the supply curve. Panel (a) in the figure applies for a more elastic supply curve: A small increase in price from P_e to P_s associated with a large increase in quantity, Q_e to Q_s. These changes keep most of the subsidy in the hands of the buyers. In contrast panel (b) illustrates that a less elastic supply curve results in a larger price increase and smaller quantity increase, thus transferring most of the subsidy to ethanol sellers.

This market model highlights the importance of the price responsiveness of market participants in determining the division of the subsidy. The reality is that little is known empirically about the relevant parameters. Considering the market situation of blenders, the buyers of fuel ethanol, it does not seem that the quantity they wish to purchase would change very much if the price of ethanol changed, implying that their demand curve has low elasticity.

The circumstances of the ethanol sellers seem quite different. Ethanol is supplied by a relatively large number of plants of different sizes and ages. Older plants are likely to be smaller and less efficient to operate than newer plants. If the plants were to be arrayed from the least efficient to the most efficient, the resulting curve would be increasing with price: as price increases less efficient plants can be profitably operated. Moreover, since ethanol plants are major users of corn, as ethanol output rises the price of corn will increase, which adds to the upward slope of the supply curve. After all plants are operating at full capacity, the supply for the given number of plants will become virtually vertical. At a later time, if more efficient and larger plants have been put into use, the supply curve will shift downward and outward as illustrated in Fig. 7 in which S_{100} and S_{200} are the supply curves for the respective number of plants. If the unit operating costs of

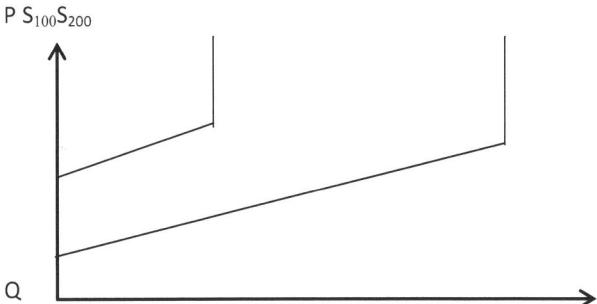

Figure 7 Hypothetical Supply Curves for Ethanol Plants.

plants of different sizes and ages are not too different, the supply curve will be relatively flat up to the capacity output except for the effect of the rising prices for the corn feedstock.

Alternative Ethanol Market Outcomes with Quantity Mandates

EISA mandated increasing use of biofuels from 9.0 billion gallons in 2008 to 36 million gallons in 2022. In the early years of the mandate corn was expected to be the main feedstock but under the legislation its use will level off at 15 billion gallons after 2015 with additional production thereafter expected to come from cellulosic ethanol plants. In terms of affecting market outcomes, an important point is whether or not the mandate is binding in a given year, where binding means that ethanol blenders are required to buy more ethanol than they would freely choose to buy under prevailing market conditions. A non-binding mandate imposes no constraints on market behavior. Two possible scenarios are analyzed in Fig. 8 under the assumption that the tariff prevents imports, and which also recognizes the government subsidy and the impact of oil prices. (Meyer et al. 2008, 2011).

In panel (a) a low oil price means that the demand for ethanol is at a low level such that buyers would choose a quantity lower than the mandated quantity, Q_m. Given the supply curve for the existing plants, Q_m will be available only at a price, P_m, that is higher than either the equilibrium price or the price that would prevail with the subsidy and without the mandate (P_s). With the mandate not only is the entire subsidy transferred to sellers but the buyers must use additional resources to pay P_m. This outcome can exist only in a transitional period while the ethanol industry is expanding in response to profit incentives resulting from the high ethanol price. In this

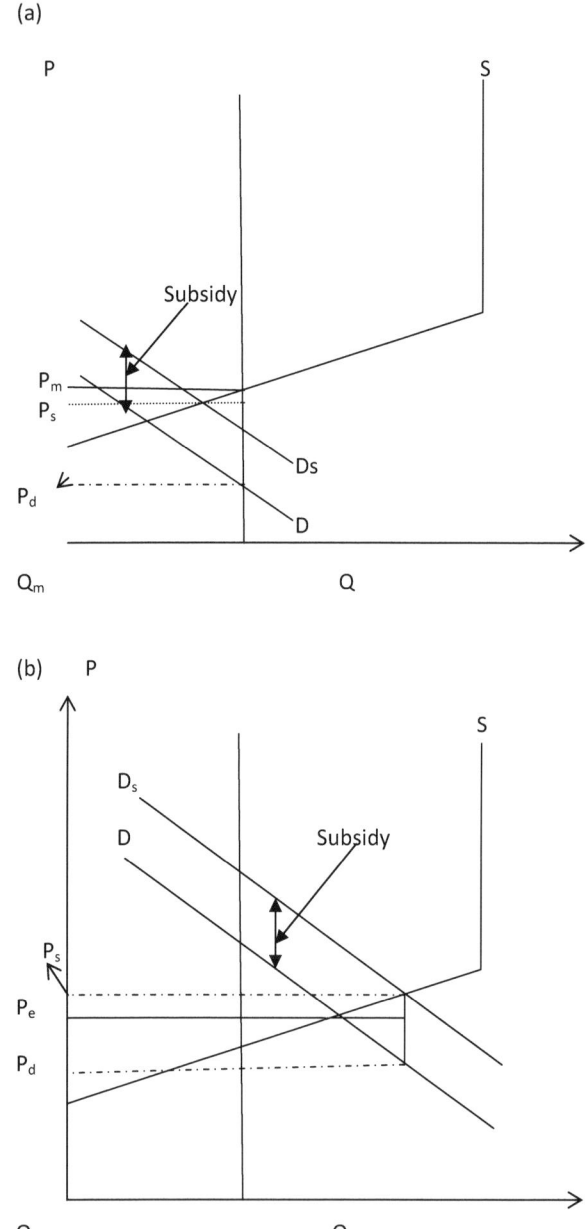

Figure 8 (a) Ethanol Market Outcome with Subsidy, Binding Mandate and Low Oil Price, (b) Ethanol Market Outcome with Subsidy, Non-binding Mandate and High Oil Price.

expansion phase with a binding mandate, the subsidy accrues completely to ethanol sellers and ethanol production is likely to be highly profitable for the operating plants.

A situation with a more mature ethanol industry is shown in panel (b). Here oil prices are higher so the demand curve for ethanol has shifted outward. The ethanol industry has expanded by increasing the number of plants and now has capacity well above a higher mandate. In this case the market will clear at P_s and Q_s as explained in the earlier discussion of the effects of a subsidy. The subsidy will be divided based on the relative elasticities of the demand and supply curves. If the demand curve has low elasticity compared to the supply curve, which seems to be the most likely, then most of the subsidy will remain with the ethanol buyers thus effectively subsidizing ethanol consumption (de Gorter et al. 2010).

How are corn producers affected by these subsidies? These effects will be transmitted through the price of corn. If the mandate or the subsidy increases ethanol quantities, then the price of corn will rise in response to the increased demand for it. In general, the higher the proportion of the corn crop used for ethanol, the greater the impact on prices, and hence the larger the price benefits passed through to corn producers.

Ethanol and Animal Feed

Corn prices more than doubled during 2007–2008 from what had been a relatively low and stable level during 2001–2006. Prices for the 2010 crop rose even higher to a historic record level. More price increases are in prospect for the 2011 marketing year given expectations that yields in the US will decline due to extreme heat in July and flooding of fields. This means that livestock producers are facing high and rising prices for one of the major animal feeds.

Higher corn prices reduce the profitability of livestock production, which is also affected by volumes of meat products produced. Livestock producers adjust output within cycles of different lengths depending on the production period for specific types of animals.

The competition between meat producers and ethanol plants is somewhat abated by the use of ethanol coproducts as animal feed. Only the starch component of corn kernels—about 50% of the kernel—is transformed into ethanol. Both wet milling and the now dominant dry milling processes produce a variety of co-products. The largest volume of these co-products, roughly a third of the corn volume, is distillers dried grains with solubles (DDGS), which is high in protein and can substitute in livestock feed for other protein sources such as soybean meal.

Some ethanol advocates add the DDGS to the feed use of corn and argue that ethanol production has not materially reduced feed supplies. The weakness in this argument is the difficulty in knowing what would have happened to corn production and the supply of animal feeds if subsidies, mandates and the import tariff had not created incentives for large-scale investment in domestic ethanol plants. While overall feed utilization may not have declined in the years of the ethanol buildup, there is no doubt that livestock producers have had to face sharply rising prices for corn and that feed costs would have been lower without the competition for corn from ethanol.

Food vs. Fuel: The "Achilles Heel" of Corn Ethanol?

The title of this section refers to the dilemma that has been created by diverting food crops to biofuel production to the detriment of the global supply of staple foods. This issue has been debated intensely since the crisis in 2007–2008 caused by rising world food prices, and is regaining prominence with the current price increases that began in 2010. It involves several major crops including corn, soybeans, vegetable oils and sugar cane, all of which can be consumed as food or feed for animals, or used to make biofuels. This discussion focuses on the connection of corn ethanol to increasing food prices.

The use of food crops for biofuels, along with rising world food needs and moderating production growth trends, has reversed the declining price patterns that prevailed for last two decades of the 20th century. International food price indices rose at an accelerated rate in 2007 and 2008, and are currently rising again. Higher food prices have several causes in addition to the use of the commodities for biofuels: growing global demand for food, especially in the high-growth developing countries; shifts in food consumption patterns, again in countries with rising per capita incomes; increased US exports in response to rising demand and a weaker dollar; and weather-related production deficits in key exporting countries.

Corn prices in 2011 reached record highs but the effects of the higher prices on food consumers have been questioned. Those effects depend on how long corn prices remain high and the extent to which higher corn prices are passed through to retail food products. The cost of the raw commodity is only a small part of the retail price for processed foods For example, the cost of corn in a box of corn flakes is about $0.05. Generally speaking, costs for transportation, processing, packaging, handling and merchandising make up 80 percent or more of the prices of retail food items. This percentage is somewhat lower for fresh meats so higher feed costs may pass through more rapidly and completely to meat consumers (Informa economics 2011).

Those on the other side of this argument point to the problems high corn prices create for import-dependent, food-insecure countries in the developing world. In these cases, higher prices serve as constraints on import capacities, which can adversely affect both food availabilities and the capabilities of poorer households to buy food.

Several recent studies of food prices have identified the increasing production of corn ethanol as only one factor affecting food prices. Other factors considered important include export demand associated with changing diets in developing countries, higher agricultural production costs primarily due to rising energy prices, underinvestment in new agricultural technology, and participation of speculators in commodity markets. These factors can continue to push up prices for corn and other food crops even if the use of corn for ethanol were to decline from its current level (Hochman et al. 2010).

Conclusions: The Future of Corn Ethanol

Corn ethanol faces an uncertain future. If support policies are reduced, or even eliminated, profitability of existing plants will fall and incentives to invest in new plants will erode. Existing plants will continue to operate as long as variable costs can be covered. If revenues fail to cover fixed costs, older and less efficient plants will close or be converted for production with other feedstock. Over time, additional plants will cease to operate. Such adjustments are characteristics of declining industries.

The import tariff is an underappreciated key determinant of the prospective decline of the corn ethanol industry. Brazil is the world's second largest producer of ethanol using sugar cane as the feedstock. This more efficient production technology is spreading to other countries that produce sugar cane, some of which may have duty-free access to the US market. With reduced, or even no, import duties, domestic producers will be hard pressed to compete with lower-cost foreign ethanol. This is an imminent threat to corn ethanol producers and an existential threat to ethanol production based on alternative feedstock.

The mandate only affects the ethanol market if it is binding. That was true in the expansion phase of the industry until capacity grew to equal or exceed the mandated annual volumes and demand for ethanol was high due to high oil prices. If the mandate is extended and oil prices were to fall, the inward shift of demand for ethanol could make the mandate binding again, thus protecting the profitability of ethanol production. Without a tariff-protected mandate a decrease in demand would hasten the closure of more corn ethanol plants.

Oil prices are expected to remain high, as well as volatile, over the medium-term future. This expectation is based on robust demand increases in high-growth developing countries (e.g., China, India), modest expansion in drilling and exploration, and vulnerability of oil supplies to OPEC policies and geopolitical instability in major exporters. This oil price outlook is favorable to ethanol producers since ethanol prices tend to closely follow gasoline prices, and high oil prices will maintain political support for curbing petroleum imports. It does not, however, assure a profitable future for US corn ethanol, which could lose its market position to ethanol produced from other feedstock as well as imports from Brazil and other countries.

References

de Gorter H and Just DR (2010) The Social Costs and Benefits of Biofuels: The Intersection of Environmental, Energy and Agricultural Policy. Appl Eco Perspect Policy 32: 4–32.

Hochman G, Rajagopal D and Zilberman D (2010) Are Biofuels the Culprit: OPEC, Food and Fuel. American Economic Review: Papers and Proceedings I00: 183–187.

Informa economics (2011) Analysis of Corn, Commodity, and Consumer Food Prices. Renew. Fuels Assoc.

Mathews JA and Tan H (2009) Biofuels and indirect land use change effects: the debate continues. Biofuels Bioprod Bioref 3: 305–317.

Meyer S, Thompson W and Westhoff P (2008) Model of the US Ethanol Market. FAPRI-MU Report 07–08 (July).

Meyer S, Thompson W and Westhoff P (2011) New Challenges in Agricultural Modeling: Relating Energy and Farm Prices. FAPRI-MU Report (May).

NCGA (2011) World of Corn. National Corn Growers Association.

Oladosu D, and Kline L (2011) Global Analysis of Biofuel: Indirect Effects and Feedstock Potential. Oakridge Natl Lab.

Taheripour F and Tyner W (2007) Ethanol Subsidies, Who Gets the Benefits? Depart Ag.l Econ, Purdue Univ.

Thompson S (2010) Biofuels Effects on Markets and Indirect Land Use for Food. J. Internl. Trade and Develop 6: 117–131.

CHAPTER 13

Future Prospects for Corn as a Biofuel Crop

Kenneth J. Moore,[1],* Douglas L. Karlen[2] and Kendall R. Lamkey[3]

ABSTRACT

Ethanol production from corn grain has increased significantly during the past ten years in the US. This increase was driven by government policy guided by the Renewable Fuel Standard (RFS) and embodied in the Volumetric Ethanol Excise Tax Credit and other legislation created to promote a biofuels industry. As corn grain ethanol production approaches the target set out in the RFS, the industry is looking to develop capacity for producing advanced biofuels, primarily from agricultural wastes and dedicated energy crops. The residues remaining following corn harvest have been identified as a voluminous and readily available feedstock for advanced biofuels. However, these residues provide important ecosystem services and their complete removal may exacerbate environmental problems associated with soil erosion, water quality, nutrient cycling, and carbon sequestration. Alternative crop management practices need to be developed and implemented to ensure that these services are maintained or enhanced for biofuel production from corn residues to be sustainable. Management practices such as reduced tillage, use of cover crops, site-specific harvest intensities, and

[1]1571 Agronomy Hall, Iowa State University, Ames, IA 50011, USA.
[2]USDA-ARS-NLAE, 2110 University Blvd, Ames, IA, 50011, USA.
[3]Department of Agronomy, 100 Osborn Drive, Iowa State University, Ames, IA 50011, USA.
*Corresponding author: kjmoore@iastate.edu

shifting marginal land currently used for corn production to perennial energy crops show potential for allowing removal of corn residue while maintaining ecosystem services.

Keywords: ethanol, soil erosion, soil quality, water quality, carbon sequestration

Current Status—A Grain Dominated System

Few people would have predicted the rapid increase in ethanol production from corn grain that occurred during the last decade. This increase was driven primarily by a "blender's credit" that subsidized the blending of ethanol with gasoline (Hoekman 2009). The credit, which helped ensure a market for corn ethanol, assured a reasonable return to capital investment and enabled the industry to respond by expanding rapidly. For both farmers and their neighbors, investment in ethanol plants created greater local demand and higher prices for corn grain while also providing an increased number of well-paying employment opportunities that helped reinvigorate the economies in many small rural communities (NAS 2009). Some argue that the number of jobs added to the local economy has been overestimated, and when the increasing corn demand for ethanol production was coupled with that for animal feed to meet increasing demands from Asia, land values and input costs have also increased (Low and Isserman 2009).

In 2006, ethanol produced from corn exceeded that produced from sugarcane worldwide (Balat and Balat 2009) and ethanol production from corn has continued to expand exponentially (Fig. 1). There were 204 plants in the US in 2011 producing 13.5 billions of gallons of ethanol (Renewable Fuels Association 2012) and the US actually exported close to a billion gallons to Brazil making the US the leading exporter of ethanol in the world (USDA 2011). The rapid rise in US grain ethanol production is an astounding accomplishment, and as we look forward to further increases in the production of it and other biofuels from sources other than corn grain, it is worthwhile to consider the factors that made corn ethanol production so successful.

Several factors have contributed to the success of the corn ethanol industry. The blender's credit, officially known as the Volumetric Ethanol Excise Tax Credit, was authorized in 2004 as part of the Jobs Creation Act (H.R. 4520). It provided manufacturers of liquid fuels with an economic incentive to blend ethanol with petroleum products. The original tax credit was 51 cents per gallon on a pure ethanol basis, but it was reduced to 45 cents per gallon in 2009 and then phased out entirely at the end of 2011. A tariff on ethanol imports was also imposed to discourage blenders from

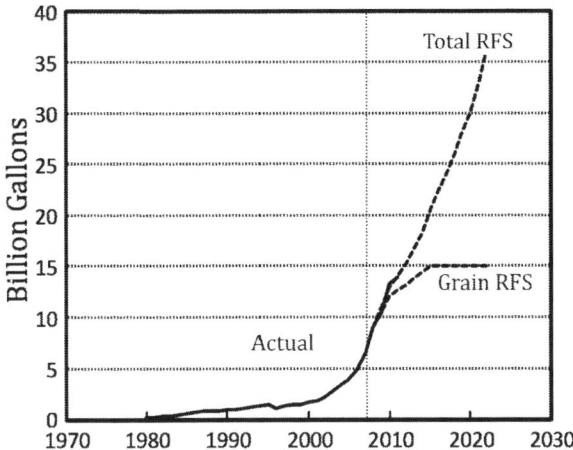

Figure 1 US ethanol production (1980–2011) and production Renewable Fuel Standard targets (2008–22). Data source, Renewable Fuels Association 2012.

using foreign sources of ethanol (Elobeid and Tokgoz 2008). This latter measure was introduced largely to discourage importation from Brazil, which had levied a protective tariff on ethanol imports. The US government also provided tax incentives to encourage investment in the development of ethanol production plants. A tax credit of ten cents per gallon was given to plants producing less than 60 million gallons per year. This contributed significantly to the proliferation of ethanol plants throughout the US Corn Belt, but some argue that such incentives and tax credits are not wise because of their impact on our national debt. However, as pointed out by Rossali-Calli (2010) the $7.7 to 11.6 billion given to the ethanol industry from 1979 to 2000 was really miniscule when compared to the $135 to 150 billion in tax breaks given to support the fossil fuel industries.

Coupled with these incentives were mandates requiring fuel manufacturers to produce increasing levels of biofuels beginning in 2006 (Hoekman 2009). The original Renewable Fuel Standard (RFS) was implemented in the Energy Policy Act of 2005 (P.L. 110-58) and was amended in the Energy Independence and Security Act of 2007 (P.L. 110-140). These acts set targets for the production of biofuels and are overseen by the EPA (De Gorter and Just 2009). The schedule set by the RFS increases the mandate from 9 billion gallons of renewable fuels in 2008 to 36 billion gallons in 2022 (Fig. 1). Ethanol produced from corn grain is capped at 15 billion gallons in 2015. The remaining 21 billion gallons are to be produced from feedstocks other than corn grain. Corn residue, the nongrain portion of the crop, is likely to contribute significantly to the production of these second generation biofuels. How much corn residue will contribute to meeting the RFS from here forward remains to be seen, but several studies evaluating the feasibility of producing fuel from biomass recognize it as a major feedstock (Perlack et al. 2005).

Another factor that contributed initially to the increase in ethanol production was the low price of corn. From 1973 through 2005, US corn grain prices averaged $2.36 ± 0.40 bu^{-1} of grain (25 kg), but for 2006 through 2010 prices averaged $3.94 ± 0.59 bu^{-1}, before spiking in 2011 to $6.01 (NASS 2011). Prior to the development of an ethanol market for corn grain, surplus production, often encouraged by federal subsidies, kept corn prices relatively low, but as the growing ethanol industry increased demand for corn grain, the price farmers received increased. Part of the rationale for pursuing an aggressive agenda in developing a corn ethanol industry was to create demand for corn that would result in increased prices and ultimately returns to farmers. More recently, the increased global demand for corn grain and the corresponding increase in its value has raised the cost of ethanol production resulting in narrower margins but also forcing increased efficiency (Babcock 2008). A recent processing change has been the increased oil extraction from the distiller's solids, but from the perspective of animal producers using DDGs for feed, this change has not been desirable.

Rural communities have benefited from construction and operation of corn ethanol production facilities (Low 2009). The industry has created new jobs in rural areas and provided local investment and marketing opportunities for corn grain. Looking forward to further expansion of ethanol produced from second-generation biomass, Ugarte et al. (2007) predicted substantial job creation in the agriculture and energy sectors. They further predicted that due to the broad geographic distribution of biomass production that many regions of the country will benefit.

As corn grain ethanol production approaches the target set out in the RFS, it is important to consider why production targets are shifting to more advanced biofuels. First, there was the realization that there is an upper limit in the amount of ethanol that can be produced from corn grain without negatively impacting food markets. Today, nearly 40% of the US corn crop is used to produce ethanol (NASS 2011), with most of the remainder being used in livestock feed. Only a small percentage contributes directly as an ingredient to foods produced for human consumption. Another factor favoring advanced biofuels is the recognition that the capacity to produce corn is constrained by land resources and even if the entire US corn crop were processed into ethanol it would only account for 12 to 15% of annual US gasoline consumption (Perlack et al. 2005).

In the intervening period since the original RFS targets were set, the specter of indirect land use change emerged and changed many perceptions regarding the sustainability of using corn grain to produce fuel (Searchinger et al. 2008). The concept behind indirect land use change is that shifting land use in the US to produce biofuels will cause a proportional conversion of land in other parts of the world to food production. Indirect land use change assumes that markets will respond in ways that result in deforestation and other practices that will have a negative impact on global carbon balance. While the theory is based on assumptions that may or may not be valid, it has nevertheless had a sobering effect on development of biofuels policies and has been a consideration in the development of revised RFS targets (EPA 2009). Furthermore, while production of biofuels on land already in cultivation could lead to a net decrease in greenhouse emissions relative to fossil fuels, tilling previously uncultivated land elsewhere could lead to increased global greenhouse gas (GHG) emissions. This could mean that instead of having a positive effect on climate change factors, using land that was previously used for food production for fuel crops could actually exacerbate the problem.

Despite all of the tangible positive benefits that have accrued from corn ethanol production concerns have been expressed about the industry and the impacts it could have on the environment and global food security (Farrell et al. 2006). Implicit in these discussions is recognition that corn ethanol production represents only a small improvement over petroleum products in terms of conversion efficiency. The energy derived from a unit of corn ethanol is on the scale of 1.4 times of that used to produce it (US DOE 2006). While it has become apparent that the efficiency of the grain to ethanol conversion can be improved, it is clear that much greater efficiencies will be realized from second-generation fuels (Hettinga 2008; Liska et al. 2009). The fossil fuel efficiency ratio for ethanol produced from biomass is predicted to be greater than 10, almost an order of magnitude over that which can be obtained in the conversion of grain to ethanol (US DOE 2006). There have also been several questions raised regarding the balance between food and fuel. Unfortunately, the issue has been portrayed very simplistically by the popular press generally neglecting the complex interactions with factors such as climate change, livelihoods, development goals, and misconceptions and misunderstandings among academics, policy makers and the public (Rosillo-Calle and Johnson 2010). Both the benefits and risks of biofuels are very context specific—a system that is sustainable in one location may not even work in another. For example potential impacts of climate change on biofuel feedstock production could be any one of several limiting factors including (1) lack of water, (2) soil erosion, (3) salinity, or (4) lack of investment (Rosillo-Calle and Johnson 2008; Wilhelm et al. 2011).

The RFS is designed to encourage a shift to second-generation biofuels in 2015. After this time, further expansion of the industry will be based on biofuels derived from agricultural and forest residues as well as energy crops. As the incentives shift to producing and using advanced biofuels, corn crop residues will play an increasingly important role in biofuel production. The Billion Ton Update (BT2), a study published by the US DOE (2011) to assess and predict potential biomass feedstock resources in the US, estimates that corn stover could account for between 65 and 140 million tons or as much as 10% of available feedstock for biofuel production in 2030. The concentration of corn production in the Midwest US and relative availability of corn residues in the region have led many analysts to predict that an industry based on cellulosic fuels will take root there and spread to other areas as technologies develop and other herbaceous and woody energy feedstocks are established.

There are obvious advantages for using corn residue as a biofuel feedstock. The practice will allow coproduction of food and fuel on the same land and therefore will not necessarily result in significant land use change. Corn residues currently represent the largest readily available supply of feedstock (DOE 2011). US production of corn grain exceeded 12.5 billion bushels in 2011 (NASS 2012), which means that by assuming a harvest index of 0.5, more than 290 million tons of residue will be produced. For several reasons, a substantial amount of this material will not be available to produce biofuels, but its sheer abundance underscores the potential. After evaluating ethanol production from corn grain and stover with respect to energy use, energy security, and resource conservation metrics, Lavigne and Powers (2007) concluded that using corn stover as a feedstock was more consistent with US national energy policy priorities than producing ethanol from grain.

Other advantages for embracing the use of corn residues for bioenergy production relate to the well-developed nature of the crop and the industry that supports it. Much of the infrastructure for producing, harvesting and transporting corn residues already exists. In terms of genetics, the corn industry has excelled at discovering fundamental knowledge about the species and translating that information into superior crop performance. The underlying knowledge for developing hybrids for coproduction of energy and grain as well as the research infrastructure for its expansion already exist. The Corn Belt has an extensive transportation infrastructure for moving agricultural products from the field, to storage, processing facilities and export markets outside the region. Time will tell, but collectively these reasons suggest that the nascent cellulosic fuel industry will likely develop around corn biomass feedstock in the US Midwest and migrate to other areas with different feedstock materials.

Concerns Regarding the Use of Corn Stover as a Biofuel Feedstock

As previously discussed, corn stover, the aboveground material left in fields after corn grain harvest, was identified by Perlack et al. (2005) as a primary biomass source in the Billion Ton Study (BTS). However, this raised concern among many soil scientists because harvesting crop residues for biofuel feedstock or any other purpose will decrease annual carbon input and may gradually diminish soil organic carbon (SOC) to a level that threatens the soil's production capacity (Johnson 2006). Concerns were accentuated knowing that for many soils artificial drainage, intensive annual tillage, and less diverse plant communities have already reduced SOC by 30 to 50% when compared to precultivation levels (Schlesinger 1985). Returning a portion of crop residues to replenish SOC was deemed essential for sustainability (Lal 2004a,b; Wilhelm et al. 2007).

With regard to advanced biofuels, cellulosic biomass has numerous advantages over corn, soybean, or other grains, including its availability from sources that do not compete with food and feed production. Biomass can be reclaimed from municipal solid waste streams and from residual products of certain forestry and farming operations (Brick 2011). It can also be grown on idle or abandoned cropland thus minimizing competition with food, feed and fiber production. Plant biomass has the potential to play an important role in the global energy future because it can be grown in a sustainable manner and converted into liquid transportation fuels using either biochemical or thermochemical conversion processes. Biofuels made from renewable feedstocks are an attractive alternative to gasoline because they can decrease the net release of greenhouse gases (GHG) from the transportation sector (Karlen et al. 2011).

Using the lessons learned from grain ethanol production, it is important to recognize that while there is sufficient rationale and scope for including corn residues as an advanced biofuel feedstock, the advantages must be balanced against potential environmental concerns. Currently, most corn residues are left in the field after harvest because they have significant impact on the cropping system. They are involved in nutrient cycling, water balance, carbon sequestration and very importantly for helping to prevent soil erosion caused by wind and water (Johnson et al. 2009, 2011). These are important ecosystem functions and harvesting corn residues will affect the performance of each (Johnson et al. 2007). Simply removing the residues without an attempt to replace the ecosystem services they provide is not an option on many sites (Johnson et al. 2010).

Harvesting corn stover as a feedstock for biofuel production could have many benefits, if the process is developed as a complete system that considers all ecosystem services provided by crop residues (Larson 1979; Karlen et al. 1984; Wilhelm et al. 2010). This includes conserving soil water, reducing surface runoff and evaporation, increasing infiltration rates, controlling soil erosion, recycling plant nutrients, providing habitat and energy for earthworms and other soil macro- and micro-organisms, improving water quality by denaturing and filtering of pollutants, improving soil structure, preserving native habitats, and maintaining biodiversity. Crop residues can also help reduce nonpoint source pollution, decrease sedimentation, minimize risks of anoxia and dead zones in coastal ecosystems, increase agronomic productivity, advance food security, and mitigate flooding by holding water on the land rather than allowing it to run off into streams and rivers (Kimble et al. 2007).

Long-term Productivity Effects

It is well recognized that excessive crop residue harvest will have negative consequences on long-term soil productivity, especially if conventional, relatively intensive (i.e., moldboard plowing, chisel plowing, multiple diskings, etc.) tillage practices are used (Larson 1979; Wilhelm et al. 2004; Blanco-Canqui and Lal 2007). However, by adopting practices such as strip-tillage, or no-tillage it may be possible to harvest a portion of the crop residues without impairing long-term soil productivity. This is especially true where improved hybrids and management are consistently resulting in grain yields exceeding 200 bu ac^{-1} and producers are actually facing "crop residue management" problems due to subsequent N immobilization or poor stand establishment due to inadequate soil-seed contact. Those latter conditions are the primary assumptions for estimates of available corn stover in the revised Billion Ton Study (BT2) (DOE 2011).

Representative corn grain yields from several ongoing field studies listed above are summarized in Table 1. Overall, they show that harvesting a portion of the corn stover for several years did not have a negative impact on subsequent grain yields. The negative response for the MN-NT95 site was caused in part by a K deficiency that developed due to long-term no-tillage (since 1995). Also, the lower yields for several of the nonremoval Iowa sites were caused by N immobilization and early-season plant N deficiencies. Exact stover harvest rates varied among locations but generally averaged 1 to 1.5 t ac^{-1} for the moderate removal rate and 2 to 2.5 t ac^{-1} for the high-removal rate.

Table 1 Representative long-term corn grain yield (bu/acre) in response to stover harvest on ARS-Renewable Energy Assessment Project (REAP) and Sun Grant Regional Partnership (RP) research sites.

Location	Tilage	Site-Years	Removal Rate		
			None	Moderate	High
MN-NT95	No-tillage	7	130	134	122
MN-NT05	No-tillage	6	152	155	150
MN-CP	Chisel plow	7	156	162	164
PA1	No-tillage	4	147	153	145
IL-NT	No-tillage	19	179	194	197
IL-CP	Chisel plow	19	205	199	202
SD1	No-tillage	4	118	120	ND[†]
IA-Conv	No-tillage	4	168	179	189
IA-Conv	Chisel plow	4	171	188	188
IA-High Pop	No-tillage	4	168	188	183
IA-High Pop	Chisel plow	4	173	189	188
IA-Biochar 1	Chisel plow	4	168	190	193
IA-Biochar 2	Chisel plow	4	177	185	191
Average	----	7	162	172	176

[†]ND – Not determined

Soil Quality Effects

Six of the most critical environmental factors that limit sustainable agricultural residue removal are: soil organic carbon, wind and water erosion, plant nutrient balances, soil water and temperature dynamics, soil compaction, and offsite environmental impacts (Wilhelm et al. 2010). One method for evaluating the impact of harvesting corn stover and other feedstock materials is to use soil quality assessment. During the past 20 years, several studies (Karlen et al. 1997, 2006; Liebig et al. 2006; Wienhold et al. 2006; Zobeck et al. 2008; Jokela et al. 2009) have used the Soil Management Assessment Framework (SMAF) developed by Andrews et al. (2004) to monitor and evaluate soil biological, chemical, and physical responses to various land uses, farming systems, and management practices. We expect that the potential land-use changes associated with development of a sustainable biofuel industry will present another opportunity to use the SMAF to guide and quantify long-term effects of such endeavors.

By focusing on soil quality impacts, the perception that crop residues are not important for modern grain production systems will hopefully be dispelled. Crop residues, both above and belowground, protect land from the ravages of wind and water erosion (Soil Conservation Society of America 1979). They also supply an annual input of carbon and replenish several of the essential plant nutrients that are assimilated during crop production

(Wilhelm et al. 2004). Traditionally, a limited amount of corn stover has been harvested for animal feed and bedding. This is usually done in a localized manner, with a substantial portion of the residues being returned to the soil, often mixed with animal manure and thus not only adding carbon but also recycling other nutrients to soils in the same location, or at least on the same farm, from whence they came.

Despite the recycling that can occur when crop residues are utilized as animal feed and then partially recycled through the manure, long-term research has conclusively shown that crop production practices often result in TOC loss (Paustian et al. 1997). Losses are often greatest where corn is produced on soils having artificial drainage, intensive annual tillage, and less diverse plant communities. Collectively, these factors have been shown to have reduced TOC by 30 to 50% when compared to precultivation levels (Schlesinger 1985). Such TOC loss can have many detrimental effects on soil productivity (Gollany et al. 1991; Mann et al. 2002) and quality (Liebig et al. 2005; Moebius-Clune et al. 2008). However, soil and crop management practices that decrease tillage and crop residue incorporation can reduce TOC losses and may even increase TOC to a limited extent (Burke et al. 1989).

Previous long-term studies, such as those by reviewed by Paustian et al. (1997), showed the importance of preventing excessive stover harvest, which was recognized in the BTS. As a result, the BTS authors limited their estimates of available feedstock in order to protect soil resources from wind and water erosion (Nelson 2002; Graham et al. 2007), but they did not account for the amount of stover required to sustain TOC levels. Soil carbon assessments are a key component of the several studies.

Soil Erosion Effects

In 1979, Larson conducted one of the first large-scale studies focused on crop residue removal and its effect on soil erosion using the Universal Soil Loss Equation. This study included the Corn Belt, the Great Plains, and the Southeast. The effect of tillage practices, i.e., conventional, conservation, and no-till and residue management were investigated with respect to rainfall and wind erosion, runoff, and potential nutrient removal. This study found that for the management practices and crop yields at the time, nearly 49 million metric tons of residue was available annually throughout the Corn Belt. Soil carbon, tilth, and productivity maintenance were not considered. As a result of limited interest in agricultural residues for energy production during the 1980s and 1990s, no additional large spatial scale assessments of residue availability were performed until more than two decades after Larsen's study. Nelson (2002) used the Revised Universal Soil Loss Equation

(RUSLE) and Wind Erosion eQuation (WEQ) to expand on Larson's analysis to develop a methodology to estimate the sustainable removal rates of corn stover and wheat straw at the soil-type level. This methodology considered rainfall and wind-induced soil erosion as a function of reduced and no-till field management practices. In 2004, Nelson et al. used the same approach to assess five other major one- and two-year cropping rotations, e.g., corn-soybean. Neither of these studies addressed soil organic matter as a function of removal. Researchers have also used the Revised Universal Soil Loss Equation, Version 2 (RUSLE2) and/or Wind Erosion Prediction System (WEPS) to address a number of erosion-based questions on crop residue removal.

Crop residue reduces water erosion primarily by dissipating rainfall energy and slowing overland flow of water so that it can infiltrate and be retained in the soil. It helps mitigate wind erosion by slowing wind speed at the soil surface—air interface and reducing opportunities for soil movement through suspension, saltation, or surface creep (Fig. 2). For controlling wind erosion it is not only important to retain an adequate amount of surface residue cover, but also to have significant vertical orientation.

In addition to the physical loss of soil particles and the nutrients or other materials attached to them through either water or wind erosion, crop residues are also important for helping to form and stabilize soil aggregates. Data from the Brookings, South Dakota show the impact of what would be considered excessive crop residue harvest based on the long-term average grain yields (Table 2). The long-term effect of excessive corn residue harvest is that the fraction of small aggregates and thus the

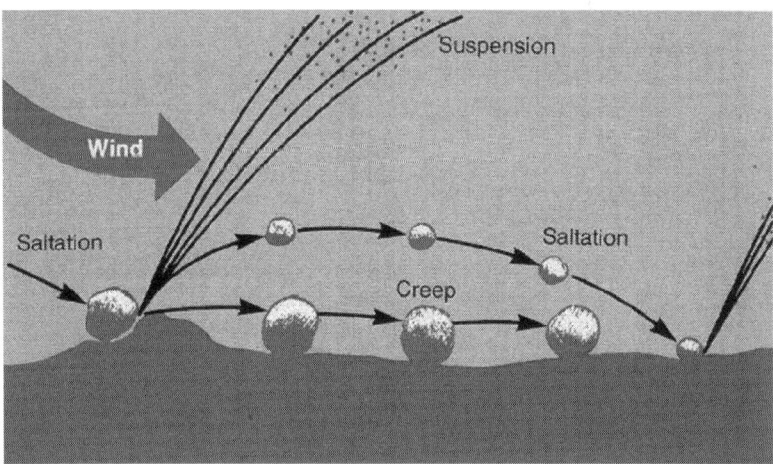

Figure 2 Erosion processes that are mathematically simulated by the WEPS model to estimate daily soil loss by wind (adapted from Hagen et al. 1996).

Table 2 Effect of corn residue removal on soil aggregate size distribution after four, two-year corn-soybean production cycles near Brookings, SD (adapted from Hammerbeck et al. 2012).

Residue Removal	Dry Aggregate Size Classification (mm)					Erodible Fraction[†]
	< 0.4	0.4–0.8	0.8–2.0	2–19	> 19	g kg^{-1}
None	59 C[‡]	34 C	75 B	416	416 A	93 C
Moderate	91 B	54 B	105 A	403	347 B	145 B
High	116 A	68 A	118 A	398	300 B	184 A
P > F	0.0004	<0.0001	0.0006	NS	0.0378	0.0048

[†]The wind erodible fraction is the mass of soil < 0.84 mm in diameter
[‡]Numbers followed by the same letter within each aggregate size group are not significantly different at α = 0.05 using the Tukey Test

highly erodible fraction increases while the fraction of large aggregates (> 19 mm) decreases, presumably due to lack of carbon input to help build more stable and therefore wind erosion resistant aggregates.

Soil Fertility Effects

Harvesting corn residue for biofuel feedstock or other bioproducts removes not only carbon needed to sustain soil biological and physical attributes, but also several essential plant nutrients. Recent studies by Karlen et al. (2011) showed as expected that nutrient removal was directly proportional to the amount of stover harvested. They showed that average N-P-K removal was increased by 26, 3 and 30 lb ac^{-1} for continuous corn and 37, 3, and 30 lb ac^{-1} for rotated corn, respectively, when compared to harvesting only the grain. There were also increases in secondary (Ca, Mg, and S) and micronutrient (Cu, Fe, Mn, Zn) removal compared to a grain-only harvest.

The increased nutrient removal associated with corn stover harvest could be beneficial if soil test values were extremely high from prior animal manure or fertilizer application rates, but if they needed to be replaced to prevent long-term nutrient depletion, the replacement cost for those nutrients is a factor that must be included when placing a monetary value on the stover. Obviously, fertilizer price is a major and everchanging factor and those values are closely associated with petroleum and transportation costs. Karlen et al. (2011) discussed how the estimated nutrient replacement costs fluctuated throughout five years and acknowledged that they will continue to do so throughout the future. However, one of the most consistent results from their work was in total nutrient replacement cost (~ $17.25 ± $1 ton^{-1}). They concluded that having documented such a consistent value makes it more feasible to determine a fair market value for both feedstock producers and consumers.

Water Quality and Quantity Effects

One of the primary environmental concerns for both grain and cellulosic ethanol production from corn continues to be the long-term effects on surface and groundwater quality (UCS 2010a, 2010b, 2011). This concern arose because as the amount of corn grown for ethanol rose from approximately 5 to 40% during the past decade, the use of fertilizer and other inputs to support the crop also increased. This raises environmental concerns because most of the US corn crop is grown in the Mississippi River watershed, which is a major contributor of nitrogen and phosphorus to the Gulf of Mexico and thus a factor creating a "dead zone" where fish cannot survive. Since corn production accounts for 42% of US N fertilizer use, more intensive corn production increases the potential for even greater N loss to surface and groundwater resources.

In addition to water quality impacts, water quantity has also been raised as a concern associated with corn ethanol. Karlen (2011) stated that to understand the complexity of predicting biofuel effects on water quantity and quality, we must first step back from biofuels production *per se* and examine the global hydrologic cycle. Currently, the US uses approximately 48% of its fresh and weakly saline water for thermoelectric power generation, but evaporation and the power generation actually consume only 2 to 3% of this water. The remaining 97 to 98% is returned in its original form and is thus potentially available for reuse. Municipal withdrawals account for another 10%, but nearly 90% of that water is returned as wastewater that can be treated and reused. Finally, industry accounts for another 21% of US freshwater withdrawals, with the quantity and quality of that being returned being industry specific and highly variable.

Agricultural water use differs from use by these other entities in that most of the water is consumed through evapotranspiration (ET) that supports plant growth and development. Water is also consumed when used to leach salts from the soil and thus manage soil salinity. With or without a biofuels industry, agriculture uses large quantities of water. Freshwater extraction ranges from less than 20% to more than 90% for different countries depending upon climate, productivity and water use efficiency (WUE) of the crops being grown. The ratio quantifying the amount of plant dry matter produced for a specific quantity of water used is defined as water use efficiency. The WUE value varies greatly depending on crop species, location, culture practice, climate and weather, and other factors. Growing plants is very water intensive because as much as 1,000 pounds of water may be required to produce just one pound of plant material. Fortunately, the transpired water is recycled in the hydrologic cycle. It falls as precipitation, and after infiltrating into the soil, running off into streams or lakes, or percolating to deeper aquifers, it is once again available for ET in support of plant growth or for other uses. Unfortunately, the groundwater portion of the cycle cannot always be replenished as fast as it is used in many drier regions and as a result groundwater is often irreversibly mined. The Ogallala Aquifer, located in the US Great Plains, is one example where water mining has occurred. A 2009 US Geological Survey (USGS) report stated that in parts of southwest Kansas and the Texas Panhandle, groundwater levels have dropped by more than 150 feet due to intensive crop irrigation and minimal aquifer recharge.

Water is also important for conversion of all feedstocks into biofuels, specifically for heating, cooling, and the chemical processes involved. For the corn-based biofuels conversion process, water consumption has decreased dramatically during the past decade, falling from an average of 5.8 gallons of water per gallon of ethanol in 1998, to 3.0 gallons/gallon or less in 2009. For comparison, the recovery and refining of crude oil requires 3.6 to 7.0 gallons of water per gallon of fuel. Water requirements for conversion of cellulosic materials will depend on the feedstock and the conversion process. These systems are not fully developed, but current estimates of water use range from 1.9 to 6.0 gallons/gallon for ethanol production or 1.0 gallon/gallon for biodiesel (Karlen 2011).

Ultimately, management practices make all the difference with regard to both water quality and quantity issues associated with biofuel production from corn or any other crop. This includes the use of nutrient management plans based on soil property measurements, replacing gullies with grass-filled channels, changing row orientation to follow the contour of the land and adding terraces and grass buffers to control water flow and reduce erosion. Cover crops can be grown from late fall to early spring to capture residual nutrients, add carbon, and protect the soil surface from wind and water erosion. Controlled drainage systems can be used to reduce short-circuiting of nutrients from the soil profile to surface waterbodies.

Prospects for Overcoming Concerns Regarding the Use of Corn for Biofuels

Using current corn production practices, removing corn residues as a feedstock for ethanol production could and will likely have negative effects on soil and water quality (Wilhelm et al. 2004). However, many of the concerns associated with harvesting corn stover might potentially be eliminated by using alternative crop production practices. The increased use of reduced tillage systems and cover crops would lessen the impact of stover removal on soil erosion and allow more carbon input and less loss from the soil (Perlack et al. 2005). Furthermore, diverting land less well suited to annual crop production to the production of perennial energy crops would address some of the more serious concerns with using corn stover as a bioenergy feedstock. Careful consideration of the ability of the land where the crop is grown to tolerate residue removal could also greatly diminish the overall impact if highly erodible and otherwise unsuitable land is excluded from the practice. Much of this land is either highly erodible or possesses other constraints that make it economically or environmentally marginal for row crop production. By creating a market for cellulosic feedstocks, using corn stover for fuel could lead to the conversion of these lands to more environmentally benign crop management systems (Brick 2011).

Alternative crop production practices to address some of the environmental concerns of harvesting corn stover have already been developed and are being used to a limited extent. The use of no and minimum tillage practices significantly reduces soil erosion and ameliorates loss of soil organic matter (West and Post 2002; Montgomery 2007). This latter effect has a positive impact on carbon sequestration when no-till practices are compared to conventional practices (Bernarcchi 2005). The benefits of no-till agriculture have been studied and known for a several years (Phillips et al. 1980). These include increased soil organic matter and therefore carbon sequestration, reduced fuel and therefore energy requirements, reduced soil erosion, decreased soil compaction, increased water infiltration, improved nutrient cycling, and improved water quality (Reicosky 2008). Despite these advantages, no till production of corn was estimated to be practiced on only 25.5 million acres of cropland in 2009 or about 29.5% of the corn acreage (Horowitz et al. 2010). The main reason no till is not practiced on more acres is the perception of lower yields (Vyn and Raimbult 1991) and returns. Lower yields from no-till corn are often related to later planting date due to slower warming of nontilled soil in spring (Fortin 1993). No-till production also creates a better environment for some crop pests and therefore requires alternative pest management practices. In addition, no-till production requires producers to adopt new production strategies and invest in new farm machinery. This said, creating a market for corn stover, may indeed encourage producers to adopt a no-till strategy in order to compensate for the loss of ecological services provided by the residue when it is left in the field. In the original BTR, production estimates were based on 100% use of no till on crop fields where the residues were being removed as a feedstock in 2030 (Perlack 2005).

The use of cover crops also has potential for reducing the environmental impacts of corn stover removal in some production areas (Fronning et al. 2008). Annual cover crops are generally planted in the fall after a row crop is harvested. They are intended to provide soil cover during the winter period when the ground is normally fallow. Thus, they intercept rainfall and reduce soil runoff reducing erosion and increasing water infiltration. They also hold the soil against wind erosion and thus improve air quality

(Hartwig and Ammon 2002). In the springtime, the actively growing cover crops take up soil nitrogen and immobilize it preventing it from leaching into the groundwater (Mitchell and Tell 1976). The primary obstacles to use of cover crops relate to cost of establishment and timing. The farther north cover crops are planted, the less the time interval is between harvest and the onset of winter weather. In some years, there is little time making their use somewhat risky. To avoid these time constraints researchers and producers have evaluated various methods for establishing cover crops in the standing grain crop using aerial seeding methods. While these methods have shown some success, they are relatively expensive to use and require time to implement. Without a reasonable return to the investment in time and other resources, their use is often hard for producers to justify. However, by allowing harvest of corn stover there may be some financial incentive for using them.

Perennial cover crops are another, although less well developed, option. Research has demonstrated that corn can be grown in the presence of a groundcover and produce yields comparable to conventional production practices (Wiggans et al. 2012). However, it has proven essential to manage competition from the cover crop in the spring with the use of contact herbicides or some other form of suppression (Echtenkamp and Moomaw 1989). Perennial cover crops avoid the establishment and time constraints associated with annual cover crops and essentially provide the same benefits. Additionally, they can have a positive influence on soil moisture because the can increase water infiltration and provide an evaporation barrier in the summer. For this reason, perennial groundcovers are sometimes referred to as living mulches. The use of perennial groundcovers for corn production is not a proven technology, but recent research indicates that it has strong potential for addressing many of the environmental issues associated with corn production and stover removal (Flynn et al. 2013).

Recent increases in the price of corn grain have led producers to alter their crop management practices. Farming land that once produced marginal returns is now profitable and has led to the conversion of this land from pasture or conservation to row-crop production. Unfortunately, much of this marginal land has a disproportionate impact on ecological services that are lost with its conversion to cropland. This is land that is subject to flooding and drought, is highly erodible, or otherwise has inherent characteristics that can lead to negative environmental effects under certain circumstances. Land of this class should not be used for grain production much less with stover removal. The environmental consequences of doing so are disproportionately large compared with the financial gain from doing so. However, development of processing facilities for converting corn stover will create an opportunity for using alternative sources of biomass. Marginal lands could thus be managed productively by growing perennial energy crops such as switchgrass thus preventing or reducing the negative environmental consequences of converting them to cropland.

Developing and implementing this landscape vision for producing biofuels feedstock could facilitate balancing the economic drivers and limiting factors needed to achieve sustainable feedstock supplies and alleviate many concerns regarding the use of corn for biofuels (Fig. 3 and Karlen et al. 2012). The landscape vision would not replace current corn and soybean production systems but rather augment them with several other potential bioenergy feedstocks such as switchgrass, *Miscanthus*, sorghum, mixed cool- and warm-season grasses, or woody species such as poplar or willow. Again, the premise for this vision is that rather than focusing solely on energy production from corn, a diversified landscape would provide multiple ecosystem services such as:

1. Sustainable grain and biomass supplies for food, feed and energy
2. Increased C sequestration
3. Protection of water quality
4. Increased productivity and profitability
5. Reduced producer and environmental risk
6. Greater biodiversity
7. Improved wildlife habitat
8. Vigorous rural community development supported by new industries and entrepreneurial opportunities

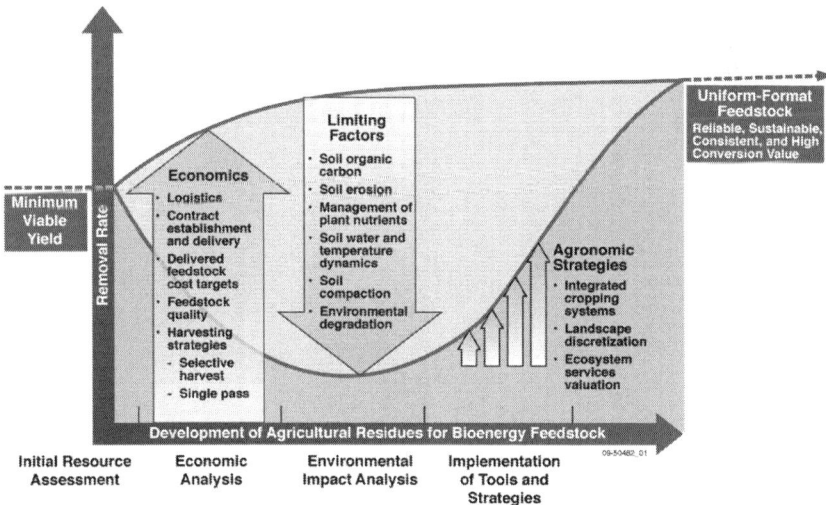

Figure 3 An illustration of competing economic drivers and limiting factors that must be balanced to achieve sustainable cellulosic feedstock supplies needed to support a transition from fossil to renewable fuels (from Wilhelm et al. 2010).

The advanced biofuel production systems associated with this landscape vision should not be viewed as limited to ethanol or any other specific fuel source. In addition to starch from the corn grain *per se*, the crop may be able to efficiently contribute additional feedstock in the form of crop residues or even by removing the pericarp before fermentation. Extraction of oil from the distiller's grain is another component of the complete system. Furthermore, the use of these additional components from the corn crop may even produce a fuel that qualifies as an advanced biofuel and, depending on the conversion process and to a large part on the soil organic carbon (SOC) dynamics associated with the entire system.

A starting point for implementing the landscape vision is to establish best management practices and standards for the entire biofuel industry. These practices should address all of the limiting factors and be supported by the improved agronomic practices that are being developed to ensure supplies for conversion facilities can be met (Fig. 3). A current example of this approach is the developmental work being done by the Council for Sustainable Biofuel Production (CSBP), which is striving to develop standards that would enable those purchasing bioenergy feedstock to appropriately compensate land owners and operators not only for the commodities *per se*, but also the ecosystem services their land provides. This could also help overcome the perception that using land for biofuel production is in direct competition with using if for food and feed production, thus confirming a point made by Rosillo-Calle and Johnson (2010) that the issue is not land availability but rather how the land is managed.

In summary, there are ample opportunities to use corn as the foundation for a viable biofuel industry in the Midwestern US. The key is management and not focusing solely on corn grain as the only feedstock. Diversity is crucial and the use of best management practices is essential. In other words, all options are open.

References

Andrews SS, Karlen DL and Cambardella CA (2004) The soil management assessment framework: a quantitative soil quality evaluation method. Soil Sci Soc Am J 68: 1945–1962.

Babcock BA (2008) Distributional implications of U.S. ethanol policy. Appl Econ Perspect Pol 30: 533–542.

Balat M and Balat H (2009) Recent trends in global production and utilization of bio-ethanol fuel. Appl Energy 86: 2273–2282.

Bernacchi CJ, Hollinger SE and Meyers T (2005) The conversion of the corn/soybean ecosystem to no-tillagriculture may result in a carbon sink. Glob Change Biol 11: 1867–1872.

Blanco-Canqui H and La R (2007) Soil and crop response to harvesting corn residues for biofuel production. Geoderma 141: 355–362.

Brick S (2011) Harnessing the popwer of biomass residuals: Opportunities and challenges for Midwestern renewable energy. Heartland Papers. Issue 4. The Chicago Council on Global Affairs, Chicago, IL, USA, 65 p.

Burke IC, Yonker CM, Parton WJ et al. (1989) Texture, climate, and cultivation effects on soil organic matter content in U.S. grasslands soils. Soil Sci Soc Am J 53: 800–805.

De Gorter H and Just DR (2009) The economics of a blend mandate for biofuels. Am J Agri Econ 9: 738–750.

Echtenkamp GW and Moomaw RS (1989) No-till corn production in a living mulch system. Weed Technol 3: 261–266 .

Elobeid, Amani and Simla Tokgoz (2008) Removing distortions in the U.S. ethanol market: What does it imply for the United States and Brazil? Am J Agri Econ 90: 918–932.

EPA (2009) EPA lifecycle analysis of greenhouse gas emissions from renewable fuels. Office of Transportation and Air Quality, EPA-420-F-09-024, Washington, DC, USA.

Farrell, Alexander E, Richard J Plevin et al. (2006) Ethanol can contribute to energy and environmental goals. Science 311: 506–508.

Flynn E Scott, Kenneth J Moore, Singer J and Kendall R Lamkey (2013) Evaluation of grass and legume species as perennial ground covers in corn production. Crop Science 53: 611–620.

Fortin MC (1993) Soil temperature, soil water, and no-till corn development following in-row residue removal. Agron J 85: 571–576.

Fronning BE, Kurt D Thelen and Doo-Hong Min (2008) Use of manure, compost, and cover crops to supplant crop residue carbon in corn stover removed cropping systems. Agron J 100: 1703–1710.

Gollany HT, Schumacher TE, Evenson P et al. (1991) Aggregate stability of an eroded and desurfaced typic Argiustoll. Soil Sci Soc Am J 55: 811–816.

Graham RL, Nelson R, Sheehan J et al. (2007) Current and potential U.S. corn stover supplies. Agron J 99: 1–11.

Hagen LJ, Wagner LE and Tatarko J (1996) Wind Erosion Prediction System (WEPS) Technical Documentation: BETA Release 95-08: http://www.weru.ksu.edu/weps/docs/weps_tech.pdf

Hahn-Hägerdal B, Galbe M, Gorwa-Grauslund MF et al. (2006) Bio-ethanol—the fuel of tomorrow from the residues of today. Trends Biotechnol 24: 549–56.

Hammerbeck AL, Stetson SJ, Osborne SL et al. (2012) Corn residue removal impact on soil aggregates in a no-till corn/soybean rotation. Soil Sci Soc Am J 76: 1390–1398.

Hartwig NL and Ammon HU (2002) Cover crops and living mulches. Weed Science 50: 688–699.

Hettinga WG, Junginger HM, Dekker SC et al. (2009) Understanding the reductions in US corn ethanol production costs: An experience curve approach. Energy Pol 37: 190–203.

Hoekman S Kent (2009) Biofuels in the U.S.—Challenges and opportunities. Renew Energy 34: 14–22.

Horowitz J, Ebel R and Ueda K (2010) "No-Till" Farming Is a Growing Practice. USDA Economic Research Service, Economic Information Bulletin No. 70, Washington DC, USA

Johnson JMF, Allmaras RR and Reicosky DC (2006) Estimating source carbon from crop residues, roots, and rhizodeposits using the national grain-yield database. Agron J 98: 622–636.

Johnson JMF, Coleman MD, Gesch RW et al. (2007) Biomass-bioenergy crops in the United States: A changing paradigm. Am J Plant Sci Biotechnol 1: 1–28.

Johnson JMF, Papiernik SK, Mikha MM et al. (2009) Soil processes and residue harvest management. In: Lal R and Stewart BA (eds) Carbon Management, Fuels, and Soil Quality. Taylor and Francis, New York, NY, USA, pp 1–44.

Johnson JMF, Karlen DL and Andrews SS (2010) Conservation considerations for sustainable bioenergy feedstock production: If, what, where, and how much? J Soil Water Conserv 65(4): 88A–91A.

Johnson JM-F, Archer DW, Karlen DL et al. (2011) Soil management implications of producing biofuel feedstock. In: Sauer TS and Hatfield JL (eds) Soil Management: Building a Stable Base for Agriculture. Am Soc Agron Soil Sci Soc Am, Madison, WI, USA, pp 371–390.

Jokela WE, Grabber JE, Karlen DL et al. (2009) Cover crop and liquid manure effects on soil quality indicators in a corn silage system. Agron J 101: 727–737.

Karlen DL, Hunt PG and Campbell RB (1984) Crop residue removal effects on corn yield and fertility of a Norfolk sandy loam. Soil Sci Soc Am J 48: 868–872.

Karlen DL, Mausbach MJ, Doran JW et al. (1997) Soil quality: A concept, definition, and framework for evaluation. Soil Sci Soc Am J 61: 4–10.

Karlen DL, Hurley EG, Andrews SS et al. (2006) Crop rotation effects on soil quality at three northern corn/soybean belt locations. Agron J 98: 484–495.

Karlen DL (2011) Unraveling water quality and quantity effects of biofuels production. In: Rudebusch L (ed) Getting into Soil and Water. Soil & Water 2011. Soil & Water Conservation Club, Iowa Water Center at Iowa State Univ, Ames, IA, USA, iowawatercenter@iaste.edu, pp 36–39.

Karlen DL, Birrell SJ and Hess JR (2011a) A Five-Year Assessment of Corn Stover Harvest in Central Iowa, USA. Soil Till Res 115–116: 47–55.

Karlen DL, Varvel GE, Johnson JMF et al. (2011b) Monitoring soil quality to assess the sustainability of harvesting corn stover. Agron J 103: 288–295.

Karlen DL, Archer D, Liska AJ et al. (2012) Energy issues affecting corn/soybean systems: Challenges for sustainable production. CAST Issue Paper No. 48. Council for Agricultural Science and Technology, Ames, IA, USA.

Kimble JM, Rice CW, Reed D et al. (eds) (2007) Soil Carbon Management, Economic, Environmental, and Societal Benefits. Taylor and Francis, Boca Raton, FL, USA.

Lal R (2004a) Soil carbon sequestration impacts on global climate change and food security. Science 304: 1623–1627.

Lal R (2004b) Is crop residue a waste? J. Soil Water Conserv 59: 136–139.

Larson WE (1979) Crop residues: Energy production or erosion control. J Soil Water Conserv 34: 74–76.

Lavigne A and Powers SE (2007) Evaluating fuel ethanol feedstocks from energy policy perspectives: A comparative energy assessment of corn and corn stover. Energy Pol 35: 5918–5930.

Liebig MA, Morgan JA, Reeder JD et al. (2005) Greenhouse gas contributions and mitigation potential of agricultural practices in northwestern USA and western Canada. Soil Till Res 83: 25–52.

Liska AJ, Yang HS, Bremer VR et al. (2009) Improvements in life cycle energy efficiency and greenhouse gas emissions of corn-ethanol. J Indust Ecol 13: 58–74.

Liebig M, Carpenter-Boggs L, Johnson JMF et al. (2006) Cropping system effects on soil biological characteristics in the Great Plains. Renew Agri Food Syst 21: 36–48.

Low SA (2009) Ethanol and the local economy industry trends, location factors, economic impacts, and risks. Econ Dev Quart 23: 71–88.

Mann L, Tolbert V and Cushman J (2002) Potential environmental effects of corn (*Zea mays* L.) stover removal with emphasis on soil organic matter and erosion. Agri Ecosyst Environ 89: 149–166.

Mitchell WH and MR Tell (1976) Winter-annual cover crops for no-tillage corn production. Agron J 69: 569–573.

Moebius-Clune BN, van Es HM, Idowu OJ et al. (2008) Long-term effects of harvesting maize stover and tillage on soil quality. Soil Sci Soc Am J 72: 960–969.

Montgomery DR (2007) Soil erosion and agricultural sustainability. Proc Natl Acad Sci USA 104: 13268–13272.

National Academy of Sciences (NAS) (2009) Liquid transportation fuels form coal and biomass: Technological status, costs, and environmental impacts. The National Academies Press, Washington DC, USA, 322 p.

Nelson RG (2002) Resource assessment and removal analysis for corn stover and wheat straw in the Eastern and Midwestern United States—rainfall and wind-induced soil erosion methodology. Biomass Bioenergy 22: 349–363.

Paustian K, Collins HP and Paul EA (1996) Management controls on soil carbon. In: Paul EA, Paustian K, Elliot ET et al. (eds) Soil Organic Matter in Temperate Agroecosystems: Long-Term Experiments in North America. CRC Press, Boca Raton, FL, USA, pp 15–49.

Perlack RD, Wright LL, Turhollow AF et al. (2005) Biomass as feedstock for a bioenergy and bioproducts industry: The technical feasibility of a billion-ton annual supply. DOE/GO-102005-2135 and ORNL/TM-2005/66: http://feedstockreview.ornl.gov/pdf/billion_ton_vision.pdf _.

Phillips RE, Thomas GW, Blevins RL et al. (1980) No-tillage agriculture. Science 208: 1108–1113.

Reicosky DC (2008) Carbon sequestration and environmental benefits from no-till systems. In: Goddard T et al. (eds) No-till Farming Systems. Special Publication No. 3. World Association of Soil and Water Conservation, Bangkok, Thailand, pp 43–58.

Renewable Fuels Association (2012) Statistics: http://www.ethanolrfa.org/pages/statistics#C.

Rosillo-Calle F and Johnson FX (eds) (2010) Food versus Fuel: An Informed Introduction to Biofuels. Palgrave Macmillian, New York, NY, USA, 217 p.

Schlesinger WH (1985) Changes in soil carbon storage and associated properties with disturbance and recovery. In: Trabalha JR and Reichle DE (eds) The Changing Carbon Cycle: A Global Analysis. Springer, New York, USA, pp 194–220.

Searchinger T, Heimlich R, Houghton RA et al. (2008) Use of U.S. croplands for biofuels increases greenhouse gases through emissions from land use change. Science 319: 1157–1268.

Soil Conservation Society of America (SCSA) (1979) Effects of tillage and crop residue removal on erosion, runoff and plant nutrients. Special Publ No 25, Ankeny, IA, USA.

Ugarte DG De La Torre, English BC and Jensen K (2007) sixty billion gallons by 2030: Economic and agricultural impacts of ethanol and biodiesel expansion. Am J Agri Econ 89: 1290–1295.

Union of Concerned Scientists (UCS) (2010a) The billion gallon challenge: Getting biofuels back on track. Washington DC, USA: http://www.ucsusa.org/smartbioenergy.

Union of Concerned Scientists (UCS) (2010b) Managing the rising tide of biofuesl: The energy-water collision. Washington DC, USA: http://www.ucsusa.org/energy-water.

Union of Concerned Scientists (UCS 2011) Corn ethanol's threat to freshwater resources. The energy-water collision. Washington DC, USA: http://www.ucsusa.org/energy-water.

USDA-National Agricultural Statistics Service (NASS) (2012) Data and Statistics [Online]. Washington DC, USA: http://www.nass.usda gov/Data_and_Statistics/Quick_Stats/index.asp.

US Department of Energy (DOE) (2006) Breaking the biological barriers to cellulosic ethanol: A joint research agenda. DOE/SC-0095, US Department of Energy Office of Energy Efficiency and Renewable Energy, Washington DC, USA.

US Department of Energy (DOE) (2011) U.S. Billion-Ton Update: Biomass supply for a bioenergy and bioproducts industry. RD Perlack and BJ Stokes (Leads), ORNL/TM-2011/224. Oak Ridge National Laboratory, Oak Ridge, TN, USA, 227 pp.

Vyn TJ and Raimbult BA (1991) Long-term effect of five tillage systems on corn response and soil structure. Agron J 85: 1074–1079.

West TO and Post WM (2002) Soil organic carbon sequestration rates by tillage and crop rotation: A global data analysis. Soil Sci Soc Am J 66: 1930–1946.

Wienhold BJ, Pikul JL Jr, Liebig MA et al. (2006) Cropping system effects on soil quality in the Great Plains: Synthesis from a regional project. Renew Agri Food Syst 21: 49–59.

Wiggans DR, Singer JW, Moore KJ et al. 2012. Maize water use in living mulch systems with stover removal. Crop Sci 52: 327–338.

Wilhelm WW, Johnson JMF, Hatfield JL et al. (2004) Crop and soil productivity response to corn residue removal: A literature review. Agron J 96: 1–17.

Wilhelm WW, Johnson JM-F, Karlen DL et al. (2007) Corn stover to sustain soil organic carbon further constrains biomass supply. Agron J 99: 1665–1667.

Wilhelm WW, Hess JR, Karlen DL et al. (2010) Balancing limiting factors and economic drivers for sustainable Midwest agricultural residue feedstock supplies. Indust Biotechnol 6 (5): 271–287.

Zobeck TM, Halvorson AD, Wienhold BJ et al. (2008) Comparison of two soil quality indexes to evaluate cropping systems in northern Colorado. J Soil Water Conserv 63: 329–338.

Index